AGL 2159

LOCATION, SCHEDULING, DESIGN and INTEGER PROGRAMMING

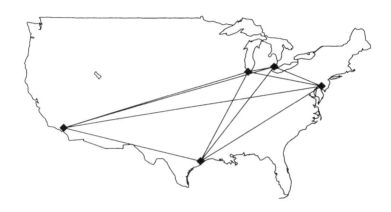

INTERNATIONAL SERIES IN
OPERATIONS RESEARCH & MANAGEMENT SCIENCE

Frederick S. Hillier, Series Editor
Department of Operations Research
Stanford University
Stanford, California

Saigal, Romesh.
The University of Michigan
  LINEAR PROGRAMMING: A Modern Integrated Analysis

Nagurney, Anna/ Zhang, Ding
University of Massachusetts @ Amherst
  PROJECTED DYNAMICAL SYSTEMS AND VARIATIONAL INEQUALITIES WITH APPLICATIONS

LOCATION, SCHEDULING, DESIGN and INTEGER PROGRAMMING

Manfred Padberg
*Professor and Research Professor
of Statistics and Operations Research
New York University*
New York, USA

Minendra P Rijal
*Lecturer
Tribhuvan University*
*Kathmandu, NEPAL,
and Visiting Assistant Professor
New York University*
New York, USA

KLUWER ACADEMIC PUBLISHERS
Boston/London/Dordrecht

Distributors for North America:
Kluwer Academic Publishers
101 Philip Drive
Assinippi Park
Norwell, Massachusetts 02061 USA

Distributors for all other countries:
Kluwer Academic Publishers Group
Distribution Centre
Post Office Box 322
3300 AH Dordrecht, THE NETHERLANDS

Library of Congress Cataloging-in-Publication Data

A C.I.P. Catalogue record for this book is available from the Library of Congress.

Copyright © 1996 by Kluwer Academic Publishers

All rights reserved. No part of this publication may be reproduced, stored in a retrieval system or transmitted in any form or by any means, mechanical, photo-copying, recording, or otherwise, without the prior written permission of the publisher, Kluwer Academic Publishers, 101 Philip Drive, Assinippi Park, Norwell, Massachusetts 02061

Printed on acid-free paper.

Printed in the United States of America

PREFACE

Location, scheduling and design problems are assignment type problems with quadratic cost functions and occur in many contexts stretching from spatial economics *via* plant and office layout planning to VLSI design and similar problems in high-technology production settings. The presence of nonlinear interaction terms in the objective function makes these, otherwise simple, problems $\mathcal{NP}$ hard. In the first two chapters of this monograph we provide a survey of models of this type and give a common framework for them as *Boolean* quadratic problems with special ordered sets (BQPSs). Special ordered sets associated with these BQPSs are of equal cardinality and either are disjoint as in *clique* partitioning problems, *graph* partitioning problems, class-room scheduling problems, operations-scheduling problems, multi-processor assignment problems and VLSI circuit layout design problems or have intersections with well defined joins as in asymmetric and symmetric Koopmans-Beckmann problems and quadratic assignment problems. Applications of these problems abound in diverse disciplines, such as anthropology, archeology, architecture, chemistry, computer science, economics, electronics, ergonomics, marketing, operations management, political science, statistical physics, zoology, etc. We then give a survey of the traditional solution approaches to BQPSs. It is an unfortunate fact that even after years of investigation into these problems, the state of algorithmic development is nowhere close to solving large-scale real-life problems exactly. In the main part of this book we follow the polyhedral approach to combinatorial problem solving because of the dramatic algorithmic successes of researchers who have pursued this approach. In particular, we define and utilize in Chapters 4 and 5 the concept of a "locally ideal" linearization to obtain improved linear programming formulations of these problems. A locally ideal linearization is a linearization that yields an ideal, i.e., minimal and complete, linear description of each pair or certain sets of pairs of variables in the quadratic interaction terms of the objective function. In a way, using this concept of formulating BQPSs is analogous to investigating thoroughly a few threads of a cobweb as a starting point for a full-fledged study of the entire cobweb. In Chapter 6 we compare alternative formulations of some scheduling problems analytically and give some results on the facial structure of their associated polytopes. Chapter 7 deals with the affine hull and the dimension of quadratic assignment polytopes and their symmetric relatives. Chapter 8 reports some very preliminary computational results.

By comparison to traveling salesman problems and other combinatorial optimization problems where we know a lot about the facial structure of the associated polytopes – knowledge that has been put to use in the actual optimization of large-scale problems – little such operational knowledge has been accumulated so far for quadratic assignment problems. We hope that this monograph will help focus interest and provoke more work along polyhedral lines of investigation into the fascinating world of location, scheduling and design problems. We are confident that following this line of work and implementing a proper branch-and-cut algorithm will push the limits of exact computation far beyond the current ones. Due to space and time limitations we have not included a survey about the polyhedral/polytopal methods that we employ in the main part of this book. There are now several texts available where the reader can find the pertaining material covered in detail. In particular, any unexplained terminology can be found in Chapters 7 and 10 of M. Padberg's *Linear Optimization and Extensions* (Springer-Verlag, Berlin, 1995).

The writing of this monograph has been made possible in part by the financial support that Professor Karla Hoffman of George Mason University and Professor Padberg have received from ONR. We would like to thank Dr. Donald Wagner of the Office of Naval Research for his continued support.

New York City

Manfred Padberg
Minendra P Rijal

CONTENTS

Preface		v
List of Figures		ix
List of Tables		xi
1	**LOCATION PROBLEMS**	1
	1.1 A Modified KB Model	5
	1.2 A Symmetric KB Model	8
	1.3 A Five-City Plant Location Example	14
	1.4 Plant and Office Layout Planning	20
	1.5 Steinberg's Wiring Problem	26
	1.6 The General Quadratic Assignment Problem	31
2	**SCHEDULING AND DESIGN PROBLEMS**	35
	2.1 Traveling Salesman Problems	35
	2.2 Triangulation Problems	36
	2.3 Linear Assignment Problems	38
	2.4 VLSI Circuit Layout Design Problems	39
	2.5 Multi-Processor Assignment Problems	44
	2.6 Scheduling Problems with Interaction Cost	47
	2.7 Operations-Scheduling Problems	50
	2.8 Graph and Clique Partitioning Problems	52
	2.9 Boolean Quadric Problems and Relatives	56
	2.10 A Classification of Boolean Quadratic Problems	57

3	**SOLUTION APPROACHES**	**59**
	3.1 Mixed zero-one formulations of QAPs	61
	3.2 Branch-and-bound algorithms for QAPs	65
	3.3 Traditional cutting plane algorithms	72
	3.4 Heuristic procedures	75
	3.5 Polynomially solvable cases	76
	3.6 Computational experience to date	77
4	**LOCALLY IDEAL LP-FORMULATIONS I**	**79**
	4.1 Graph Partitioning Problems	82
	4.2 Operations Scheduling Problems	88
	4.3 Multi-Processor Assignment Problems	95
5	**LOCALLY IDEAL LP FORMULATIONS II**	**105**
	5.1 VLSI Circuit Layout Design Problems	105
	5.2 A General Model	111
	5.3 Quadratic Assignment Problems	117
	5.4 Symmetric Quadratic Assignment Problems	122
6	**QUADRATIC SCHEDULING PROBLEMS**	**133**
	6.1 Alternative Formulations of the OSP	133
	6.2 Quadratic Scheduling Polytopes	144
7	**QUADRATIC ASSIGNMENT POLYTOPES**	**151**
	7.1 The Affine Hull and Dimension of QAP_n	151
	7.2 Some Valid Inequalities for QAP_n	157
	7.3 The Affine Hull and Dimension of SQP_n	161
8	**SOLVING SMALL QAPs**	**167**
A	**FORTRAN PROGRAMS FOR SMALL SQPs**	**173**
	REFERENCES	**205**
	INDEX	**217**

LIST OF FIGURES

1.1	A 5×5 plant-location assignment example	2
1.2	United States plant-location assignment example	15
1.3	Reduction of **T** in the U.S. example to increase sparsity	17
1.4	Section of the backboard of a Univac Solid-State Computer	27
2.1	A layout of a small condition-code circuit made up completely of standard cells	39
2.2	Example of a sea-of-cells master	40
2.3	A 5×3 circuit layout design example	41
2.4	Cell placement in the sea-of-cells technology	42
2.5	A 5×3 task-processor assignment example	45
2.6	A 5×3 activity-facility assignment example	48
2.7	A 5×3 work-center assignment of operations example	51
2.8	A 6-node graph and its 2-partition	53
2.9	A 6-node complete graph and its 2-partition	55
2.10	A classification of BQPSs	58
4.1	Traditional and locally ideal linearizations of the BQP	80
5.1	The locally ideal linearization of CLDPs	107
5.2	The locally ideal linearization of the general model	113
5.3	The locally ideal linearization of QAPs	118
6.1	The partitioning of the inequalities $(6.10),\ldots,(6.12)$	137
7.1	The matrix **F** used in the proof of Proposition 7.3	163
7.2	Summary of the construction of the proof of Proposition 7.3	165

LIST OF TABLES

1.1	Data for a Koopmans-Beckmann problem with $n = 5$ U.S. cities	16
1.2	The equation system (1.9) of size 44×95 for the U.S. example with $a = 6$	18
1.3	The inequality system (1.11) of size 50×95 for the U.S. example with $a = 6$	19
1.4	Reduction in problem size and LP values for the U.S. example	20
1.5	The 19 facilities, their functions and optimal locations	22
1.6	Distance and flow matrix for 19 facilities	23
1.7	Reduction in problem size and LP values for the hospital layout example	24
1.8	Connection matrix and distance matrix in Manhattan-norm for the wiring problem	29
1.9	Reduction in problem size and LP values for the wiring problem	30
2.1	Data for a circuit layout design problem with $m = 5, n = 3$	43
2.2	Data for a multi-processor problem with $m = 5, n = 3$	46
2.3	Data for a class-room scheduling problem with $m = 5, n = 3$	49
2.4	Data for an operations-scheduling problem with $m = 5, n = 3$	52
4.1	The feasible 0-1 vectors of the local polytope P of GPP	83
4.2	The feasible 0-1 vectors of the local polytope P of OSP	90
4.3	The feasible 0-1 vectors of the local polytope P of MPP	96
5.1	The feasible 0-1 vectors of the local polytope P of SQP	123
7.1	All cut inequalities needed for a complete description of QAP_4	162
8.1	Computational results for super sparse QAPLIB problems	169
8.2	Computational results for some selected QAPLIB problems	170

1
LOCATION PROBLEMS

This monograph analyzes various classes of *Boolean quadratic problems with special ordered set constraints* (BQPSs) in order to develop a practical approach to solving these problems. The BQPS provides a framework of mathematical abstraction for a variety of scheduling, design and assignment problems with a combination of linear assignment and quadratic interaction cost, not necessarily nonnegative, that arise in a wide variety of real-life contexts. We start with a detailed discussion of quadratic assignment problems which appear to have their roots in three separate spheres of scientific interest – in spatial economics which has a long history of its own, see e.g. Weber [1909], and in industrial engineering and computer science, both of which are comparatively young disciplines.

Koopmans and Beckmann [1957] introduced the classical quadratic assignment problem in the context of analyzing the problem of locating economic activities in an exchange economy. The problem of assigning indivisible economic activities to locations is essentially a matching of a set of n economic activities to a set of n locations so as to maximize the benefits of locating the respective economic activities. Given a set $N = \{1, \ldots, n\}$ of economic activities and their possible locations, the assignment of an activity $i \in N$ to a location $j \in N$ accrues a benefit while the interaction between every two activity-location pairs (i, j) and (k, ℓ) for $i \neq k \in N$ and $j \neq \ell \in N$ results in an interaction cost; see Figure 1.1. Koopmans and Beckmann [1957] describe a variation of the plant location problem of maximizing the total assignment benefits net of the interaction cost as an example of the problem of locating economic activities.

The plant location problem represents an idealization of a variety of practical decision problems. The quadratic terms in the cost (revenue) function arise due to circumstances which make the profitability of locating a plant at a certain

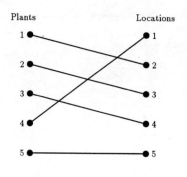
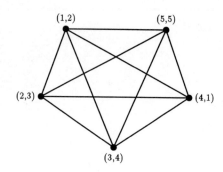

A feasible 5 × 5 plant-location pairing   Edges with quadratic cost of the pairing

Figure 1.1 A 5 × 5 plant-location assignment example

location dependent on the configuration in which the remaining plant-location pairs are matched. A typical example of a "direct" interaction cost is the cost of transportation for the flow of commodities (or bundles of commodities) between plants; more generally, the benefits of improvements in one location that extend to adjacent locations or the detrimental effects of noise, vibration or pollutants stemming from the surrounding plants can also be viewed as the interaction cost of a given set of plant-location matchings. The cost of interplant transportation considered in Koopmans and Beckmann [1957] gives rise to the quadratic terms in the cost function. This interplant transportation cost comprises two components: a location independent amount of flow between plants and a plant assignment independent transportation cost between locations. Defining two $n \times n$ matrices $\mathbf{T} = (t_{ik})$ and $\mathbf{D} = (d_{j\ell})$ where

$$t_{ik} = \text{total amount to be transported from plant } i \text{ to plant } k \text{ and}$$
$$d_{j\ell} = \text{unit transportation cost from location } j \text{ to location } \ell,$$

for $i, k, j, \ell \in N$, the interaction cost of interplant transportation, i.e. the quadratic part of the objective function, are given by $t_{ik} d_{j\ell}$ with $t_{ii} = 0$ and $d_{jj} = 0$ for all $i, j \in N$. On the other hand, the semi-net revenue $-c_{ij}$, the revenue before subtracting the interplant transportation cost that is generated from the operation of a plant $i \in N$ at a given location $j \in N$, gives rise to linear assignment terms in the revenue (cost) function. Note that the matrix $\mathbf{T}$ need not be symmetric. Koopmans and Beckmann [1957] assume that the unit transportation cost satisfy a triangular inequality

$$d_{ij} \leq d_{ik} + d_{kj} \quad \text{for } 1 \leq i, j, k \leq n,$$

which means that transportation from location i to location j via a third location k is at least as expensive as direct transportation. Moreover, it is assumed that flows and distances are nonnegative, i.e. $t_{ik}, d_{j\ell} \geq 0$ for all $1 \leq i, j, k, \ell \leq n$.

Denoting the plant-location pairings by an $n \times n$ matrix $\mathbf{X} = (x_{ij})$ where

$$x_{ij} = \begin{cases} 1 & \text{if economic activity } i \in N \text{ is located at location } j \in N, \\ 0 & \text{otherwise,} \end{cases}$$

the Koopmans-Beckmann location allocation problem (KBP) can be stated as the following zero-one quadratic optimization problem.

$$\min \sum_{i,j \in N} c_{ij} x_{ij} + \sum_{i,k \in N} \sum_{j,\ell \in N} t_{ik} d_{j\ell} x_{ij} x_{k\ell}$$

subject to
$$\sum_{i \in N} x_{ij} = 1 \quad \text{for } j \in N \quad (1.1)$$

$$\sum_{j \in N} x_{ij} = 1 \quad \text{for } i \in N \quad (1.2)$$

$$x_{ij} \in \{0, 1\} \quad \text{for } i, j \in N. \quad (1.3)$$

The equalities (1.1), (1.2), (1.3) model the requirement that each plant is indivisible and has to be matched with exactly one location in the KBP. Denote the set of all feasible exact matchings of these indivisible plants to locations by

$$\mathcal{X}_n = \left\{ \mathbf{X} \in \mathbb{R}^{n \times n} : \mathbf{X} = (x_{ij}) \text{ where } x_{ij} \text{ satisfies } (1.1), (1.2), (1.3) \right\}.$$

The KBP can then be stated, in matrix notation, also as

$$\min \left\{ tr(\mathbf{T}(\mathbf{X}\mathbf{D}\mathbf{X}^T)) : \mathbf{X} \in \mathcal{X}_n \right\},$$

where $tr(\cdot)$ denotes trace of a square matrix, i.e. the sum of its diagonal elements.

Koopmans and Beckmann [1957] formulate this location allocation problem as the following mixed zero-one linear programming problem.

$$\min \sum_{i,j \in N} c_{ij} x_{ij} + \sum_{i,k \in N} \sum_{j,\ell \in N} d_{j\ell} z_{ij}^{k\ell}$$

subject to
$$\mathbf{X} \in \mathcal{X}_n \quad (1.4)$$

$$t_{ik} x_{ij} + \sum_{\ell \in N} z_{i\ell}^{kj} - t_{ik} x_{kj} - \sum_{\ell \in N} z_{ij}^{k\ell} = 0 \quad \text{for } i, j, k \in N \quad (1.5)$$

$$z_{ij}^{k\ell} \geq 0 \quad \text{for } i, j, k, \ell \in N \quad (1.6)$$

$$z_{ij}^{i\ell} = 0 \quad \text{for } i, j, \ell \in N. \quad (1.7)$$

The new variables $z_{ij}^{k\ell}$ correspond to the quadratic terms $t_{ik}x_{ij}x_{k\ell}$ of the objective function of the KBP and model the flow from location j to location ℓ of the commodity supplied by plant i to plant k. The constraints (1.5) express the fact that the production of the commodity supplied by plant i to plant k from location j plus the total inflow of that commodity into location j must equal the consumption of the same commodity plus its total outflow from location j, i.e. these constraints are the usual flow conservation constraints of network theory. The constraints (1.7) express the fact that there is no flow from plant i to itself. We note that

$$z_{ij}^{kj} = 0 \quad \text{for all } i,j,k \in N \qquad (1.7a)$$

holds as well since there is no intralocational transport (case $i = k$) and since no two plants can be at the same location (case $i \neq k$), but these constraints are not stated explicitly in the original article. Besides the nonnegativity conditions on the flow variables, the remaining constraints are the assignment constraints (1.1), (1.2) and (1.3). The correctness of the formulation follows since by the triangular inequality for the transportation cost we do not need to consider any transshipments. Thus for every feasible assignment of plants to locations the remaining flow problem decomposes into n^2 trivial flow problems that assure that each plant i supplies each plant k directly with t_{ik} units of the required commodity. Dropping the variables (1.7), (1.7a) from the formulation it follows that we have n^2 zero-one variables, $n^2(n-1)^2$ flow variables and $n^2(n-1) + 2n$ equations.

Let us now briefly summarize some of the characteristics of *optimal* solutions to this mixed zero-one formulation of the KBP and its straight-forward linear relaxation obtained by relaxing the assumption of indivisibilities of plants, i.e. by replacing the constraint set (1.3) by $0 \leq x_{ij} \leq 1$ for all $1 \leq i,j \leq n$, as detailed in Koopmans and Beckmann [1957]. Assuming that $d_{j\ell} > 0$ for all $1 \leq j \neq \ell \leq n$ and that the semi-net revenue terms c_{ij} are location independent, i.e. $c_{ij} = c_i$ for all $1 \leq i,j \leq n$, an optimal solution to this linear relaxation problem is to distribute each plant in equal fraction $1/n$ over all locations in which case there is no need for transportation, i.e. $z_{ij}^{k\ell} = 0$ for all $i,j,k,\ell \in N$. Moreover, if the flow coefficients t_{ik} for all $1 \leq i \neq k \leq n$ are positive, this fractional solution is the unique optimal solution. If some of the flows are equal to zero then alternate optima exist. Presence of at least one positive flow coefficient t_{ik} for some $1 \leq i \neq k \leq n$ is sufficient to preclude the existence of any integral optimal solution. In contrast, in the absence of the quadratic terms we retrieve the famous *linear assignment* or *marriage* problem (we will have more about this problem in Chapter 2) which always has an integer optimal solution that can be found easily using one of the various network flow algorithms; see Ahuja *et al.* [1993]. In addition, such an integer optimal solution is always

Location Problems

stable, in other words, there is no incentive for any plant owner to relocate his plant in some location other than the one prescribed by the overall optimal integral solution. Thus, this optimal plant assignment is sustainable in an exchange economy governed solely by a market mechanism operating through a profit-maximizing response of each and every plant owner. This is not the case when there are quadratic terms in the cost function; see Koopmans and Beckmann [1957] for a more detailed discussion.

The particular linearization $z_{ij}^{k\ell} = t_{ik} x_{ij} x_{k\ell}$ used by Koopmans and Beckmann [1957] shifts some of the data from the objective function into the constraint set. The resulting problem formulation is data-dependent, it has an interesting interpretation, but it looses the property of having only zero-one variables, since the flow variables of the formulation take on the discrete values of 0 or t_{ik}. To stay in a pure zero-one environment – which has its advantages and disadvantages – we use a different linearization later on because it will permit us to integrate the KBP and various other quadratic zero-one problems into a unifying framework.

1.1 A Modified KB Model

Instead of accepting the historical formulation of the problem at face value, let us play with it and examine different aspects of the underlying real problem. To remove the assumption about the triangularity of transportation cost, which may be unrealistic, we note that the flow conservation constraints (1.5) can be replaced by transportation-type constraints

$$-t_{ik} x_{ij} + \sum_{\ell \in N} z_{ij}^{k\ell} = 0, \quad -t_{ik} x_{kj} + \sum_{\ell \in N} z_{i\ell}^{kj} = 0 \quad \text{for } i \neq k, j \in N. \quad (1.5a)$$

It follows that for every feasible assignment of plants to locations the resulting transportation problem decomposes into n^2 trivial transportation problems and thus we have the correctness of the changed formulation. Indeed, every feasible solution to the changed formulation is feasible for the Koopmans-Beckmann formulation, but not *vice versa*. The changed formulation has n^2 zero-one variables, $n^2(n-1)^2$ flow variables and $2n^2(n-1) + 2n$ equations.

Inspecting the changed formulation we can draw several conclusions. First, we can derive a trivial lower bound on the quadratic part of the objective function of the KBP as follows. Let $d_j = min\{d_{j\ell} : 1 \leq j \neq \ell \leq n\}$ for $1 \leq j \leq n$. Then from the first part of (1.5a) we find that

$$t_{ik}d_{j\ell}x_{ij}x_{k\ell} \geq d_j x_{k\ell} \sum_{h \in N} z_{ij}^{kh}$$

for all $1 \leq i \neq k \leq n$ and $1 \leq j \neq \ell \leq n$. From (1.7) it follows that this inequality holds also for all $1 \leq i = k \leq n$. Moreover, since $t_{ii} = 0$ and $x_{kj}x_{ij} = 0$ for all $k \neq i$, $1 \leq j \leq n$ and $\mathbf{X} \in \mathcal{X}_n$, the inequality holds as well for all $1 \leq j = \ell \leq n$. Consequently, summing over all $i, k, j, \ell \in N$ and using the Koopmans-Beckmann linearization $z_{ij}^{kh} = t_{ik}x_{ij}x_{kh}$ again we find that

$$\sum_{i,k \in N} \sum_{j,\ell \in N} t_{ik}d_{j\ell}x_{ij}x_{k\ell} \geq \sum_{i,j \in N} (d_j \sum_{k \in N} t_{ik})x_{ij}.$$

So the optimal objective function value of the KBP is greater than or equal to

$$min \left\{ \sum_{i,j \in N} (c_{ij} + d_j \sum_{k \in N} t_{ik})x_{ij} : \mathbf{X} \in \mathcal{X}_n \right\}. \qquad (LWB)$$

Thus by solving the linear assignment problem (LWB) we get a lower bound on the KBP. Moreover, if $d_j = d$ for $1 \leq j \leq n$ and $c_{ij} = 0$ for all i and j, then the minimization problem is trivial and its objective function value equals $d \sum_{i,k \in N} t_{ik}$. Surprising as it may seem, (LWB) is sometimes sharp for the linear programming relaxation of the changed formulation; see Chapter 1.5. Second, from (1.5a) and (1.6) we find immediately that

$$z_{ij}^{k\ell} = 0 \quad \text{for all } \ell \neq j \in N \quad \text{if } t_{ik} = 0, \qquad (1.5b)$$

no matter what $i \neq k \in N$. Thus we can drop all corresponding flow variables and constraints from the formulation since we need not fool our computer into believing that these flow variables or constraints exist. Third, suppose that $t_{ik} = t_{ki} \neq 0$ for some $i, k \in N$. From the Koopmans-Beckmann linearization it follows that

$$z_{ij}^{k\ell} = t_{ik}x_{ij}x_{k\ell} = t_{ki}x_{k\ell}x_{ij} = z_{k\ell}^{ij} \quad \text{for all } \ell \neq j \in N \quad \text{if } t_{ik} = t_{ki}, \qquad (1.5c)$$

i.e. the flow between plants i and k is symmetric irrespective of their location. Knowing these identities we will, of course, reduce the necessary number of variables and change the objective function of our model accordingly; but the identities (1.5c) also affect the number of equations (1.5a). For if you look at the constraints (1.5a) for $i < k$ and assume that $t_{ik} = t_{ki}$ where $i, k \in N$ then you find that the constraint pair corresponding to t_{ki} is identical to the one

Location Problems

for t_{ik} when you use the identities (1.5c). Thus for each $j \in N$ we need only one pair of the constraints (1.5a) with $i < k$, say, if $t_{ik} = t_{ki} \neq 0$. Consequently, if

$$a = \text{number of off-diagonal elements } t_{ik} = 0 \text{ and}$$
$$b = \text{number of elements } t_{ik} = t_{ki} \neq 0 \text{ with } i \neq k \in N,$$

it follows that we can formulate the KBP with n^2 zero-one variables, $n^2(n-1)^2 - (a+b)n(n-1)$ flow variables and $2n^2(n-1) - 2(a+b)n + 2n$ equations. So if the matrix $\mathbf{T}$ of interplant shipments is symmetric with $a = 0$, then $b = n(n-1)/2$ and the number of flow variables equals $n^2(n-1)^2/2$, i.e. it is half the original number of flow variables, and the number of equations (1.5a) is reduced to about half the original number, i.e. to $n^2(n-1) + 2n$ equations.

We know from the assignment problem that the rank of the constraint matrix given by (1.2) and (1.3) equals $2n - 1$. So we can expect the rank of the system (1.5a) to be deficient as well. Indeed, adding the first part of (1.5a) for all $j \in N$ and subtracting from it the sum of the second part of (1.5a) over all $j \in N$ as well we create the trivial equation $0 = 0$ where we have used equation (1.3). Consequently, from these elementary rank considerations we find that we can drop additional $b + 1$ equations from the formulation, which now has $2n^2(n-1) - 2n(a+b) - b + 2n - 1$ equations in the symmetric case. Of course, this is only a *preliminary* investigation into the rank of the *required* equation system. We will deal fully with the issue of finding a minimal equation system of full rank in Chapter 7.

If n becomes large, then it can be expected that a substantial number of the interplant shipments t_{ik} equals zero since it is realistic to assume that plants exchange goods with only a small subset of the other plants. Thus the above changes should bring about a substantial reduction in the size of the model. Indeed, by a simple observation one can always create zero elements in the flow matrix $\mathbf{T}$ even if initially there are none. Let $p \in N$, define

$$\alpha_p = min\{t_{ip} : 1 \leq i \neq p \leq n\}, \quad \beta_p = min\{t_{pk} : 1 \leq k \neq p \leq n\},$$

and suppose $\alpha_p > 0$ or $\beta_p > 0$. We define new objective function coefficients

$$c'_{ij} = \begin{cases} c_{pj} + \alpha_p \sum_{\ell \in N} d_{\ell j} + \beta_p \sum_{\ell \in N} d_{j\ell} & \text{for } i = p, 1 \leq j \leq n \\ c_{ij} & \text{otherwise,} \end{cases}$$

$$t'_{ik} = \begin{cases} t_{ip} - \alpha_p & \text{for } k = p, 1 \leq i \neq p \leq n \\ t_{pk} - \beta_p & \text{for } i = p, 1 \leq k \neq p \leq n \\ t_{ik} & \text{otherwise.} \end{cases}$$

It follows from a straight-forward calculation, using $d_{jj} = 0$ for $1 \leq j \leq n$ and the fact that $\mathbf{X} \in \mathcal{X}_n$ implies that $x_{pj} x_{p\ell} = 0$ for any $p \in N$ and $1 \leq j \neq \ell \leq n$, that for all $\mathbf{X} \in \mathcal{X}_n$

$$\sum_{i,j \in N} c'_{ij} x_{ij} + \sum_{i,k \in N} \sum_{j,\ell \in N} t'_{ik} d_{j\ell} x_{ij} x_{k\ell} = \sum_{i,j \in N} c_{ij} x_{ij} + \sum_{i,k \in N} \sum_{j,\ell \in N} t_{ik} d_{j\ell} x_{ij} x_{k\ell}.$$

Now we have created at least one zero element in the flow matrix, we can reapply the reasoning and iterate until the correspondingly recalculated $\alpha_p = \beta_p = 0$ for all $p \in N$, i.e. every row and every column of $\mathbf{T}$ has at least one off-diagonal element equal to zero. Note that if $\mathbf{T}$ is a symmetric matrix then the new flow matrix that results is symmetric as well. In case that $\mathbf{T}$ has symmetric as well as asymmetric elements, then in order to preserve symmetric elements we use $\alpha = min\{\alpha_p, \beta_p\}$ in the updating formulas for c'_{ij} and t'_{ik} instead of α_p and β_p.

The preceding goes by the name of "reduction procedures" in the literature and we will have more on that in Chapter 3. Since the sparsity of the flow matrix $\mathbf{T}$ gives rise to a formulation of the KBP having fewer flow variables and fewer equations (1.5a) we will assume that the matrix $\mathbf{T}$ has been reduced accordingly. We shall call the formulation of the KBP that results from the changes that we have just discussed the *modified Koopmans-Beckmann* formulation.

1.2 A Symmetric KB Model

In the modified formulation of the KBP we have utilized the symmetry of possibly only few elements of the flow matrix $\mathbf{T}$. Let us assume now that both matrices $\mathbf{T}$ and $\mathbf{D}$ are entirely symmetric. Using the symmetry of the elements t_{ik} and $d_{j\ell}$ as well as $t_{ii} = d_{ii} = 0$ for $1 \leq i \leq n$ *you* prove e.g. by induction on $n \geq 2$ that

$$\sum_{i,k \in N} \sum_{j,\ell \in N} t_{ik} d_{j\ell} x_{ij} x_{k\ell} = 2 \sum_{i<k \in N} \sum_{j<\ell \in N} t_{ik} d_{j\ell} (x_{ij} x_{k\ell} + x_{i\ell} x_{kj}).$$

For all $\mathbf{X} \in \mathcal{X}_n$ it follows that $x_{ij} x_{k\ell} + x_{i\ell} x_{kj} \in \{0,1\}$ for all $1 \leq i < k \leq n$ and $1 \leq j < \ell \leq n$. We will use this fact in Chapter 4 when we linearize symmetric quadratic terms in a zero-one framework. At present let us linearize the quadratic terms in the spirit of Koopmans and Beckmann and introduce new variables

$$\xi_{ij}^{k\ell} = t_{ik}(x_{ij} x_{k\ell} + x_{i\ell} x_{kj}) \quad \text{for } 1 \leq i < k \leq n, 1 \leq j < \ell \leq n.$$

Like in the general Koopmans-Beckmann case the symmetric flow variables $\xi_{ij}^{k\ell}$ assume the discrete values of 0 or t_{ik} for every $\mathbf{X} \in \mathcal{X}_n$. Adapting an old "trick" to linearize quadratic zero-one terms, see Padberg [1976], we can write down linear relations as follows.

$$\begin{aligned}
-t_{ik}(x_{ij} + x_{i\ell}) + \xi_{ij}^{k\ell} &\leq 0 \\
-t_{ik}(x_{k\ell} + x_{kj}) + \xi_{ij}^{k\ell} &\leq 0 \\
t_{ik}(x_{ij} + x_{i\ell} + x_{k\ell} + x_{kj}) - \xi_{ij}^{k\ell} &\leq t_{ik} \\
x_{ij}, x_{i\ell}, x_{k\ell}, x_{kj}, \xi_{ij}^{k\ell} &\geq 0.
\end{aligned}$$

This gives $3n^2(n-1)^2/4$ inequalities in $n^2 + n^2(n-1)^2/4$ nonnegative variables. When we intersect this constraint set with the requirement that $\mathbf{X} \in \mathcal{X}_n$ and make the appropriate substitutions in the objective function, we get a mixed zero-one linear program that models the symmetric KBP correctly; see also the introduction to Chapter 4 where we discuss the linearization in the context of zero-one variables in greater detail. Now we calculate

$$\sum_{j=1}^{\ell-1} \xi_{ij}^{k\ell} + \sum_{j=\ell+1}^{n} \xi_{i\ell}^{kj} = t_{ik} \sum_{j=1}^{\ell-1}(x_{ij}x_{k\ell} + x_{i\ell}x_{kj}) + t_{ik}\sum_{j=\ell+1}^{n}(x_{i\ell}x_{kj} + x_{ij}x_{k\ell})$$

$$= t_{ik}\left(x_{k\ell}(\sum_{j=1}^{\ell-1} x_{ij} + \sum_{j=\ell+1}^{n} x_{ij}) + x_{i\ell}(\sum_{j=1}^{\ell-1} x_{kj} + \sum_{j=\ell+1}^{n} x_{kj})\right)$$

$$= t_{ik}(x_{k\ell} + x_{i\ell} - 2x_{i\ell}x_{k\ell}),$$

where we have used that $\mathbf{X} \in \mathcal{X}_n$. But $x_{i\ell}x_{k\ell} = 0$ for all $\mathbf{X} \in \mathcal{X}_n$, $1 \leq i < k \leq n$ and $1 \leq \ell \leq n$. Consequently every feasible solution to the mixed zero-one program satisfies the linear equation that results from dropping the term $2x_{i\ell}x_{k\ell}$. Using the new equations, the equations (1.2) and the nonnegativity of the flow variables we show next that the third set of the $3n^2(n-1)^2/4$ inequalities above is redundant. For let $1 \leq r < s \leq n$ and $1 \leq g < h \leq n$. From (1.2) and the new equations we calculate

$$\begin{aligned}
2t_{rs} &= t_{rs}(\sum_{j=1}^{n} x_{rj} + \sum_{j=1}^{n} x_{sj}) + t_{rs}(x_{rg} + x_{sg}) - \sum_{\ell=1}^{g-1} \xi_{r\ell}^{sg} - \sum_{\ell=g+1}^{n} \xi_{rg}^{s\ell} + t_{rs}(x_{rh} + x_{sh}) \\
&\quad - \sum_{\ell=1}^{h-1} \xi_{r\ell}^{sh} - \sum_{\ell=h+1}^{n} \xi_{rh}^{s\ell} + \sum_{\{g,h\}\neq j=1}^{n}\left(-t_{rs}(x_{rj} + x_{sj}) + \sum_{\ell=1}^{j-1} \xi_{r\ell}^{sj} + \sum_{\ell=j+1}^{n} \xi_{rj}^{s\ell}\right) \\
&= 2\left(t_{rs}(x_{rg} + x_{rh} + x_{sg} + x_{sh}) - \xi_{rg}^{sh} + \sum_{\{g,h\}\neq j=1}^{n-1} \sum_{\{g,h\}\neq \ell=j+1}^{n} \xi_{rj}^{s\ell}\right).
\end{aligned}$$

Consequently, since $\xi_{rj}^{s\ell} \geq 0$ we find that the constraints

$$t_{rs}(x_{rg} + x_{rh} + x_{sg} + x_{sh}) - \xi_{rg}^{sh} \leq t_{rs} \quad \text{for } 1 \leq r < s \leq n, 1 \leq g < h \leq n,$$

are superfluous and can be dropped from the formulation. Like we did above we calculate next

$$\sum_{\substack{i=1 \\ t_{ik} \neq 0}}^{k-1} \frac{1}{t_{ik}} \xi_{ij}^{k\ell} + \sum_{\substack{i=k+1 \\ t_{ik} \neq 0}}^{n} \frac{1}{t_{ik}} \xi_{kj}^{i\ell} = \sum_{\substack{i=1 \\ t_{ik} \neq 0}}^{k-1} (x_{ij}x_{k\ell} + x_{i\ell}x_{kj}) + \sum_{\substack{i=k+1 \\ t_{ik} \neq 0}}^{n} (x_{kj}x_{i\ell} + x_{k\ell}x_{ij})$$

$$\begin{cases} = x_{k\ell} + x_{kj} - 2x_{k\ell}x_{kj} & \text{if } t_{ik} \neq 0 \text{ for all } i \in N - k, \\ \leq x_{k\ell} + x_{kj} & \text{if } t_{ik} = 0 \text{ for some } i \in N - k, \end{cases}$$

where we have used that $\mathbf{X} \in \mathcal{X}_n$ implies $\sum_{\text{some } j} x_{ij} \leq 1$ and the nonnegativity of the x_{ij}. We can drop the quadratic term $2x_{k\ell}x_{kj}$ like we did before. Thus we get linear equations that must be satisfied by every feasible solution to the mixed zero-one program corresponding to $k \in N$ with $t_{ik} \neq 0$ for all $i \in N - k$ (the "dense" columns of the matrix $\mathbf{T}$) and the corresponding less-than-or-equal-to linear inequalities for the "sparse" columns of $\mathbf{T}$, i.e. those columns that have at least one off-diagonal element equal to zero. Using the nonnegativity of the flow variables, the new set of equations/inequalities implies

$$\xi_{ij}^{k\ell} \leq t_{ik}(x_{k\ell} + x_{kj}) \text{ for } i < k \in N, \qquad \xi_{kj}^{i\ell} \leq t_{ik}(x_{k\ell} + x_{kj}) \text{ for } k < i \in N.$$

Consequently the first two sets of the $3n^2(n-1)^2/4$ inequalities are implied by the new equations/inequalities. Thus they can all be dropped from the formulation. Let us denote

$$D = \{k \in N : t_{ik} > 0 \text{ for all } i \in N - k\}, \quad S = \{k \in N : k \notin D\},$$

i.e. D is the index set of all dense columns of the matrix $\mathbf{T}$ and $S = N - D$ its complement in N. Summarizing we get the following mixed zero-one linear program for the *symmetric Koopmans-Beckmann* problem or SKP, for short.

$$\min \sum_{i,j \in N} c_{ij} x_{ij} + 2 \sum_{i<k \in N} \sum_{j<\ell \in N} d_{j\ell} \xi_{ij}^{k\ell}$$

subject to $\qquad \mathbf{X} \in \mathcal{X}_n \qquad (1.8)$

$$-t_{ik}(x_{i\ell} + x_{k\ell}) + \sum_{j=1}^{\ell-1} \xi_{ij}^{k\ell} + \sum_{j=\ell+1}^{n} \xi_{i\ell}^{kj} = 0 \qquad \text{for } i < k \in N, \ell \in N \qquad (1.9)$$

$$-(x_{k\ell} + x_{kj}) + \sum_{i=1}^{k-1} \frac{1}{t_{ik}} \xi_{ij}^{k\ell} + \sum_{i=k+1}^{n} \frac{1}{t_{ik}} \xi_{kj}^{i\ell} = 0 \qquad \text{for } j < \ell \in N \text{ and } k \in D \qquad (1.10)$$

$$-(x_{k\ell} + x_{kj}) + \sum_{\substack{i=1 \\ t_{ik} \neq 0}}^{k-1} \frac{1}{t_{ik}} \xi_{ij}^{k\ell} + \sum_{\substack{i=k+1 \\ t_{ik} \neq 0}}^{n} \frac{1}{t_{ik}} \xi_{kj}^{i\ell} \leq 0 \qquad \text{for } j < \ell \in N \text{ and } k \in S \qquad (1.11)$$

$$\xi_{ij}^{k\ell} \geq 0 \qquad \text{for } i < k \in N, j < \ell \in N. \qquad (1.12)$$

If all nondiagonal elements t_{ik} of **T** are positive, then the formulation of the symmetric KBP has $n^2 + n^2(n-1)^2/4$ nonnegative variables and $2n + n^2(n-1)$ equations.

Like we did above let us now discuss the effect that nondiagonal elements $t_{ik} = 0$ have on the size of the formulation. So if a denotes as before the number of off-diagonal zero elements of the matrix **T**, then from the equations (1.9) of the formulation it follows that $an(n-1)/4$ variables $\xi_{ij}^{k\ell}$ must all equal zero. Thus there is no need to introduce them nor their corresponding equations into the model. Assuming that $\mathbf{D} = \emptyset$, i.e. that the flow matrix **T** is in reduced form, it follows that $n^2 + n^2(n-1)^2/4 - an(n-1)/4$ nonnegative variables, $2n - 1 + n^2(n-1)/2 - an/2$ equations and at most $n^2(n-1)/2$ inequalities (1.11) suffice to model the symmetric KBP correctly. In particular, there are no equations of the type (1.10). The number of inequalities does not bother us; we can generate them "on the fly" as needed by a *dynamic* simplex algorithm, see e.g. Padberg [1995]. Indeed, scrutinizing the derivation of (1.11) we can find more *valid* inequalities since from (1.1) we calculate in fact

$$\sum_{\substack{i=1 \\ t_{ik} \neq 0}}^{k-1} \frac{1}{t_{ik}} \xi_{ij}^{k\ell} + \sum_{\substack{i=k+1 \\ t_{ik} \neq 0}}^{n} \frac{1}{t_{ik}} \xi_{kj}^{i\ell} = x_{kj} + x_{k\ell} - 2x_{kj}x_{k\ell} - x_{kj} \sum_{\substack{i \in N-k \\ t_{ik}=0}} x_{i\ell} - x_{k\ell} \sum_{\substack{i \in N-k \\ t_{ik}=0}} x_{ij}.$$

Consequently using $\mathbf{X} \in \mathcal{X}_n$ again we find that in addition to (1.11) the inequalities

$$\sum_{\substack{i \in N-k \\ t_{ik}=0}} x_{ij} + \sum_{\substack{i=1 \\ t_{ik} \neq 0}}^{k-1} \frac{1}{t_{ik}} \xi_{ij}^{k\ell} + \sum_{\substack{i=k+1 \\ t_{ik} \neq 0}}^{n} \frac{1}{t_{ik}} \xi_{kj}^{i\ell} \leq 1 \quad \text{for } j < \ell \in N \text{ and } k \in S, \quad (1.11a)$$

$$\sum_{\substack{i \in N-k \\ t_{ik}=0}} x_{i\ell} + \sum_{\substack{i=1 \\ t_{ik} \neq 0}}^{k-1} \frac{1}{t_{ik}} \xi_{ij}^{k\ell} + \sum_{\substack{i=k+1 \\ t_{ik} \neq 0}}^{n} \frac{1}{t_{ik}} \xi_{kj}^{i\ell} \leq 1 \quad \text{for } j < \ell \in N \text{ and } k \in S, \quad (1.11b)$$

are satisfied by the feasible solutions of the mixed zero-one program corresponding to the SKP. To *formulate* the SKP the system (1.8),...,(1.12) suffices, but (1.11a) and (1.11b) may be needed for a complete linear description of the *convexification* of the mixed-discrete solution set of the SKP, i.e. for the underlying *polytope* in the space of dimension $n^2 + n^2(n-1)^2/4 - an(n-1)/4$. There are at most $n^2(n-1)$ inequalities (1.11a) and (1.11b) i.e. polynomially many in terms of the parameter n, and thus the inequalities (1.11), (1.11a) and (1.11b) can be *checked* in a reasonable amount of time. Of course, like (1.11) the inequalities (1.11a) and (1.11b) are *not* needed if for some $k \in S$ $t_{ik} = 0$ for all $i \in N$.

Reading the constraints (1.9) carefully we find the following. For each pair (i, k) with $1 \leq i < k \leq n$ and $t_{ik} > 0$ the submatrix or "block," formed by the $n(n-1)/2$ columns corresponding to the flow variables $\xi_{ij}^{k\ell}$ with $1 \leq j < \ell \leq n$ and the n rows corresponding to the terms $-t_{ik}(x_{i\ell} + x_{k\ell})$ for $1 \leq \ell \leq n$, is the incidence matrix of an *undirected complete graph* K_n having n nodes. Moreover, distinct pairs (i, k) with $t_{ik} > 0$ gives rise to blocks that are disjoint in the overall constraint matrix given by (1.9). Since the incidence matrix of K_n has rank n for all $n \geq 3$, we can now calculate the rank of the equation system of symmetric KBPs with $D = \emptyset$. The rank of (1.1) and (1.2) equals $2n - 1$ and the corresponding submatrix is disjoint from the above blocks. There are $n^2 - n - a$ nonzero elements in $\mathbf{T}$ and thus the rank of the entire equation system equals $2n - 1 + n(n^2 - n - a)/2$, i.e. after dropping one of the constraints (1.1) the system of equations of symmetric KBPs with $D = \emptyset$ has full row rank; see also Table 1.2 for an illustration when $n = 5$ and $a = 6$.

The preceding rank consideration has shown that all equations (1.9) are required in the formulation of the SKP. Assuming that $D = \emptyset$, i.e. that the flow matrix has been reduced so that every column contains an off-diagonal zero entry, the question is whether or not there are any additional equations that must be taken into consideration. Equations are important because they determine and are determined by the *dimension* of the set of feasible solutions. As it turns out there are in general more equations required for sparse SKPs. To find more *valid* equations for this problem we use the following identity for $\mathbf{X} \in \mathcal{X}_n$ and $U \subseteq N$ which is readily verified e.g. by induction on $|U| \geq 2$.

$$\sum_{i<k\in U}(x_{ij}x_{k\ell} + x_{i\ell}x_{kj}) = (\sum_{i\in U}x_{ij})(\sum_{k\in U}x_{k\ell}) \text{ for } 1 \leq j < \ell \leq n.$$

Let $U \subseteq N$ be such that $t_{ik} > 0$ for all $i \neq k \in U$ and assume that $N - U$ satisfies $|N - U| \geq 2$ and $t_{ik} > 0$ for all $i \neq k \in N - U$ as well. In other words, we take any partitioning of the set of plants into U and $N - U$ so that every plant exchanges goods with every other plant in U and likewise for all plants in $N - U$. Such a partitioning of N may, of course, not exist. If it does not exist we *conjecture* that for $D = \emptyset$ the equations (1.9) are a minimal and complete description of the *affine hull* of the polytope of the feasible solutions to the SKP. So suppose U and $N - U$ exist. Then we calculate for arbitrary $1 \leq j < \ell \leq n$ as follows.

$$-\sum_{i<k\in U}\frac{1}{t_{ik}}\xi_{ij}^{k\ell} + \sum_{i<k\in N-U}\frac{1}{t_{ik}}\xi_{ij}^{k\ell}$$
$$= -\sum_{i<k\in U}(x_{ij}x_{k\ell} + x_{i\ell}x_{kj}) + \sum_{i<k\in N-U}(x_{ij}x_{k\ell} + x_{i\ell}x_{kj})$$

Location Problems

$$= -(\sum_{i \in U} x_{ij})(\sum_{k \in U} x_{k\ell}) + (1 - \sum_{i \in U} x_{i\ell})(1 - \sum_{k \in U} x_{kj})$$

$$= 1 - \sum_{i \in U} x_{ij} - \sum_{k \in U} x_{k\ell}.$$

Thus we have for all $U \subseteq N$ that qualify the additional equations

$$\sum_{i \in U} x_{ij} - \sum_{i \in N-U} x_{i\ell} - \sum_{i<k \in U} y_{ij}^{k\ell} + \sum_{i<k \in N-U} y_{ij}^{k\ell} = 0, \qquad (1.11c)$$

for all $1 \leq \ell < j \leq n$ where we have set $y_{ij}^{k\ell} = \frac{1}{t_{ik}} \xi_{ij}^{k\ell}$. Note that (1.11c) is symmetric in U and $N - U$. Consequently, only half the number of all possible equations (1.11c) matters. If the flow matrix **T** is reduced, but relatively "dense", then there are potentially many such additional equations that have to be taken into consideration. The question that ensues is the one of the *minimality* of the system of equations that is necessary to describe the affine hull of the polytope given by the *convex hull* of the feasible solutions to (1.8),...,(1.12). Nothing is known about such a minimal system at present. In Chapter 7 we discuss what we know about the case of a dense matrix **T**. From a numerical problem-solving point-of-view it is desirable, if not imperative, to study the question of the minimality of the equation system since most problems tend to be sparse, unless they are randomly generated. Randomly generated problems are hardly ever representative of what the practitioner of combinatorial optimization needs to solve.

Similarly to what we did to derive (1.11a) and (1.11b) we can derive additional *valid* inequalities for the SKP polytope from the last observations, i.e. new inequalities that all mixed-zero-one solutions to (1.8),...(1.12) must satisfy. Like in the case of the additional equations the question that ensues is simply where to stop and/or to look for new inequalities that truly "matter". To this end one distinguishes between valid inequalities that define *facets* of the SKP polytope and those that do not. Facet-defining inequalities are inequalities that are required in an ideal, i.e. minimal and complete, linear description of the SKP polytope and moreover, such a description is *quasi-unique*. So in principle we know what we have to look for when we wish to describe symmetric Koopmans-Beckmann or related combinatorial optimization problems by the way of linear equations and inequalities. It is perhaps ironic to note the fact that the study of quadratic assignment problems from this polyhedral point-of-view is roughly where pertaining studies of the notorious *traveling salesman* problem were over twenty years ago. You will find any unexplained terminology used in this section in Chapters 7 and 10 of Padberg [1995].

Having obtained the reduced formulation utilizing the sparsity of the flow matrix $\mathbf{T}$, we can scale the remaining flow variables and write the entire program as a pure zero-one programming problem. This is done by introducing new variables $y_{ij}^{k\ell} = (1/t_{ik})\xi_{ij}^{k\ell}$, like we did in (1.11c). As a consequence, the objective function changes and clearing the t_{ik} in (1.9), the elements of the constraint matrix are $0, +1$ or -1. These details are left to the reader.

Every feasible zero-one solution to symmetric KBPs has exactly $n + (n^2 - n - a)/2$ variables equal to one. From the rank consideration it thus follows that we have at least $n!$ highly degenerate *bases* for the relaxed linear program. Massive *primal* degeneracy can cause problems for most simplex-based computer software. In addition, many of the cost coefficients of the objective function are equal in value and due to the structure of the constraint set we can expect a high degree of *dual* degeneracy as well. One kind of degeneracy of a linear program can usually be dealt with by solving e.g. the associated dual linear program. To have both primal and dual degeneracy in a linear program, frequently, spells unmitigated numerical disaster. It would therefore be naive to expect that large scale KBP-type linear programs can be solved easily by "off-the-shelf" simplex algorithms. Rather – and this is the case with most other difficult combinatorial optimization problems as well – *advanced* pivot strategies and *creative* use of simplex-based software are an absolute necessity for numerical success, unless it so happens that n is fairly small or a very close to its maximum of $n(n-1)$. Alternatively, non-simplex-type algorithms must be utilized for the resolution of the linear programs.

1.3 A Five-City Plant Location Example

We now illustrate by way of a small example how the KBP arises in a real-life situation. We will also illustrate the effect on the size and the "goodness" of the formulations that result from the various formulation devices that we have discussed above. Suppose a company is faced with a decision to open 5 new plants in 5 major cities of the United States: Chicago, Detroit, Houston, Los Angeles and Philadelphia. This simple decision scenario is complicated by the fact that the output of a plant is an input to the production process of another plant. Hence, a certain number of units of the output of a plant located at one of the potential sites has to be transported to another plant located at some other potential site. Such interplant shipments result in cost which depend on both the interacting plants and their locations; see Figure 1.2; these cost are represented as $t_{ik}d_{j\ell}$ in the formulation of the KBP. The cost of

Location Problems

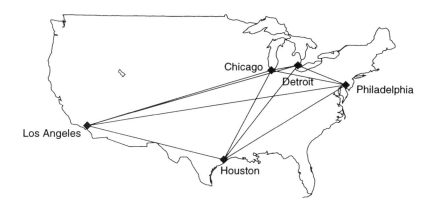

Figure 1.2 United States plant-location assignment example

assigning plants to locations is represented by c_{ij} in the formulation. To keep our framework general, some c_{ij} may be zero or negative. Table 1.1 summarizes the information on how many units t_{ik} of products have to be transported from plant i to plant k, the distance $d_{j\ell}$ between every pair of potential locations and the cost c_{ij} of locating these plants at different potential locations. The entries in the intercity distance table are the actual aerial distances between the cities of Figure 1.2 expressed in units of 100 miles which we take to equal the unit transportation cost. The interplant shipments and the linear assignment cost (third table) have been chosen by us arbitrarily.

The unique optimal solution to this example is to locate plant 1 in Detroit, plant 2 in Chicago, plant 3 in Philadelphia, plant 4 in Los Angeles and plant 5 in Houston with a total cost of 1,812. You can verify this by enumerating all $5! = 120$ assignments of plants to locations that are possible for $n = 5$, by evaluating their cost and choosing the minimum cost assignment. Of course, enumeration becomes impossible – even on the fastest computers that will ever be built – if n becomes large, where "large" means – today – about $n = 15$.

The relaxation of Koopmans and Beckmann's mixed zero-one linear programming formulation (1.4), ..., (1.7) gives a linear program with 650 nonnegative variables and 260 equations. As we have discussed above, the optimum linear programming solution to this problem equals $x_{ij} = 0.2$ for $1 \leq i, j \leq 5$ and all flow variables have a value of zero. The corresponding optimum objective

Interplant Shipments					
Plants	1	2	3	4	5
1	0	8	8	4	2
2	8	0	7	6	4
3	8	7	0	0	6
4	4	6	0	0	9
5	2	4	6	9	0

Intercity Distances					
Cities	L	H	D	P	C
L	0	22	32	39	28
H	22	0	18	22	15
D	32	18	0	7	4
P	39	22	7	0	11
C	28	15	4	11	0

	Cities				
Plants	L	H	D	P	C
1	26	44	4	24	54
2	0	0	44	26	28
3	0	134	2	0	10
4	6	28	18	2	134
5	46	0	36	82	0

Table 1.1 Data for a Koopmans-Beckmann problem with $n = 5$ U.S. cities

function value equals 149.6, which is a truly *bad* lower bound on the minimum cost of 1,812 for the mixed zero-one problem.

The relaxation of the changed Koopmans-Beckmann formulation (1.4), (1.5a) and (1.6) gives a linear program with 425 nonnegative variables and 210 equations. Its optimum solution is not integer, indeed it has many "fractional" variables, but its optimum objective function value equals 1,511.6 which is a *far better* lower bound on the true minimum of 1,812 than the previous one. This is not surprising as the replacement of the "aggregated" equations (1.5) by their "disaggregated" form (1.5a) forces many flow variables to become positive.

Observing that the interplant shipments matrix **T** is symmetric and as $a = 2$ zero off-diagonal entries we can write down the modified Koopmans-Beckmann formulation. This would give us a linear program with 205 nonnegative variables and 90 equations. Since the transshipment matrix has dense columns we can apply the reduction procedure described in Chapter 1.1. In Figure 1.3 we show how the reduction procedure transforms the flow matrix **T** and how it changes the linear assignment cost matrix $\mathbf{C} = (c_{ij})$. The encircled elements are used in the reduction and the reduced matrix **T'** has $a = 6$ zero off-diagonal elements. This gives a linear program with only 165 nonnegative variables and 72 equations. Solving the linear program we find a solution with an objective function value of 1,714.0, which is better than the previous one. Note that a substantially *smaller* linear program was sufficient to get this improved result. While for a small problem like this one the problem size reduction may not be impressive, the size of the linear program *does* matter when n grows larger. The

Location Problems

$$T = \begin{pmatrix} 0 & 8 & 8 & 4 & \boxed{2} \\ 8 & 0 & 7 & 6 & 4 \\ 8 & 7 & 0 & 0 & 6 \\ 4 & 6 & 0 & 0 & 9 \\ \boxed{2} & 4 & 6 & 9 & 0 \end{pmatrix} \rightarrow \begin{pmatrix} 0 & 6 & 6 & 2 & 0 \\ 6 & 0 & 7 & 6 & \boxed{4} \\ 6 & 7 & 0 & 0 & 6 \\ 2 & 6 & 0 & 0 & 9 \\ 0 & \boxed{4} & 6 & 9 & 0 \end{pmatrix} \rightarrow \begin{pmatrix} 0 & 2 & 6 & 2 & 0 \\ 2 & 0 & 3 & 2 & 0 \\ 6 & 3 & 0 & 0 & 6 \\ 2 & 2 & 0 & 0 & 9 \\ 0 & 0 & 6 & 9 & 0 \end{pmatrix} = T'$$

$$C = \begin{pmatrix} 26 & 44 & 4 & 24 & 54 \\ 0 & 0 & 44 & 26 & 28 \\ 0 & 134 & 2 & 0 & 10 \\ 6 & 28 & 18 & 2 & 134 \\ 46 & 0 & 36 & 82 & 0 \end{pmatrix} \rightarrow \begin{pmatrix} 510 & 352 & 248 & 340 & 286 \\ 0 & 0 & 44 & 26 & 28 \\ 0 & 134 & 2 & 0 & 10 \\ 6 & 28 & 18 & 2 & 134 \\ 46 & 0 & 36 & 82 & 0 \end{pmatrix} \rightarrow \begin{pmatrix} 510 & 352 & 248 & 340 & 286 \\ 968 & 616 & 532 & 658 & 492 \\ 0 & 134 & 2 & 0 & 10 \\ 6 & 28 & 18 & 2 & 134 \\ 46 & 0 & 36 & 82 & 0 \end{pmatrix} = C'$$

Figure 1.3 Reduction of **T** in the U.S. example to increase sparsity

relaxation of the symmetric KBP (1.8),...,(1.12) using the reduced matrix T' gives – in terms of variables – an even smaller linear program. It has 95 non-negative variables, 44 equations of the type (1.9) and 50 inequalities (1.11). For your convenience, we have displayed the entire constraint matrix in Tables 1.2 and 1.3, except the equations (1.11c) of which there are ten in this case. As you can see the linear program that we wish to solve is highly "structured." Moreover, remember that after *scaling*, see Chapter 1.2, all nonzero entries of the matrix are either one (+) or minus one (−). Solving first the linear program with 95 variables and 44 equations we find an objective function value of 1,700.0. Now 8 inequalities of type (1.11) are violated by the optimum solution to the linear program. We add them to the existing linear program, reoptimize and we get the optimal integer solution with an objective function value of 1,812. Thus our linear programming relaxation has a relative error of 0% in this particular, small instance of the SKP. In Table 1.4 we summarize the reduction in size and the corresponding linear programming solution values.

In Chapter 7.1 we shall give a data-independent formulation for the Koopmans-Beckmann problem with $n^2 + n^2(n-1)^2/2$ nonnegative variables and $2n(n-1)^2 - (n-1)(n-2)$ equations. Moreover, we show that this is a minimal system of equations of full rank. For our example problem with $n = 5$ this gives a linear program with 225 nonnegative variables and 148 equations. Solving this linear program we find a zero-one valued solution with an objective function value of 1,812, i.e the optimal solution to the problem.

In Chapter 7.3 we utilize the symmetry of the data and give a data-independent formulation of the symmetric Koopmans-Beckmann problem having $n^2+n^2(n-1)^2/4$ nonnegative variables and $2n-1+n^2(n-2)$ equations, which we will also show to be a minimal system of equations of full rank for the problem. For our

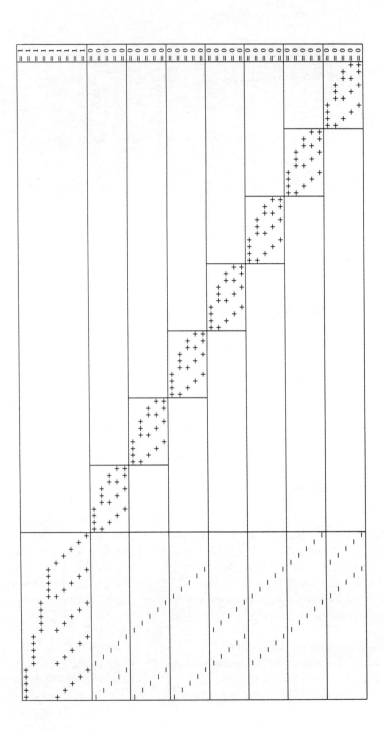

Table 1.2 The equation system (1.9) of size 44 × 95 for the U.S. example with $a = 6$

Location Problems

Table 1.3 The inequality system (1.11) of size 50 × 95 for the U.S. example with $a = 6$

	No of vars	No of equns	Value z_{LP}
Original KBP	650	260	149.6
Changed KBP	425	210	1,511.6
Modified KBP	165	72	1,714.0
Symmetric KBP	95	44 (8)	1,812.0

Table 1.4   Reduction in problem size and LP values for the U.S. example

example problem with $n = 5$ this gives a linear program with 125 nonnegative variables and 84 equations. Solving this linear program you find the optimal zero-one solution to the problem as well.

Since $n = 5$ is very small it is not surprising that small linear programs provide optimal zero-one solutions; for large n *many* more inequalities are needed to assure this outcome. Yet the preceding should have convinced *you* that elementary tricks *and* mathematics can be used to bring the *size* of KBP-type mixed zero-one optimization problems "down" substantially and that the chances of finding optimal solutions are improved dramatically by a thorough analysis of the problem. In Chapters 4-7 we study some of the required additional inequalities for the Koopmans-Beckmann and related problems.

1.4   Plant and Office Layout Planning

Rational factory planning and plant layout was recognized by industrial engineers of the 1940s and 1950s as a topic of immense practical and theoretical interest. Many articles – mostly in the *Journal of Industrial Engineering* – attest to this fact, see e.g. Apple [1950], Armour and Buffa [1963], Buffa [1955], Cameron [1952], Hillier [1963], Hillier and Connors [1966] among others for further historical references. The problem remains of paramount interest for the 1990s *and beyond* as regards the design of automated storage/retrieval systems and mechanized production units as well as the determination of the most functional layout of e.g. private and public office buildings. The general problem here is the location of work centers, storage bins, departments, etc. in relation to each other so as to produce a best layout in terms of material flow, communication flow, accessibility and so forth. We shall illustrate this general problem by a hospital layout problem from the 1970s. In this case several clinics of a public hospital are to be located relative to one another so as to minimize the

Location Problems 21

total distance in meters that its patients must walk to receive treatment in the hospital's clinics.

Alwalid N. Elshafei, who worked at the time at the Institute of National Planning in Cairo (Egypt), describes his problem as follows:

> "... *The hospital concerned (the Ahmed Maher Hospital) is located in a rather densely populated part of Cairo. It is composed of six major departments: Out-patient, In-patient, Dental Research, Accident and Emergency, Physiotherapy and Housekeeping and Maintenance, each department occupying a separate building. In recent years the center of gravity of activity within the hospital has been moving steadily from the wards towards the Out-patient department. As a result, this latter department has been becoming more and more overcrowded with the average daily number of patients now exceeding 700, and with these patients having to move along 17 clinics in the department. The location of the clinics relative to each other has been criticized for causing too much traveling for patients and for causing bottlenecks and serious delays. It was therefore decided to conduct a study aimed at an improvement in the layout of the department leading to a reduction in the total distance traveled by patients and hence in the frequency of bottlenecks and congestions ...*"; see Elshafei [1977].

Like in the Koopmans-Beckmann problem we have thus a number of plants (clinics) and a number of possible locations for them. These locations are at certain distances from each other that can be measured and/or estimated reasonably well. Patients travel between the clinics and their respective numbers constitute the "interplant flows" of the KBP. These flow numbers can be estimated in a representative way by conducting a patient count for each pair of clinics over a reasonable time period, e.g. over a year's time. Thus we have, in principle at least, the same problem as in the KBP: we wish to assign the clinics to locations so as to minimize the total distance in meters travelled by the patients of this hospital per year.

In plant layout planning there is, however, an additional complication. The departments or clinics may have different space requirements in terms of the square meters occupied by them. If this is the case, a "trick" that usually works is to split the bigger departments in "dummy" smaller departments which are all of equal size. By assigning "infinite" flows among the dummy departments that result from splitting a big department, one can usually capture most of the location problem adequately. We note, however, that differences in space requirements certainly deserve further attempts at the modeling level to get

	Facility's Function	Opt Loc		Facility's Function	Opt Loc
1	Receiving and Recording	17	11	X-Ray	10
2	General Practitioner	18	12	Orthopedic	13
3	Pharmacy	19	13	Psychiatric	7
4	Gynecological and Obstretric	11	14	Squint	5
5	Medicine	12	15	Minor Operations	15
6	Paediatric	9	16	Minor Operations	16
7	Surgery	3	17	Dental	8
8	Ear, Nose and Throat	14	18	Dental Surgery	4
9	Urology	1	19	Dental Prosthetic	6
10	Laboratory	2			

Table 1.5   The 19 facilities, their functions and optimal locations

a better formulation of it. However, in the case of Elshafei's hospital layout problem this idea worked and we quote from his paper:

"... The outpatient department is composed of a receiving and recording room, a waiting room and 17 clinics. There is also an administration section, a lecture room, a staff housing facility and stairs between floors. The flow of patients is, however, confined between the receiving and recording room and the 17 clinics, i.e. 18 facilities in total. Thus it was decided to fix the other sections at their original location and investigate the relative location of the 18 facilities. All the facilities needed roughly the same area with the exception of the Minor Operation section which occupied nearly double the space necessary for any other facility. Thus it was split in two pseudo facilities which have to exist beside each other. As a result, the total number of facilities is 19 ..."; see Elshafei [1977].

In Table 1.5 we have reproduced the 19 facilities that result, their respective functions and optimal locations. We are faced now with the problem of determining the data for the problem. Data collection and/or estimation is frequently a hairy problem and it is instructive to see how it was done in this case:

"... Estimates of the patient flows between clinics were available on a yearly basis. Entries in the flow matrix were obtained by averaging the flow between each pair of clinics, thus generating a symmetric matrix. The distances between locations were actually measured by tracing the paths taken by patients while moving from one location to another. Whenever the movement involved a change in floors, the corresponding vertical distance was multiplied by a subjective factor of 3. It was noticed that a patient, after being through a sequence of visits to more than one clinic, must return to the first clinic he visited to mark off his card. In doing so he traces, more or less, the same

Location Problems

	C0001	C0002	C0003	C0004	C0005	C0006	C0007	C0008	C0009	C0010	C0011	C0012	C0013	C0014	C0015	C0016	C0017	C0018	C0019
L0001		76687		415	545	819	135	1368	819	5630		3432	9082	1503			13732	1368	1783
L0002	12		40951	4118	5767	2055	1917	2746	1097	5712				268		1373	268		
L0003	36	24		3848	2524	3213	2072	4225	566			404	9372		972		13538	1368	
L0004	28	75	47		256					829	128								
L0005	52	82	71	42					47	1655	287		42				226		
L0006	44	75	47	34	42				926	161									
L0007	110	108	110	148	125	148			196	1538	196								
L0008	126	70	73	111	136	111	46				301								
L0009	94	124	126	160	102	162	46	69		1954	418								
L0010	63	86	71	52	22	52	136	141	102			282							
L0011	130	93	95	94	73	96	47	63	34	64		1686					226		
L0012	102	106	110	148	125	148	30	46	45	118	47								
L0013	65	58	46	49	32	49	108	119	84	29	56	100							
L0014	98	124	127	117	94	117	51	68	23	95	54	51	77						
L0015	132	161	163	104	130	152	79	121	80	131	94	89	113	79		99999			
L0016	132	161	163	109	130	152	79	121	80	131	94	89	113	79	10				
L0017	126	70	73	111	136	111	46	27	69	141	63	46	119	68	113	113			
L0018	120	64	67	105	130	105	47	24	64	135	46	40	113	62	107	107	6		
L0019	126	70	73	111	136	111	41	36	51	141	24	36	119	51	119	119	24	12	

Table 1.6   Distance and flow matrix for 19 facilities

path he has taken in his forward trip because all the clinics are in the same building and there is only one main corridor per floor. Thus the distance matrix can also be taken to be symmetric even for pairs of locations on two different floors ... The flow between pseudo facilities 15 and 16 is put equal to an extremely large number so as to force them to be in two adjacent locations ..."; see Elshafei [1977].

In Table 1.6 we have reproduced in the upper triangular part the flows between the clinics and in the lower triangular part the distances of the respective locations. Of course, there is no flow from any clinic to itself and the distance from any location to itself equals zero. Thus we can formulate the problem as a symmetric Koopmans-Beckmann problem.

To solve the problem Elshafei [1977] devised, jointly with Mokhtar S. Bazaraa, a heuristic or suboptimal algorithm and found an "acceptable" solution to the quadratic assignment problem for the Ahmed Maher Hospital in reasonable computation time. The solution that the heuristic produced had a total value of 11,281,887 patient meters per year as opposed to the 13,973,298 patient meters per year that the existing layout of the hospital required. Thus a decrease of roughly 19.2% in meters to be walked on an annual basis was achieved, a substantial *expected* gain for the patients of Cairo's hospital.

The question that remained open until 1993 was simply: how "good" was the solution produced by the heuristic algorithm and more importantly, how much more potential was there to improve the walking burden of Ahmed Maher Hospital's patients? Of course, we do not have a floor plan of the hospital and its physical shape today may very well have changed from what it was in the 1970s. An optimal solution to the SKP with the data of Table 1.6 was

	No of vars	No of equns	Value z_{LP}
Original KBP	130,682	13,756	0.0
Changed KBP	117,325	13,034	NA
Modified KBP	19,513	2,109	5,059,178.5
Symmetric KBP	9,937	1,101 (751)	8,138,457.5

Table 1.7   Reduction in problem size and LP values for the hospital layout example

calculated by T. Mautor [1993]. The solution was actually *found* by Bazaraa and Sherali [1980] and to quote from their paper they wrote "... *We also obtained a significant improvement over the best known solution to Elshafei's hospital layout problem* ..." With hindsight – because Mautor showed it in 1993 – their statement was too modest. But it took 13 years to prove that fact, i.e. the optimality of their solution.

It turns out that an optimal assignment of the various clinics to locations produces 8,606,274 patient meters per year which indicates that the improved layout due to Elshafei's "acceptable" solution could itself be improved by roughly 23.7%. In terms of the original situation this means that a reduction of about 38.4% in annual patient meters walked was achievable by a more functional layout of the Ahmed Maher Hospital. Evidently, the patients of this Cairo hospital had a very good reason to complain about the location of its clinics.

In Table 1.7 we show the sizes of the various mixed-integer programming formulations that we have discussed in the previous sections when applied to the data of Table 1.6. Given the sheer size of the original and the changed KBP formulations we did not solve the linear programming relaxation of either problem. Indeed, from our discussion in Chapter 1.1 we know that the optimum solution to the linear programming relaxation of the original KBP formulation equals $x_{ij} = 1/19$ for $1 \leq i,j \leq 19$, all flow variables being equal to zero. Since the linear part of the objective function has all $c_{ij} = 0$, we thus get an optimal objective function value of 0.0 which is the most trivial bound for this problem. The linear programming relaxation of the modified KBP formulation gives an optimal objective function value of 5,059,178.5. We computed it by generating the entire linear program of size 2,109 × 19,513 and solving it directly using the CPLEX routine *dualopt* of CPLEX Optimization Inc, with the steepest-edge pricing option. To do so required about 3 minutes of elapsed CPU time on our computer; see below.

The flow matrix of Table 1.6 has 112 nonzero entries which gives a density of 31% and as you can verify from the table, the flow matrix is already in reduced form. The resulting symmetric KBP has thus 1,101 equations in 9,937 nonnegative variables, of which there are 361 zero-one variables. But there are also the inequalities (1.11) that have to be taken into account, as well as the inequalities (1.11a) and (1.11b) which we can use to improve the lower bound on the quadratic assignment problem. So we wrote a FORTRAN program to solve the associated symmetric KBP including the inequalities (1.11), (1.11a) and (1.11b), but not any of the possible equations (1.11c). To do so required about 7 days of intense work by one of the authors. The program implements the dynamic simplex algorithm, see Padberg [1995], where constraints and variables are both dropped and added dynamically. In Chapter 8 we describe the various components of the computer program in greater detail and a complete listing is contained in Appendix A.

There are 9,747 inequalities (1.11), (1.11a) and (1.11b) to be considered in this case. To solve the various linear programming relaxations we used the program package CPLEX Callable Library of CPLEX Optimization Inc, as subroutines. To optimize the entire linear programming problem took about 45 minutes of elapsed CPU time on a Solaris 2.4 computer running on a single dedicated processor of this machine – which makes our computer comparable to a Sun SPARC workstation 20. Its objective function value equaled 8,138,457.5, which is thus a lower bound on the optimum objective function value of the quadratic assignment problem. To find the lower bound, the biggest linear programming problem ever solved had at most 5,934 variables and 1,852 constraints, i.e. all remaining variables and constraints were checked outside of the LP solver properly speaking.

Our procedure also incorporates a heuristic algorithm and the fixing of certain variables which is mathematically correct using the linear programming reduced cost and a heuristically obtained upper bound. Our heuristic found a best value of 9,806,342 which is about 13% better than Elshafei's solution value. The computation time to find twenty "acceptable" solutions was negligible and took less than 1 second. (Their respective solution values range from 9,806,342 to 13,617,354 with a mean value of 12,084,204.6.) Due to the relatively large gap between the linear programming lower bound and the heuristic upper bound, the program fixed only 391 variables to zero. This left a mixed zero-one problem with 9,546 variables. The missing constraints of the type (1.11) – they are necessary for a *formulation* of the problem – were then added automatically. The resulting problem had 9,546 variables of which 357 must be zero-one and 4,366 equations and/or inequalities. This problem was fed into CPLEX's branch-and-bound routine *mipoptimize* and an optimal solution to it was computed. All calculations were done automatically and to solve this problem from scratch took about two hours of elapsed CPU time on our computer until the program stopped and the optimal solution displayed in Table 1.5 was obtained. The optimal objective function value of the mixed zero-one problem is thus about 5.75% above the optimum value of its linear programming relaxation and the optimum solution to the problem agrees with Mautor's [1993] solution. We note that all numerical cost values of this section have to be multiplied by a factor of two to make them comparable to the value published in an updated version of QAPLIB [1991].

1.5 Steinberg's Wiring Problem

The late 1950s were marked not only by the emergence of *rock'n roll*, but also by the advent of the computer age. Computer production had become commercialized and as a result, engineers working in the computer industry began to pose themselves questions as to how to mechanize the layout of a computer, see e.g. Glaser [1959], Kodres [1959], Loberman and Weinberger [1957] and Steinberg [1961]. Young, hopeful academics – like Paul Gilmore [1962], Donald Knuth [1961] and the late Eugene Lawler [1960, 1963] – also got involved, formulated problems arising in the computer industry and proposed methods for their solution. Computing power was, of course, insufficient in the fifties and the amount of core memory much too limited to permit the optimization of most of the proposed formulations because of sheer problem size. Moreover, suitable algorithms for the resolution of the resulting combinatorial optimization problems were simply not available and, in the rush of things happening, the underlying mathematics of the proposed formulations were frequently not studied in sufficient detail.

Leon Steinberg, who worked for Remington Rand Univac, describes one of these problems – the computer backboard wiring problem – in the following words, see Steinberg [1961]:

> "... *Let us suppose that we are given a set* $E = \{E_1, E_2, \ldots, E_n\}$ *of n [computer] elements and we are told that* E_i *is connected to* E_j *by* t_{ij} *wires. If we set* $t_{ii} = 0$, *we obtain the symmetric connection matrix* $\mathbf{T} = (t_{ij})_{j=1,\ldots,n}^{i=1,\ldots,n}$. *In addition, let r points* $P_1, P_2, \ldots, P_r$ *be given, where* $r \geq n$. *If d is some metric and* $d_{\alpha\beta} = d(P_\alpha, P_\beta)$, *the matrix will also be symmetric, with zeroes down the diagonal* ..."

We have taken the liberty of changing Steinberg's C_{ij}'s to t_{ij}'s. The optimization problem that arises is, of course, the optimal placing of the computer

Location Problems

P01 •	P02 •	P03 •	P04 •	P05 •	P06 •	P07 •	P08 •	P09 •
P10 •	P11 •	P12 •	P13 •	P14 •	P15 •	P16 •	P17 •	P18 •
P19 •	P20 •	P21 •	P22 •	P23 •	P24 •	P25 •	P26 •	P27 •
P28 •	P29 •	P30 •	P31 •	P32 •	P33 •	P34 •	P35 •	P36 •

Figure 1.4 Section of the backboard of a Univac Solid-State Computer

elements on the backboard so as to minimize some weighted measure of the total wire length. Here "length" is the length given by the metric that we choose to work with. This is typically the Euclidean norm or the Manhattan norm. That is, if (x_α, y_α) and (x_β, y_β) are the Cartesian coordinates of the points P_α and P_β in the plane, then

$$d_2(P_\alpha, P_\beta) = \sqrt{(x_\alpha - x_\beta)^2 + (y_\alpha - y_\beta)^2}$$

is the Euclidean distance of P_α and P_β and their Manhattan distance is

$$d_1(P_\alpha, P_\beta) = |x_\alpha - x_\beta| + |y_\alpha - y_\beta|,$$

i.e. $d_2(P_\alpha, P_\beta)$ is the ℓ_2-norm and $d_1(P_\alpha, P_\beta)$ the ℓ_1-norm in $\mathbb{R}^2$. Introducing $r - n$ "fictitious" elements $E_{n+1}, \ldots, E_r$ with no wires running to them or between them, i.e. $t_{ij} = 0$ for $1 \leq i \leq r, n+1 \leq j \leq r$, we get r elements and r positions that have to be paired, where the objective is to minimize the weighted total wire length of the assignment. Suppose that the elements E_i and E_j are assigned to positions $P_{s(i)}$ and $P_{s(j)}$, respectively. Since t_{ij} wires connect E_i and E_j, the required wire length of the connection equals $t_{ij} d(P_{s(i)}, P_{s(j)})$ in the metric d. Thus adding over all $1 \leq i < j \leq n$ we obtain a measure of the required total wire length which we wish to minimize. Evidently, every element E_i (including the fictitious ones) must be assigned to some position P_α and every position P_α must be assigned to some element E_i.

As you must have guessed already, Steinberg's problem is another instance of the symmetric Koopmans-Beckmann problem and thus we know how to formulate the problem as a mixed zero-one linear program. Steinberg, a computer engineer, devised a heuristic algorithm to find an "acceptable" solution to the problem. Of course, he must have been, at the time, quite unaware of the Koopmans-Beckmann problem which had more or less just been published in the journal *Econometrica*, a journal that a computer engineer would have hardly read in those days (and, most probably, would not consult even today).

Rather than giving a contemporary application, we shall illustrate the fundamental usefulness of combinatorial optimization in computer design by the backboard wiring problem from Steinberg's article of 1961. About 35 years have passed since its inception and an optimal solution to this problem – for which we can choose several norms to measure the distances in the objective function – is still elusive today, despite innumerous attempts at its solution.

Figure 1.4 shows a section of the backboard of a modified Univac Solid-State Computer – a computer dinosaur of the fifties that you find today, perhaps, in a museum. The dots $P_1, P_2, \ldots, P_{36}$ indicate the possible positions where the electronic elements must be placed. As you see from the picture, the positions form a regular grid in the plane and any two adjacent dots are at a distance of 1 unit, both vertically and horizontally.

In Table 1.8 we state the upper-triangular part of the connection matrix $\mathbf{T}$ and the lower-triangular part of the distance matrix $\mathbf{D}$ in the Manhattan norm. Thus there are, for instance, 29 wires connecting elements E_4 and E_5 and 316 wires connecting elements E_{11} and E_{12}. There are indeed only 34 elements that have to be placed in 36 possible positions; so E_{35} and E_{36} are "fictitious" elements as discussed above and thus two positions will be empty in every assignment. From the bottom part of Table 1.8 we see that point P_3 is 2 distance units away from P_1, while P_{36} is 11 distance units away from P_1, etc.

While the distance matrix – except for the diagonal elements – is full of nonzero elements, the connection matrix $\mathbf{T}$ is comparatively sparse: there are $36 \times 36 = 1296$ possible entries and only 344 nonzero entries, which gives a density of $\mathbf{T}$ of 26.5%. Indeed, Steinberg [1961] writes "... *an average connection matrix contains over 60 per cent zeroes* ..." We know from our discussion in the previous sections that the density of the matrix $\mathbf{T}$ impacts the problem size of the resulting mixed zero-one optimization problem tremendously. Yet reading the contemporary literature on solution attempts to solve Koopmans-Beckmann problems one gets the feeling that the matrix $\mathbf{T}$ is assumed to be dense – just like the distance matrix – and no one seems to have tried to exploit the sparsity of the matrix $\mathbf{T}$ that was already noted by Steinberg in 1961 in any systematic effort at the formulation stage of the problem. Of course, attempts to utilize sparsity in the design of heuristics for the problem exist.

Table 1.9 shows the number of variables and equations of the four formulations that we have discussed in this chapter when applied to Steinberg's wiring problem. As you can verify from Table 1.8, the flow matrix $\mathbf{T}$ is already in reduced form. Since E_{35} and E_{36} are fictitious elements and since there are no other isolated elements, there are thus 21,420 inequalities (1.11) from the symmetric KBP formulation that have to be taken into account as well.

Table 1.8 Connection matrix and distance matrix in Manhattan-norm for the wiring problem

Formulation	No of vars	No of equns	Value z_{LP}
Original KBP	1,680,912	93,384	0.0
Changed KBP	1,588,896	90,792	NA
Modified KBP	218,016	12,283	5,250.0
Symmetric KBP	109,656	6,263 (6,233)	7,793.96

Table 1.9 Reduction in problem size and LP values for the wiring problem

The solution of the modified KBP took about 181 hours or $7\frac{1}{2}$ days of elapsed CPU time on our computer (see Chapter 1.4). The length of the linear programming calculations is of lesser concern to us than the lack of the goodness of the bound that is obtained. It turns out that the simple lower bound (LWB) of Chapter 1.1 gives precisely the same value of 5,250 which equals the total flow of the problem because $d_j = 1$ for $1 \leq j \leq 36$. LWB can, of course, be computed in a split second.

The solution of the linear programming relaxation of (1.8),..., (1.12) including the automatic generation of 6,233 inequalities (1.11), (1.11a) and (1.11b) produced a lower bound of 7,793.96. In view of the best known solution value of 9,526, see Skorin-Kapov [1990], this can be taken either way: either it is a bad lower bound – which is possible – or the best known solution is not good – which is also possible. This is indeed so because if we assume that the optimal objective function value is 10% above the optimal LP value, then we expect the mixed integer optimum to have an objective function value of about 8,574. On the other hand, 10% *may* be too optimistic.

To calculate the lower bound took roughly one month of elapsed CPU time on our machine. There are several ways to explain the seemingly long duration of the linear programming calculations. One is the slowness of the computer utilized – which is a fact. Far faster machines exist and it was impossible for us to utilize the "parallelization" devices that the Solaris 2.4 computer offers for particularly simply structured FORTRAN programs. We are using only one processor of this machine and at 50MHertz this makes our computer considerably slower than most *laptop* computers presently available. Secondly, our LP solver appeared to have particularly unusual numerical difficulties with the linear programs, especially in the "endgame" of the optimization, i.e. when it was pinpointing down the exact LP optimum. The numerical difficulties may be explained by the fact that the developer did not encounter linear programs similar to our ones in the code development phase – a hypothesis that can be tested by running our problems using e.g. IBM's OSL routines. As OSL was not

Location Problems 31

available to us we could not pursue this avenue. There is another explanation for the unexpected numerical behaviour of the problem. It might just be that the "right" cuts are missing from (1.11), (1.11a), (1.11b), i.e. those facet-defining inequalities for the SKP polytope that move the objective function into the neighborhood of the optimum mixed zero-one objective function value. See Chapter 9.5 of Padberg [1995] for more detail.

In any case, we are confident that – possibly by using LP algorithms other than simplex algorithms – these difficulties can be overcome. The important question concerns the *goodness* of the lower bound. Our calculations have improved the best known lower bound of 7,480, see Chakrapani and Skorin-Kapov [1994], somewhat to 7,794. This bound was obtained through an essentially minimal development effort of only about 7 days after which the computer was set to run. It is clear that much more effort is needed and should be expended to *solve* this interesting riddle posed to combinatorial optimizers well over 35 years ago.

1.6 The General Quadratic Assignment Problem

Lawler [1963] proposed a generalization of the Koopmans-Beckmann-Steinberg problem called the *quadratic assignment* problem (QAP) and stated the problem as follows

$$min\left\{\sum_{i=1}^{n}\sum_{j=1}^{n}\sum_{k=1}^{n}\sum_{\ell=1}^{n} a_{ijk\ell}x_{ij}x_{k\ell} : \mathbf{X} \in \mathcal{X}_n\right\}, \qquad (1.13)$$

where $\mathcal{X}_n$ is as defined before and $a_{ijk\ell}$ are n^4 arbitrary cost coefficients for $i,j,k,\ell \in N$. Because $x_{ij}x_{ij} = x_{ij}$ for all $x_{ij} \in \{0,1\}$ and $1 \leq i,j \leq n$, we can define $c_{ij} = a_{ijij}$ and write the objective function as the sum of a linear part and a quadratic part as in (1.13), but with $a_{ijij} = 0$ for all $1 \leq i,j \leq n$. Since for $\mathbf{X} \in \mathcal{X}_n$ it follows that $x_{ij}x_{kj} = x_{ji}x_{jk} = 0$ for all $1 \leq i \neq k \leq n$ and $1 \leq j \leq n$, the corresponding objective function coefficients are irrelevant. Thus we can assume without loss of generality that the objective function of (1.13) satisfies

$$a_{ijkj} = a_{jijk} = 0 \quad \text{for all } 1 \leq i,k \leq n, 1 \leq j \leq n.$$

Now observe that $x_{ij}x_{k\ell} = x_{k\ell}x_{ij}$. Hence with our conventions we can write the objective function of a quadratic assignment problem as

$$\sum_{i,j \in N} c_{ij}x_{ij} + \sum_{i<k\in N}\sum_{j<\ell\in N} (a_{ijk\ell} + a_{k\ell ij})x_{ij}x_{k\ell},$$

where the $a_{ijk\ell}$ satisfy the stated conditions. To write this in matrix form, denote by $\mathbf{x} \in \mathbb{R}^{n^2}$ vector formed by "stringing" out the rows of the matrix $\mathbf{X} \in \mathcal{X}_n$; i.e. the components of $\mathbf{x}$ are ordered as $(x_{11}, \ldots, x_{1n}, x_{21}, \ldots, x_{2n}, \ldots, x_{n1}, \ldots, x_{nn})$. Let

$$AP_n = \left\{ \mathbf{x} \in \mathbb{R}^{n^2} : \mathbf{x} \text{ satisfies } (1.1), (1.2) \text{ and } (1.3) \right\}. \tag{1.14}$$

Define $\mathbf{Q} \in \mathbb{R}^{n^2 \times n^2}$ to be the upper triangular matrix with zero-diagonal

$$\mathbf{Q} = \begin{pmatrix} \mathbf{O} & \mathbf{Q}_{12} & \mathbf{Q}_{13} & \cdots & \mathbf{Q}_{1n} \\ \mathbf{O} & \mathbf{O} & \mathbf{Q}_{23} & \cdots & \mathbf{Q}_{2n} \\ \vdots & \vdots & \vdots & \ddots & \vdots \\ \mathbf{O} & \mathbf{O} & \mathbf{O} & \cdots & \mathbf{Q}_{n-1,n} \\ \mathbf{O} & \mathbf{O} & \mathbf{O} & \cdots & \mathbf{O} \end{pmatrix}, \tag{1.15}$$

where $\mathbf{O} \in \mathbb{R}^{n \times n}$ consists of zeroes only and $\mathbf{Q}_{ik} \in \mathbb{R}^{n \times n}$ for $1 \le i < k \le n$ is

$$\mathbf{Q}_{ik} = \begin{pmatrix} 0 & a_{i1k2} + a_{k2i1} & \cdots & a_{i1kn} + a_{kni1} \\ a_{i2k1} + a_{k1i2} & 0 & \cdots & a_{i2kn} + a_{kni2} \\ \vdots & \vdots & \ddots & \vdots \\ a_{ink1} + a_{k1in} & a_{ink2} + a_{k2in} & \cdots & 0 \end{pmatrix}.$$

The QAP can then alternatively be stated as follows

$$min\{\mathbf{c}\mathbf{x} + \mathbf{x}^T \mathbf{Q} \mathbf{x} : \mathbf{x} \in AP_n\}, \tag{QAP}$$

where $\mathbf{Q} \in \mathbb{R}^{n^2 \times n^2}$ is of the form (1.15) and $\mathbf{c} \in \mathbb{R}^{n^2}$ is the vector of the c_{ij}'s arranged like $\mathbf{x}$.

Letting $a_{ijk\ell} = t_{ik} d_{j\ell}$ the KBP can also be stated in the form of a QAP. Since in the KBP we assume always that $t_{ii} = d_{ii} = 0$ for all $1 \le i \le n$, we have the above assumptions about the $a_{ijk\ell}$ automatically satisfied. The entry of row j and column ℓ of the matrix $\mathbf{Q}_{ik}$ is in this case given by $t_{ik} d_{j\ell} + t_{ki} d_{\ell j}$ where $1 \le i < k \le n$ and $1 \le j, \ell \le n$.

In the general case of a QAP the submatrices $\mathbf{Q}_{ik}$ of $\mathbf{Q}$ will be asymmetric. Whenever all $\mathbf{Q}_{ik}$ for $1 \le i < k \le n$ are symmetric, we call the resulting problem the *symmetric quadratic assignment* problem or SQP, for short. Symmetry of $\mathbf{Q}_{ik}$ means that $a_{ijk\ell} + a_{k\ell ij} = a_{i\ell kj} + a_{kji\ell}$ for all $1 \le i < k \le n$ and $1 \le j, \ell \le n$. Consequently, if $a_{ijk\ell} = a_{i\ell kj}$ or $a_{ijk\ell} = a_{kji\ell}$ for all $1 \le i, j, k, \ell \le n$ in (1.13), then the QAP is symmetric. In the case of the Koopmans-Beckmann problem, we get a SQP if *either* the interplant shipment matrix **T** *or* the distance matrix **D** is symmetric; see also Chapter 1.2. Like in the case of the KBP it follows that the objective function of the SQP can be written as

$$\sum_{i,j \in N} c_{ij} x_{ij} + \sum_{i<k \in N} \sum_{j<\ell \in N} (a_{ijk\ell} + a_{k\ell ij})(x_{ij} x_{k\ell} + x_{i\ell} x_{kj}). \qquad (1.16)$$

In the SKP we have assumed symmetry of **T** and of **D** and thus $a_{ijk\ell} = a_{k\ell ij}$ follows, which explains the factor of two in the objective function of the SKP.

A wide variety of applications of the QAP and the KBP has been reported in the literature; some of the major applications are:

- in electronics, the backboard wiring problem, the problem of minimizing the "latency" in magnetic drums and the synthesis of sequential switching circuits; see Glaser [1959], Knuth [1961], Kodres [1959], Lawler [1960, 1963], Steinberg [1961];

- in chemistry, the analysis of chemical reactors for organic compounds; see Ugi *et al.* [1979];

- in ergonomics, the design of control panels and typewriter keyboards; see Burkard and Offerman [1977], Land [1963], Pollatschek *et al.* [1976];

- in sports, the ranking of teams in a relay race; see Heffley [1977];

- in architecture, the computer aided design of facility layout; see Elshafei [1977], Krarup and Pruzan [1978];

- in the ranking of archeological data; see Grötschel and Wakabayashi [1989], Opitz and Schader [1984], Tüshaus [1983];

- in the balancing of turbine runners; see Laporte and Mercure [1988], Schlegel [1987];

- in scheduling, the problem of minimizing mean completion time; see Burkard [1990];

- in information retrieval, the optimal ordering of interrelated data on a magnetic tape; see Burkard [1990];

- in contemporary computer manufacturing, the design of computer chips and of very large integrated systems (VLSI design); see Grötschel [1992], Jünger *et al.* [1994], Korte *et al.* [1990], Lengauer [1990], Martin [1992], Weissmantel [1992].

Since its introduction in the late 1950s, a steady stream of literature has flowed on the theory and applications of the QAP and the computation of exact and approximate solutions of it. Many well known combinatorial optimization problems can be modeled as special cases of the QAP. The traveling salesman problems and the triangulation problems are two important examples of the so-called $\mathcal{NP}$-hard problems, see e.g. Garey and Johnson [1979] for definitions of various terms of complexity theory that we employ), which occur as special cases of the QAP; and hence, the QAP itself is $\mathcal{NP}$-hard. Simply put, this means that the existence of a polynomial-time (or technically good) algorithm for the QAP would imply the same for a whole host of other difficult combinatorial optimization problems, i.e. that the class $\mathcal{P}$ of polynomial-time solvable combinatorial problems coincides with the problem class $\mathcal{NP}$ for which only *non-deterministic* polynomial-time methods are known. Most researchers in our field believe that $\mathcal{P} \neq \mathcal{NP}$, but at present this is an article of faith. For the QAP even the problem of finding a feasible solution which is *guaranteed* to approximate the optimal objective function value by some $\varepsilon > 0$ is $\mathcal{NP}$-hard; see Sahni and Gonzales [1976]. Moreover, Dyer *et al.* [1986] show that solving the average case takes exponential time, when the objective function coefficients of QAPs are taken from some simple sample space of random numbers. Thus QAP is by all known measures a truly difficult combinatorial optimization problem.

Allowing only quadratic terms in the cost function may still be a restrictive assumption for a real-life situation. Some of the commodity bundles flowing between pairs of plants might have, for example, one or more commodities in common. A reassignment of plants to locations in such a situation leads to some reshuffling of the flows of intermediate commodities between plants. In addition, the production process can always be adjusted to input availabilities. To capture interactions of an order greater than two, cubic, quartic, ..., or, even n-adic assignment problems may have to be taken into account. They can be modeled using higher-order polynomials; see Padberg and Wilczak [1993] for the linearization of general polynomials in zero-one variables.

2
SCHEDULING AND DESIGN PROBLEMS

In addition to location problems, a truly *amazing* variety of scheduling and design problems has been formulated by numerous professionals in industrial engineering, management science, computer science and the social sciences as Boolean quadratic problems with special ordered set constraints (BQPSs). These include notorious problems such as the traveling salesman problem and seemingly innocuous, but $\mathcal{NP}$-hard optimization problems such as the unconstrained quadratic zero-one optimization problem. In this chapter we collect a representative number of these problems with the aim of classifying them into a schema that will permit us to detect commonalities and differences for further in-depth study of the essential problem classes. Right from the outset, we wish, however, to make clear that we do not advocate the exclusive treatment of every zero-one optimization problem that fits into our framework within the classes of BQPSs that we consider. Additional structural properties of a combinatorial optimization problem – if present – must be exploited fully in order to achieve numerical success and while we subscribe to the often heard maxim "...*as global as possible, as local as necessary* ...", we do it with the right amount of caution.

2.1 Traveling Salesman Problems

Given a set of cities and traveling cost between these cities, the traveling salesman problem (TSP) seeks to find a least cost tour starting from a home-city, visiting each of these cities exactly once and finally returning to the home-city. The TSP can be stated as a special case of the KBP; see Koopmans and Beckmann [1957]. If we define the elements of the matrix $\mathbf{D}$ as the cost of travel between the cities and the matrix **T** to be a fixed cyclic permutation matrix of the following form

$$\mathbf{T} = (t_{ik}) = \begin{pmatrix} 0 & 1 & 0 & \cdots & 0 \\ 0 & 0 & 1 & \cdots & 0 \\ \vdots & \vdots & \vdots & \ddots & \vdots \\ 0 & 0 & 0 & \cdots & 1 \\ 1 & 0 & 0 & \cdots & 0 \end{pmatrix},$$

then the resultant KBP given by $min\{tr(\mathbf{T}(\mathbf{XDX}^T)) : \mathbf{X} \in \mathcal{X}_n\}$ is the TSP. That is, defining AP_n as in (1.14), the TSP can be formulated as

$$min\{\mathbf{x}^T \mathbf{Q} \mathbf{x} : \mathbf{x} \in AP_n\}, \qquad (TSP)$$

where $\mathbf{Q} \in \mathbb{R}^{n^2 \times n^2}$ is an upper triangular matrix partitioned as in (1.15) and

$$\mathbf{Q} = \begin{pmatrix} \mathbf{O} & \mathbf{D} & \mathbf{O} & \cdots & \mathbf{D}^T \\ \mathbf{O} & \mathbf{O} & \mathbf{D} & \cdots & \mathbf{O} \\ \vdots & \vdots & \vdots & \ddots & \vdots \\ \mathbf{O} & \mathbf{O} & \mathbf{O} & \cdots & \mathbf{D} \\ \mathbf{O} & \mathbf{O} & \mathbf{O} & \cdots & \mathbf{O} \end{pmatrix},$$

with $\mathbf{O} \in \mathbb{R}^{n \times n}$ as defined before. Moreover, if the distance matrix **D** is symmetric, then the resultant TSP is a symmetric TSP while an asymmetric distance matrix results in the case of an asymmetric TSP. For a proof that the formulation (TSP) is correct see e.g. Burkard [1990].

The fact that the TSP can be formulated as a KBP is a mathematical curiosity that has had – at least so far – no consequence for the numerical side of problem solving for this problem. Indeed, the study of the TSP in its "natural" formulation due to Dantzig, Fulkerson and Johnson [1954] has progressed to the point where TSPs with 10,000 cities can be optimized today; see Jünger, Reinelt and Rinaldi [1995] for an excellent recent overview.

2.2 Triangulation Problems

Given an $n \times n$ *input-output matrix* of an economy divided into n sectors, the triangulation problem (TP) seeks to permute the rows and columns of this input-output matrix *simultaneously* so as to minimize the sum of the entries above the main diagonal in the permuted matrix; see Leontief [1951], Leontief [1963], Leontief [1966] and e.g. Hoffman and Padberg [1985] for more detail. The TP can also be stated as a special case of the KBP; see Korte and Oberhofer [1968, 1969] and Burkard [1990]. If we define $\mathbf{D}$ as the input-output matrix, i.e. if the $d_{j\ell}$ denote the amount of flow from sector j to sector ℓ of the economy for $1 \leq j, \ell \leq n$, and $\mathbf{T}$ as an upper triangular matrix with $t_{ik} = 1$ if $i < k$, 0 otherwise for $1 \leq i, k \leq n$, then the resultant KBP given by $min\{tr(\mathbf{T}(\mathbf{XDX}^T)) : \mathbf{X} \in \mathcal{X}_n\}$ is the TP. Defining AP_n as in (1.14), the TP can then be formulated as

$$min\{\mathbf{x}^T \mathbf{Q} \mathbf{x} : \mathbf{x} \in AP_n\}, \qquad (TP)$$

where $\mathbf{Q} \in \mathbb{R}^{n^2 \times n^2}$ is an upper triangular matrix partitioned as in (1.15) and

$$\mathbf{Q} = \begin{pmatrix} \mathbf{O} & \mathbf{D} & \mathbf{D} & \cdots & \mathbf{D} \\ \mathbf{O} & \mathbf{O} & \mathbf{D} & \cdots & \mathbf{D} \\ \vdots & \vdots & \vdots & \ddots & \vdots \\ \mathbf{O} & \mathbf{O} & \mathbf{O} & \cdots & \mathbf{D} \\ \mathbf{O} & \mathbf{O} & \mathbf{O} & \cdots & \mathbf{O} \end{pmatrix},$$

with $\mathbf{O} \in \mathbb{R}^{n \times n}$; see also Burkard [1990]. If the input-output matrix $\mathbf{D}$ is symmetric, then interchanging rows and columns simultaneously does not decrease the sum of entries above the main diagonal and all $n!$ permutations are equally good (or bad) in terms of the objective. From an applied point of view, economies are hardly symmetric in this sense and so the problem of finding an optimal triangulation is a real one when $\mathbf{D}$ is not symmetric.

In numerical analysis the same problem arises when one attempts to reorder the rows and columns of a sparse nonsymmetric matrix *simultaneously* so as to produce as few non-zero entries above the main diagonal as possible. To achieve the objective all that has to be done is to replace the non-zero elements of the matrix by ones, whereas the zero elements remain zeros. The related problem of reordering the rows and columns of a sparse nonsymmetric matrix *independently of each other* leads to a similar, but different mixed zero-one formulation.

Like in the case of the travelling salesman problem, the triangulation problem and its relatives can be formulated and studied more directly than via the QAP – which has produced substantially better computational results than what one might expect from the computational record of QAPs to date. Grötschel *et al.* [1984, 1985b] formulate the TP as a *linear ordering* problem (LOP) defined in a *digraph*. A *linear ordering* (or, *permutation*) of a finite set V with $|V| = m$

is a bijective mapping $\sigma : \{1, \ldots, m\} \mapsto V$. Given a complete digraph $D_n = (V, A_n)$ with arc weights d_{ij} for all $(i,j) \in A_n$, a *tournament* is a sub-digraph $D = (V, A)$ of D_n such that for every two nodes u and v it has exactly one arc with endnodes in u and v. A linear ordering of the nodes of D is an arc-set $\{(u,v) : \sigma^{-1}(u) < \sigma^{-1}(v)\}$ that induces an acyclic tournament and *vice versa*. The LOP seeks to find a maximum weight spanning acyclic tournament in the digraph D_n; see also Reinelt [1985] for an excellent treatment of LOP.

The TP, also called *permutation* problem (see Young [1979]), can also be formulated as a *feedback arc set* problem (or, *dicycle covering*) and an *acyclic subgraph* problem as shown by Grötschel *et al.* [1984]. Given a digraph $D = (V, A)$ with arc weights d_{ij} for all $(i,j) \in A$, the acyclic subgraph problem seeks to find an acyclic subdigraph $D' = (V, A')$ of D with $A' \subseteq A$ such that $\sum_{(i,j) \in A'} d_{ij}$ is maximized. Given a digraph $D = (V, A)$ with arc weights d_{ij} for all $(i,j) \in A$, the feedback arc set problem seeks to find an arc set $A' \subseteq A$ such that every dicycle in D contains at least one arc of A' and $\sum_{(i,j) \in A'} d_{ij}$ is minimized. A minimum weight feedback arc set induces a maximum weight acyclic subdigraph and *vice versa*; see also Jünger [1985] for an excellent treatment.

2.3 Linear Assignment Problems

Given two sets of n items and some cost of pairing any two items drawn one each from these two sets, the linear assignment problem (LAP) seeks to find a minimum cost of pairing of these $2n$ items such that every pair consists of an item drawn from each of these two sets. Given a LAP with cost coefficients c_{ij} of pairing an item i from the first set with an item j from the second set for $1 \leq i, j \leq n$, if we redefine the quadratic cost coefficients of the QAP as follows

$$a_{ijkl} = \begin{cases} c_{ij} & \text{if } (i,j) = (k,\ell), \\ 0 & \text{otherwise,} \end{cases}$$

then we obtain the LAP as a special case of the QAP. Of course, to do so is from a computational point of view disadvantageous, because the LAP, also called the *personnel assignment* problem (see Thorndike [1950]), can be solved very efficiently and in polynomial ($\mathcal{O}(n^3)$) time in the worst case; see e.g. Ahuja *et al.* [1993].

Scheduling and Design Problems 39

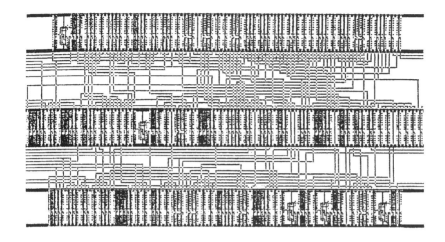

Figure 2.1 A layout of a small condition-code circuit made up completely of standard cells (Source: Lengauer [1990])

2.4 VLSI Circuit Layout Design Problems

In the design of electronic circuits of modern computers, very large scale integration (VLSI) has made it possible that hundreds of thousands of transistors, integrated on few square centimeters of a silicon chip, perform an enormous number of operations at an incredible speed. An electronic circuit is most often described as a *netlist* of a collection of components and their connecting wires. These components may be transistors, *gates* or more complicated subcircuits or *cell blocks* described recursively by the same mechanism. An instance of a cell block is described by the *pins* at which wires connect to it, a name identifying the type of the cell block and a name identifying the cell block instance. The *circuit layout* problem that arises in VLSI design (see Figure 2.1) is the problem of finding an assignment of the geometric co-ordinates of the *netlists* in the plane or in one of a few planar layers such that the requirements of the fabrication technology are met and the associated cost is minimized; see Lengauer [1990], Grötschel [1992], Jünger *et al.* [1994] and Müller [1993] for excellent accounts on this problem.

On the lowest level, the layout is a set of *masks* that guide the fabrication process of the circuit. Different sets of *design rules*, which are much alike in structure, specify the requirements that each mask has to meet in isolation and as a collection of mutually consistent entities. The circuits are usually

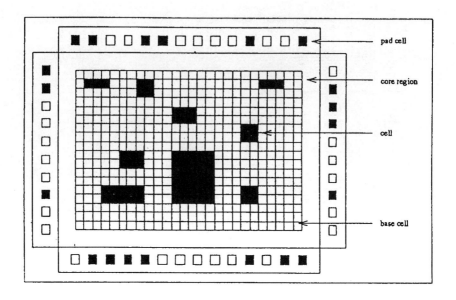

Figure 2.2 Example of a sea-of-cells master (Source: Grötschel [1992])

iso-oriented rectangles but are sometimes polygonal. They are circular only in analog circuitry. However, the circuit layout, today, is not carried out on the mask data level. It is composed *topologically* as a set of rectangular or connected rectangular regions of *grids* connected by wire paths running along the edges of the grid.

Even with the presently available technology, the circuit layout problem cannot be addressed from a total system's point of view. Instead it is carried out in a hierarchical fashion starting with large blocks of circuit components, which are themselves laid out recursively in a similar fashion. Moreover, at each stage in the hierarchy, the process of circuit layout is broken down into subproblems of component *placement* and *routing*, usually with a stage or two of *compaction* in between them. More often than not, the placement does not assign cells to locations on a fixed grid but rather yields a *floorplan*. A floorplan is a tiling of rectangular cells representing the circuit. During the general cell placement phase following the determination of the logic that will perform the full task of a circuit, this logic is cast in silicon, i.e. placed onto the substrate surface, so that certain cost criteria, e.g. the area necessary for wiring, is minimized.

Scheduling and Design Problems

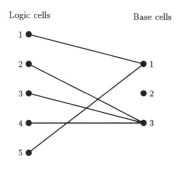
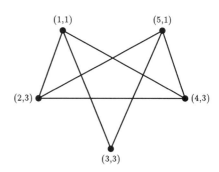

A feasible 5 × 3 logic-base cells pairing   Edges with quadratic cost of the pairing

Figure 2.3 A 5 × 3 circuit layout design example

Since the placement phase uses rough estimates of the necessary wiring area in the cost function, it is beneficial to reiterate the placement as soon as the global routing is done whereby these cost estimates can be refined.

There are two types of layout methodologies: *full-custom* layout and *semi-custom* layout. In full-custom layout, the designer starts with an empty silicon while in semi-custom layout he usually has a prefabricated silicon that already contains all switching elements or *gate arrays*. However, the technology that is currently in wide use falls somewhere in the boundary between full-custom and semi-custom layout. This technology is known as the *sea-of-gates* technology.

In the sea-of-gates layout style, see Figure 2.2 and Figure 2.4, a rectangular *master* chip filled with transistors is given. The layout procedure is carried out to decide whether channels should be routed and if routed, how they should configured. Only a fraction among a large number of transistors can be used since the connection areas of the remaining ones are occupied by wires, thus rendering them unusable. Among the feasible masters, a master, as small as possible, is chosen such that the given circuit can be realized on it. This master consists of a set $N = \{1, \ldots, n\}$ of base cells where a set of logic cells $M = \{1, \ldots, m\}$ with $m \geq n$ are to be assigned such that all logic cells fit without any two logic cells overlapping each other and all nets are routed. The circuit layout problem seeks to accomplish such an assignment with a smallest possible total net length.

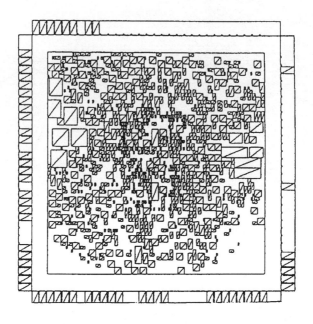

Figure 2.4 Cell placement in the sea-of-cells technology (Source: Grötschel [1992])

Defining

$$x_{ij} = \begin{cases} 1 & \text{if logic cell } i \in M \text{ is assigned to base cell } j \in N, \\ 0 & \text{otherwise,} \end{cases}$$

the VLSI circuit layout design problem (CLDP), ignoring the routing problem, can be formulated as the following zero-one program; see Grötschel [1992].

$$\begin{aligned} \min \quad & \sum_{i,k \in M} \sum_{j \neq \ell \in N} a_{ijk\ell} x_{ij} x_{k\ell} \\ \text{subject to} \quad & \sum_{j \in N} x_{ij} = 1 \quad \text{for } i \in M & (2.1) \\ & x_{ij} \in \{0,1\} \quad \text{for } i \in M, j \in N, & (2.2) \end{aligned}$$

where $a_{ijk\ell} = t_{ik} d_{j\ell} + \lambda o_{ijk\ell}$ for all $i, k \in M$ and $j \neq \ell \in N$. Here t_{ik} denotes the number of nets between logic cells i and k, $d_{j\ell}$ denotes the distance between the base cells j and ℓ, $o_{ijk\ell}$ denotes the number of overlapping base cells, if logic cells i and k are assigned to base cells j and ℓ, and λ is a penalty parameter for such overlaps; see Figure 2.3. The CLDP does not explicitly model the requirement that no two logic cells may overlap each other, but the model *penalizes* such occurrences.

Scheduling and Design Problems 43

No. of nets between L. cells					
L. Cells	1	2	3	4	5
1	0	2	2	1	1
2	2	0	1	2	1
3	2	1	0	1	2
4	1	2	1	0	1
5	1	1	2	1	0

Distance between B. cells			
B. Cells	1	2	3
1	0	1	2
2	1	0	1
3	2	1	0

Table 2.1 Data for a circuit layout design problem with $m = 5, n = 3$

Example. We illustrate the CLDP with a small example where we want to minimize the total wire-length required to assign five logic cells to three available base cells. Table 2.1 summarizes the information on the number t_{ik} of nets between logic cells and the distance $d_{j\ell}$ between base cells. In addition, we assume a penalty for overlap of 10 for each pair of logic cells assigned to the same base cell to formulate this problem as a CLDP. An optimal solution to this example is to assign logic cells 1 and 2 to base cell 1, logic cells 3 and 5 to base cell 2 and logic cell 4 to base cell 3 with a total cost of 44. This problem has four alternative optimal solutions. □

The CLDP is related to the QAP in the sense that the CLDP has quadratic terms in the objective function like the QAP, but it is different from the latter since it has *one* instead of two sets of assignment type constraints. In addition, the CLDP does not have linear terms in the objective function. The CLDP is $\mathcal{NP}$-hard in general; see Grötschel [1992]. Let $\mathbf{Q} \in \mathbb{R}^{mn \times mn}$ be the upper triangular matrix with zero-diagonal given by

$$\mathbf{Q} = \begin{pmatrix} \mathbf{O} & \mathbf{Q}_{12} & \mathbf{Q}_{13} & \cdots & \mathbf{Q}_{1m} \\ \mathbf{O} & \mathbf{O} & \mathbf{Q}_{23} & \cdots & \mathbf{Q}_{2m} \\ \vdots & \vdots & \vdots & \ddots & \vdots \\ \mathbf{O} & \mathbf{O} & \mathbf{O} & \cdots & \mathbf{Q}_{m-1,m} \\ \mathbf{O} & \mathbf{O} & \mathbf{O} & \cdots & \mathbf{O} \end{pmatrix}, \qquad (2.3)$$

where the submatrices $\mathbf{Q}_{ik} \in \mathbb{R}^{n \times n}$ for $1 \leq i < k \leq m$ are

$$\mathbf{Q}_{ik} = \begin{pmatrix} 0 & a_{i1k2} & \cdots & a_{i1kn} \\ a_{i2k1} & 0 & \cdots & a_{i2kn} \\ \vdots & \vdots & \ddots & \vdots \\ a_{ink1} & a_{ink2} & \cdots & 0 \end{pmatrix}.$$

Then the CLDP can alternatively be stated as

$$min\{\mathbf{x}^T\mathbf{Q}\mathbf{x} : \mathbf{x} \text{ satisfies (2.1) and (2.2)}\}. \qquad (CLDP)$$

2.5 Multi-Processor Assignment Problems

The multi-processor assignment problem (MPP) arises as a problem of allocating the tasks of a software system to the processors in a distributed computing environment; see Stone [1977]. In a distributed computing environment, the task modules in a working set of a modular program may be assigned to different processors at load time or/and be allowed to float from one processor to another processor during program-execution. This leads to two types of mutually conflicting cost: interprocessor communication cost and computational cost of the program. Interprocessor communication cost is reduced if all the program modules in a working set are co-resident in a single processor during the execution of the whole working set. Computational cost, on the other hand, is reduced if program modules are assigned to the processors on which they run most efficiently. In a typical multi-processor environment, memory, control and arithmetic capability constitute a processor unit, two or more of which are connected through a data link or high-speed bus. Concurrent execution of different task modules is allowed, while a task can be executed by only one processor at any particular moment. Some modules may have a fixed assignment reflecting the capability of the computing environment while many others are free to float between processors during execution to improve program execution speed. Interprocessor communication cost is very expensive and hence program modules assigned to the same processor are assumed to incur no additional overhead cost of communication.

Given a modular program consisting of a set of tasks $M = \{1, \ldots, m\}$ and a set of processors $N = \{1, \ldots, n\}$ with different processing speeds, the multi-processor assignment problem seeks to minimize the sum of the total task processing and communication time at any given interval. Each task has to be assigned to a processor but each processor can process any number of tasks and typically $m \geq n$. Due to variable speeds of the processors, c_{ij} time units are required to process a task $i \in M$ by a processor $j \in N$. If a task i is assigned to a processor j and a task k is assigned to a processor ℓ for $i, k \in M$ and $j \neq \ell \in N$ a communication time of $t_{ik}d_{j\ell}$ is required where t_{ik} is the number of units of data to be transferred between tasks i and k and $d_{j\ell} = d_{\ell j}$ is the time required to transfer one unit of data between a pair of processors j and ℓ. Moreover, a time $f_{j\ell} = f_{\ell j}$ is required for set up if the processors

Scheduling and Design Problems

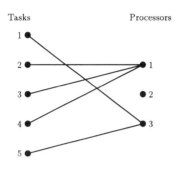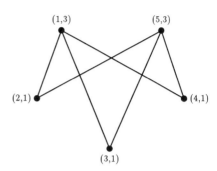

A feasible 5 × 3 task-processor pairing   Edges with quadratic cost of the pairing

Figure 2.5 A 5 × 3 task-processor assignment example

$j \neq \ell \in N$ communicate. The total communication time $a_{ijk\ell}$ is given by $a_{ijk\ell} = t_{ik}d_{j\ell} + f_{j\ell}$ for $i, k \in M$ and $j \neq \ell \in N$, see Figure 2.5 and moreover, $a_{ijk\ell} = a_{i\ell k j}$ for all i, j, k, ℓ.

Defining

$$x_{ij} = \begin{cases} 1 & \text{if task } i \in M \text{ is assigned to processor } j \in N, \\ 0 & \text{otherwise}, \end{cases}$$

the MPP can be formulated as the following zero-one program; see Magirou and Milis [1989].

$$\begin{array}{ll} \min & \sum_{i \in M} \sum_{j \in N} c_{ij} x_{ij} + \sum_{i,k \in M} \sum_{j \neq \ell \in N} a_{ijk\ell} x_{ij} x_{k\ell} \\ \text{subject to} & \sum_{j \in N} x_{ij} = 1 \quad \text{for } i \in M \hfill (2.4) \\ & x_{ij} \in \{0, 1\} \quad \text{for } i \in M, j \in N. \hfill (2.5) \end{array}$$

Example. We illustrate the MPP with a small example where we want to minimize the total communication time required to process a modular program consisting of five tasks on three processors. Table 2.2 summarizes the information on the task/processor speeds c_{ij}, the number of units t_{ik} of data transferred between tasks, the time units $d_{j\ell}$ required to transfer one unit of data between pairs of processors and the set-up time $f_{j\ell}$ if processors communicate. Setting $a_{ijk\ell} = t_{ik}d_{j\ell} + f_{j\ell}$ for $i, k \in M$ and $j \neq \ell \in N$ we formulate this problem as a MPP. The unique optimal solution to this example is to assign task 2 to

Tasks	Processors		
	1	2	3
1	145	82	89
2	20	93	134
3	79	46	169
4	68	117	5
5	123	134	116

Amount of data-transfers					
Tasks	1	2	3	4	5
1	0	2	2	1	2
2	2	0	2	2	1
3	2	2	0	0	3
4	1	2	0	0	1
5	2	1	3	1	0

Transfer time per data-unit			
Processors	1	2	3
1	0	1	2
2	1	0	1
3	2	1	0

Communication set-up time			
Processors	1	2	3
1	0	6	11
2	6	0	7
3	11	7	0

Table 2.2 Data for a multi-processor problem with $m = 5, n = 3$

processor 1, tasks 1, 3 and 5 to processor 2 and task 4 to processor 3 with a total cost of 409. □

Although the MPP has quadratic terms in the objective function and one set of assignment type constraints like the CLDP, it is different from the latter since the quadratic terms in the MPP are symmetric in the sense that $a_{ijk\ell} = a_{i\ell k j}$ for $i, k \in M$ and $j \neq \ell \in N$ in the MPP (while quadratic terms in the CLDP may be asymmetric) and also that the MPP, unlike the CLDP, has linear terms in the objective function. For $n \geq 3$, the MPP can be shown to be equivalent to the *multi-way cut* problem in a graph, see Stone [1977], and hence the MPP is $\mathcal{NP}$-hard in general; see Magirou and Milis [1989].

If the communication cost between a pair of tasks is independent of the processors they are assigned to, i.e. if $a_{ijk\ell} = a_{ik}$ for all $j \neq \ell \in N$, then the minimand of the objective function of the MPP can also be expressed as follows

$$\sum_{i \in M} \sum_{j \in N} c_{ij} x_{ij} + \sum_{i,k \in M} \sum_{j \neq \ell \in N} a_{ijk\ell} x_{ij} x_{k\ell}$$
$$= \sum_{i \in M} \sum_{j \in N} c_{ij} x_{ij} + \sum_{i,k \in M} \sum_{j \neq \ell \in N} a_{ij} x_{ij} x_{k\ell}$$
$$= \sum_{i \in M} \sum_{j \in N} c_{ij} x_{ij} + \sum_{i,k \in M} \sum_{j \in N} a_{ik} x_{ij} (1 - x_{kj})$$
$$= \sum_{i \in M} \sum_{j \in N} c_{ij} x_{ij} + \sum_{i,k \in M} a_{ik} \sum_{j \in N} x_{ij} - \sum_{i,k \in M} \sum_{j \in N} a_{ik} x_{ij} x_{kj}$$
$$= \sum_{i \in M} \sum_{j \in N} c_{ij} x_{ij} + \sum_{i,k \in M} a_{ik} - \sum_{i,k \in M} \sum_{j \in N} a_{ik} x_{ij} x_{kj}.$$

This variation of the MPP is similar to the *graph partitioning* problem described in Chapter 2.8, except that the direction of optimization is reversed, which is, however, immaterial if no sign restrictions are imposed on the objective function coefficients.

Scheduling and Design Problems

Let $\mathbf{Q} \in \mathbb{R}^{mn \times mn}$ be partitioned as in (2.3), $\mathbf{O} \in \mathbb{R}^{n \times n}$ be as defined before and redefine the submatrices $\mathbf{Q}_{ik} \in \mathbb{R}^{n \times n}$ $1 \leq i < k \leq m$ to be

$$\mathbf{Q}_{ik} = \begin{pmatrix} 0 & a_{i1k2} & \cdots & a_{i1kn} \\ a_{i1k2} & 0 & \cdots & a_{i2kn} \\ \vdots & \vdots & \ddots & \vdots \\ a_{i1kn} & a_{i2kn} & \cdots & 0 \end{pmatrix}.$$

The MPP can then alternatively be stated as

$$min\{\mathbf{cx} + \mathbf{x}^T \mathbf{Q}\mathbf{x} : \mathbf{x} \text{ satisfies (2.4) and (2.5)}\}. \qquad (MPP)$$

In accordance with a given situation, the objective function of minimizing the total running time can be appropriately modified. For example, one may wish to minimize the total dollar value of program execution. In this case the intermodular reference cost is measured in dollars per transfer and the processor assignment cost is measured in dollar amounts by taking into account the relative processor speeds and the relative processor cost per computation.

2.6 Scheduling Problems with Interaction Cost

Scheduling of operations to work-centers is a common decision problem faced by operations managers of modern manufacturing and service organizations alike. There exists a rich variety of scheduling problems according to different performance measures. A scheduling problem with particular interaction cost is considered by Carlson and Nemhauser [1966]. This type of problem arises when several activities are competing for the simultaneous use of a limited number of homogeneous facilities. For example, when scheduling courses in a university there may be several courses competing to be scheduled in the same time periods. An "interaction cost" or "cost of conflict" arises when students find two or more desired courses scheduled during the same time period. A course-schedule is feasible if every course is scheduled in exactly one time-period. On the other hand, any number of courses can be scheduled during the same time-period. A course-schedule is optimal if the total cost of conflict is minimal. Since the problem of scheduling activities with interaction cost arises in various contexts besides course-scheduling, we give a general mathematical statement of it.

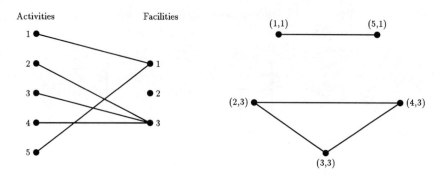

Figure 2.6 A 5 × 3 activity-facility assignment example

Given a set of activities $M = \{1, \ldots, m\}$, a set of facilities $N = \{1, \ldots, n\}$ with $m \geq n$ and corresponding interaction cost a_{ij} define

$$x_{ij} = \begin{cases} 1 & \text{if activity } i \in M \text{ is scheduled in facility } j \in N, \\ 0 & \text{otherwise.} \end{cases}$$

The scheduling problem of minimizing the interaction cost, which we call CSP hereafter, can now be stated as follows; see Figure 2.6.

$$\begin{aligned} \min \quad & \sum_{i,k \in M} \sum_{j \in N} a_{ij} x_{ij} x_{kj} \\ \text{subject to} \quad & \sum_{j \in N} x_{ij} = 1 \quad \text{for } i \in M & (2.6) \\ & x_{ij} \in \{0, 1\} \quad \text{for } i \in M, j \in N. & (2.7) \end{aligned}$$

Example. We illustrate the CSP with a small example where we want to minimize the total quadratic cost of interaction resulting from assigning five activities to three facilities. A pair of activities assigned to the same facility gives rise to a quadratic interaction cost that is independent of the facility where this pair of activities is assigned to. Table 2.3 summarizes the interaction cost a_{ij} between every pair of activities. An optimal solution to this example is to assign activity 1 to facility 1, activities 2 and 5 to facility 2 and activities 3 and 4 to facility 3 with a total cost of 42. There are six alternative optimal solutions to this problem. □

The CSP is related to the MPP in the sense that both of them have quadratic terms in the objective function and one set of assignment type constraints.

Scheduling and Design Problems

Job interaction cost					
Jobs	1	2	3	4	5
1	0	22	32	39	28
2	22	0	18	22	14
3	32	18	0	7	4
4	39	22	7	0	11
5	28	14	4	11	0

Table 2.3 Data for a class-room scheduling problem with $m = 5, n = 3$

However, these two problems are different because the quadratic cost of interaction occurs in the CSP between a pair of jobs assigned to the same machine while only those tasks that are assigned to different processors incur quadratic cost in the MPP. Moreover, the quadratic terms in the CSP are independent of the facility to which we assign an interacting pair of activities. The CSP, unlike the MPP, does not have linear terms in the objective function.

Let $\mathbf{Q} \in \mathbb{R}^{mn \times mn}$ be partitioned as in (2.3), $\mathbf{O} \in \mathbb{R}^{n \times n}$ be defined before and redefine the submatrices $\mathbf{Q}_{ik} \in \mathbb{R}^{n \times n}$ for $1 \leq i < k \leq m$ to be

$$\mathbf{Q}_{ik} = \begin{pmatrix} a_{ik} & 0 & \cdots & 0 \\ 0 & a_{ik} & \cdots & \\ \vdots & \vdots & \ddots & \vdots \\ 0 & 0 & \cdots & a_{ik} \end{pmatrix}.$$

The CSP can then alternatively be stated as

$$min\{\mathbf{x}^T \mathbf{Q} \mathbf{x} : \mathbf{x} \text{ satisfies (2.6) and (2.7)}\}. \quad (CSP)$$

Carlson and Nemhauser [1966] outline a heuristic utilizing Dorn's [1961] results on Lagrangian multipliers to obtain a local minimum of the CSP. The local minimum obtained by their procedure is a global minimum if the cost function is convex, or equivalently, if the matrix $\mathbf{A} = (a_{ik})$ for $i \in M$ and $k \in N$ is positive semidefinite. Since by definition of the problem, the matrix $\mathbf{A}$ is symmetric, nonzero and has $a_{ii} = 0$ for all $i \in M$, the matrix $\mathbf{A}$ is, however, always indefinite; but the objective function value corresponding to the fractional solution obtained by this procedure can be used as a lower bound for the original problem; see Carlson and Nemhauser [1966] for detail. We will show in Chapter 4 that the zero-one formulation of the CSP yields a variation of the *clique partitioning* problem, which we describe later in this chapter. Thus the CSP is an $\mathcal{NP}$-hard problem in general and has a variety of applications:

- in zoology, economics, marketing, political science, anthropology etc., as a clustering problem or as the problem of partitioning a given set of objects into homogeneous disjoint classes, see Grötschel and Wakabayashi [1989],

- in computer science, as a subproblem in VLSI design for the placement of cells and routing of nets in a silicon chip; see Kernighan and Lin [1970].

2.7 Operations-Scheduling Problems

We now consider a class of scheduling problems which generalize the CSP. In the class-room scheduling problem, the interaction cost terms are independent of the work-center to which a pair of activities giving rise to interaction cost is scheduled. However, the interaction cost in the OSP is a function of the interacting pair of activities as well as the work-center where they are scheduled. Moreover, the OSP, unlike the CSP, also has linear assignment cost in the objective function. We call this problem the operations-scheduling problem (OSP) for its applications in the problem of scheduling operations to work-centers. Given a set $M = \{1, 2, \ldots, m\}$ of operations competing to be scheduled in a set $N = \{1, 2, \ldots, n\}$ of work-centers with $|M| \geq |N| \geq 2$, the cost of assigning an operation $i \in M$ to a work-center $j \in N$ gives rise to the linear cost c_{ij} while assigning a pair of operations $i, k \in M$ to the same work-center $j \in N$ gives rise to quadratic interaction cost a_{ikj}. A feasible operations-schedule is an assignment such that each operation is scheduled in exactly one work-center. On the other hand, any number of operations can be scheduled in a work-center; see Figure 2.7.

Defining

$$x_{ij} = \begin{cases} 1 & \text{if operation } i \in M \text{ is scheduled in work-center } j \in N, \\ 0 & \text{otherwise,} \end{cases}$$

the operations-scheduling problem of minimizing the total assignment and interaction cost can be stated as follows

$$\min \quad \sum_{i \in M} \sum_{j \in N} c_{ij} x_{ij} + \sum_{i,k \in M} \sum_{j \in N} a_{ikj} x_{ij} x_{kj}$$
$$\text{subject to} \quad \sum_{j \in N} x_{ij} = 1 \quad \text{for } i \in M \tag{2.8}$$
$$x_{ij} \in \{0, 1\} \quad \text{for } i \in M, j \in N. \tag{2.9}$$

Example. We illustrate the OSP with a small example where we want to minimize the total quadratic cost of interaction resulting from assigning five

Scheduling and Design Problems 51

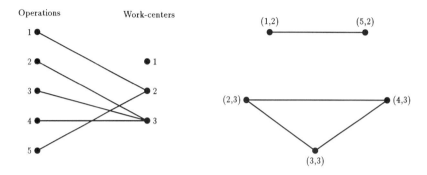

A feasible 5 × 3 operations-work-center pairing   Edges with quadratic cost of the pairing

Figure 2.7 A 5 × 3 work-center assignment of operations example

operation to three work-centers. A pair of operations assigned to the same work-center gives rise to a quadratic interaction cost that is dependent on the work-center where this pair of operations is assigned to. Tables 2.4 summarizes the information on operations processing times c_{ij} and interaction cost a_{ikj} for each of these three work-centers. The unique optimal solution to this example is to assign operations 3 and 4 to work-center 1, operation 1 to work-center 2, operations 2 and 5 to work-center 3 with a total cost of 147. □

Let $\mathbf{Q} \in \mathbb{R}^{mn \times mn}$ be partitioned as in (2.3), $\mathbf{O} \in \mathbb{R}^{n \times n}$ be as before and redefine the submatrices $\mathbf{Q}_{ik} \in \mathbb{R}^{n \times n}$ for $1 \leq i < k \leq m$ to be

$$\mathbf{Q}_{ik} = \begin{pmatrix} a_{ik1} & 0 & \cdots & 0 \\ 0 & a_{ik2} & \cdots & \\ \vdots & \vdots & \ddots & \vdots \\ 0 & 0 & \cdots & a_{ikn} \end{pmatrix}.$$

The OSP can then alternatively be stated as

$$min\{\mathbf{cx} + \mathbf{x}^T \mathbf{Q} \mathbf{x} : \mathbf{x} \text{ satisfies } (2.8) \text{ and } (2.9)\}. \qquad (OSP)$$

The OSP generalizes a number of *combinatorial* optimization problem, e.g. the *graph partitioning* problem, the *clique partitioning* problem, the *max cut* problem and the *Boolean quadric* problem that we describe later in this chapter. Hence, a wide range of the applications of these problems arising as special cases of the OSP are subsumed as the applications of the OSP. Thus, the

Jobs	Machines		
	1	2	3
1	29	17	38
2	14	19	27
3	16	29	14
4	34	23	21
5	38	39	13

Interaction cost (Machine 1)					
Jobs	1	2	3	4	5
1	0	14	24	35	24
2	14	0	25	16	19
3	24	25	0	9	4
4	35	16	9	0	13
5	24	19	4	13	0

Interaction cost (Machine 2)					
Jobs	1	2	3	4	5
1	0	22	32	39	28
2	22	0	18	22	15
3	32	18	0	7	4
4	39	22	7	0	11
5	28	15	4	11	0

Interaction cost (Machine 3)					
Jobs	1	2	3	4	5
1	0	18	30	29	28
2	18	0	7	18	11
3	30	7	0	21	14
4	29	18	21	0	15
5	28	11	14	15	0

Table 2.4 Data for an operations-scheduling problem with $m = 5, n = 3$

OSP is an $\mathcal{NP}$-hard problem in general and represents an idealization of a variety of practical decision problems ranging from clustering problems, i.e. the partitioning of a given set of objects into homogeneous disjoint classes, to electronic circuit layout problems that arise in VLSI design in the context of computer chip manufacturing.

2.8 Graph and Clique Partitioning Problems

A set F of edges in a graph $G = (V, E)$ is called a *n-partitioning* of G if there exists a partition $\{W_1, \ldots, W_n\}$ of the set V of the nodes of G such that $V = W_1 \cup \cdots \cup W_n$, $W_i \cap W_j = \emptyset$ for $1 \le i < k \le n$, $W_i \ne \emptyset$ for $1 \le i \le n$ and $F = \cup_{i=1}^n E(W_i)$, where $E(W_i) = \{e \in E : e \text{ has both endpoints in } W_i\}$. Given a weighted connected graph $G = (V, E)$ with edge weights a_{ij} for all $e = (i, j) \in E$, the graph partitioning problem (GPP) seeks to partition the nodes of G into $n \le m = |V|$ subsets so as to minimize the total weight of the edges with end nodes in two different subsets, i.e. the edges that are cut as a result of the partitioning $\{W_1, \ldots, W_n\}$ of the graph G; see Figure 2.8.

If we require that each partition be such that the subgraph $G[W_i]$ induced by W_i for $1 \le i \le n$ is a *clique*, i.e. a complete (but not necessarily maximal) subgraph of G, then the resultant partitioning is called a *clique partitioning*. The associated optimization problem is the *clique partitioning* problem (CPP).

Scheduling and Design Problems

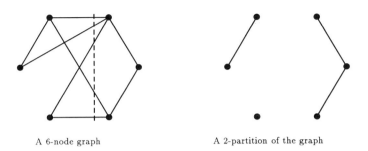

A 6-node graph   A 2-partition of the graph

Figure 2.8 A 6-node graph and its 2-partition

If G is a complete graph, then every partition of the node set of G induces a clique partitioning; see Figure 2.9. Hence, the GPP and the CPP are exactly the same in this case. The clique partitioning problem in a general *sparse* graph can be reduced to that one on a complete graph by assigning edge weights of $-\infty$ to the missing edges of the graph and changing the objective function; see below our discussion of the optimization problem.

The GPP arises in various contexts ranging from clustering of qualitative and quantitative data to VLSI layout design. For example, one important application (Kernighan and Lin [1970]) of the GPP is the placing of components of an electronic circuit onto printed circuit cards or substrates, so as to minimize the number of connections between cards. The objective of minimizing the number of interconnections between cards is justified because connections between cards have high cost when compared to connections within a board. Another application (Kernighan and Lin [1970]) consists of the problem of improving the paging properties of programs for use in computers with paged memory organization. A program is a set of connected entities, such as subroutines, procedure blocks, or single instructions and data items. Possible flow, transfer of control or reference from one entity to another represent the connections between entities. The problem is to assign entities to "pages" of a given size such that the total number of references between the objects lying in different pages is minimized.

The CPP also has a wide range of applications. For example, the so-called problem of *aggregation of binary relations into equivalence relations*, which is basically the clustering problem of finding a "best" partition of a set of given objects into non-overlapping classes of homogeneous objects, can be modeled as the CPP; see Grötschel and Wakabayashi [1989] for details. Other interesting applications of the CPP in a wide range of disciplines are reported in

Barthélemy and Monjardet [1981], Grötschel and Wakabayashi [1989], Marcotorchino and Michaud [1980, 1981a, 1981b], Opitz and Schader [1984], Tüshaus [1983].

Defining
$$x_{ij} = \begin{cases} 1 & \text{if node } i \in V \text{ belongs to set } W_j, \\ 0 & \text{otherwise,} \end{cases}$$
where $n \leq m = |V|$, the problem GPP of partitioning the nodes of V into n classes of nodes achieving the stated objective can be stated mathematically as

$$\begin{aligned} max \quad & \sum_{(i,k) \in E} \sum_{j \in N} a_{ik} x_{ij} x_{kj} \\ subject\ to \quad & \sum_{j \in N} x_{ij} = 1 \quad \text{for } i \in V \quad (2.10) \\ & x_{ij} \in \{0,1\} \quad \text{for } i \in V, j \in N, \quad (2.11) \end{aligned}$$

where $N = \{1, \ldots, n\}$. We note that given any solution x_{ij} to (2.10) and (2.11)
$$W_j = \{i \in V : x_{ij} = 1\}$$
for $1 \leq j \leq n$. It follows that $W_j \neq \emptyset$ and $W_i \cap W_j = \emptyset$ for $1 \leq i < j \leq n$. Consequently, $F = \cup_{i=1}^{n} E(W_i)$ is an n-partitioning of G. On the other hand, it is straightforward to show that every n-partitioning gives rise to a feasible solution to (2.10) and (2.11). Secondly, we note that the objective function accounts for the total weight of all edges with both ends in the sets W_j for $j \in N$ and it is maximized. We calculate using (2.10)

$$\begin{aligned} \sum_{(i,k) \in E} \sum_{j \in N} a_{ik} x_{ij} x_{kj} &= \sum_{(i,k) \in E} \sum_{j \in N} a_{ik} x_{ij} (1 - \sum_{\ell \in V-k} x_{\ell j}) \\ &= \sum_{(i,k) \in E} a_{ik} - \sum_{(i,k) \in E} \sum_{j \in N} a_{ik} x_{ij} \sum_{\ell \in V-k} x_{\ell j}. \end{aligned}$$

Thus the objective function of GPP achieves the minimization of the total weight of all edges that are cut by the partitioning. This follows because for every feasible solution x_{ij} to (2.10) and (2.11) $\sum_{\ell \in V-k} x_{\ell j} \in \{0, 1\}$ and $\sum_{\ell \in V-k} x_{\ell j} = 1$ if and only if $x_{kj} = 0$, i.e. $k \notin W_j$, for any $j \in N$ and $k \in V$.

To find a clique partitioning in a sparse graph $G = (V, E)$ with edge weights a_{ik} for all $(i, k) \in E$, define weights $\hat{a}_{ik} = a_{ik}$ for all $(i, k) \in E$, $\hat{a}_{ik} = -\infty$ otherwise. Let $E^\star$ denote the set of all possible edges on the node set V of G. We replace the weights a_{ik} in the objective function of GPP by $\hat{a}_{ik}$, replace E by $E^\star$ and solve the corresponding problem. If the optimum solution to this problem has an objective function value of $-\infty$, then the clique partitioning problem in G has no feasible solution for the given value of n. Otherwise, let the sets W_j

Scheduling and Design Problems

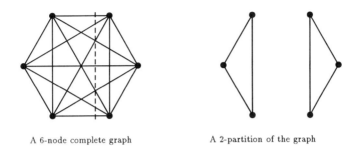

A 6-node complete graph   A 2-partition of the graph

Figure 2.9   A 6-node complete graph and its 2-partition

for $1 \leq j \leq n$ be defined as before from an optimal solution to the problem. It follows that $E(W_j)$ is a clique in G for $1 \leq j \leq n$ and by construction, $F = \cup_{j=1}^{n} E(W_j)$ is a clique-partitioning maximizing the objective function of GPP when the original weights a_{ik} of the sparse graph G are used. But then it follows from the previous reasoning that the clique-partitioning that we have found is optimal. We note for completeness that the assignment of weights of $-\infty$ to the "missing" edges of G corresponds to requiring that $x_{ij}x_{kj} = 0$ for $1 \leq j \leq n$ and all $(i,k) \in E^* - E$. This has implications for the linearization of this particular quadratic programming problem.

Let the nodes of the graph associated with the GPP represent jobs in the CSP, then it follows that the GPP is a generalization of the CSP where a pair of jobs can be assigned to an identical machine only if there is an edge joining the nodes representing these jobs. Hence, the GPP and the CPP over a complete graph are of same general form as the CSP.

Let $\mathbf{Q} \in \mathbb{R}^{mn \times mn}$ be partitioned as in (2.3), $\mathbf{O} \in \mathbb{R}^{n \times n}$ be as defined before and redefine the submatrices $\mathbf{Q}_{ik} \in \mathbb{R}^{n \times n}$ for $1 \leq i < k \leq m$ to be

$$\mathbf{Q}_{ik} = \begin{pmatrix} a_{ik}(I) & 0 & \cdots & 0 \\ 0 & a_{ik}(I) & \cdots & \\ \vdots & \vdots & \ddots & \vdots \\ 0 & 0 & \cdots & a_{ik}(I) \end{pmatrix},$$

where $a_{ik}(I) = a_{ik}$ if $(i,k) \in E$, 0 otherwise. Then the GPP can be stated as

$$max\{\mathbf{x}^T \mathbf{Q} \mathbf{x} : \mathbf{x} \text{ satisfies (2.10) and (2.11)}\}. \quad (GPP)$$

Both the GPP and the CPP are $\mathcal{NP}$-hard in general; see e.g. Garey and Johnson [1979].

2.9 Boolean Quadric Problems and Relatives

Given a set $M = \{1, \ldots, m\}$ and a vector $\mathbf{x} \in \mathbb{R}^m$ with components $x_1, \ldots, x_m$, the unconstrained Boolean quadric problem (BQP) studied by Padberg [1989] is the quadratic zero-one optimization problem

$$\begin{aligned} max \quad & \mathbf{c}\mathbf{x} + \mathbf{x}^T \mathbf{Q} \mathbf{x} \\ subject\ to \quad & x_i \in \{0, 1\} \quad \text{for } 1 \leq i \leq m, \end{aligned} \quad (2.12)$$

where $\mathbf{c} \in \mathbb{R}^m$ is a vector of rational numbers and $\mathbf{Q} \in \mathbb{R}^{m \times m}$ is an upper triangular matrix with zero-diagonal. Many problems arising in network and graph theory, such as the *min cut* problem, the *stable set* (or independent set) problem, etc., have been formulated as BQPs; see e.g. Hammer (Ivănescu) [1965].

A close relative of the BQP is a combinatorial optimization problem called the *equi-partitioning* problem (EQP) and has been studied by Conforti *et al.* [1990]. Given a weighted connected graph $G = (V, E)$ with edge weights a_{ik} for $(i, k) \in E$, the EQP seeks to partition of the node set V into two subsets S and $V - S$ with $|S| = \lfloor |V|/2 \rfloor$ or $|S| = \lceil |V|/2 \rceil$ so as to minimize the total weight of the cut edges with one endpoint in each subset. Like the BQP the EQP is $\mathcal{NP}$-hard in general and arises in the study of the ground state of spin glasses having zero magnetization; see Barahona and Casari [1987]. Defining

$$x_i = \begin{cases} 1 & \text{if node } i \in V \text{ is in set } S, \\ 0 & \text{otherwise,} \end{cases}$$

the EQP is the quadratic zero-one optimization problem

$$min \left\{ \sum_{(i,k) \in E} a_{ik} x_i (1 - x_k) : \sum_{i \in V} x_i = \lfloor |V|/2 \rfloor,\ x_i \in \{0, 1\} \text{ for all } i \in V \right\}.$$

Setting $S = \{i \in V : x_i = 1\}$ for an optimal solution $\mathbf{x} \in \mathbb{R}^v$ to this problem we have the desired equi-partition of G into two "almost equal" halves, where $v = |V|$. Since $\sum_{(i,k) \in E} a_{ik} x_i (1 - x_k) = \sum_{(i,k) \in E} a_{ik} x_i - \sum_{(i,k) \in E} a_{ik} x_i x_k$ we can find a vector $\mathbf{c} \in \mathbb{R}^v$ and an upper triangular matrix $\mathbf{Q} \in \mathbb{R}^{v \times v}$ with zero-diagonal by simply supplying $a_{ik} = 0$ for all edges $(i, k) \notin E$ on the node set V of G. Consequently we can write the equi-partioning problem in the form

$$\begin{aligned} max \quad & \mathbf{c}\mathbf{x} + \mathbf{x}^T \mathbf{Q} \mathbf{x} \\ subject\ to \quad & \sum_{i \in V} x_i = \lfloor |V|/2 \rfloor & (2.13) \\ & x_i \in \{0, 1\} \quad \text{for } 1 \leq i \leq v. & (2.14) \end{aligned}$$

Scheduling and Design Problems

The BQP, a $\mathcal{NP}$-hard problem in general, is equivalent to a combinatorial optimization problem called the *max cut* problem on the complete graph $K_{m+1} = (V', E')$ where $V' = V \cup \{m+1\}$ and $E' = E \cup \{(i, m+1) : i \in V\}$; see Padberg [1989] and Barahona et al. [1989]. Given a weighted complete graph $G = (V', E')$ with edge weights a_e for all $e \in E'$, the max cut problem (MCP) seeks to find a partition of the node set V' into two subsets such that the total weight of the cut edges with one endpoint in each subset is maximized. If all the edge weights are nonpositive (or equivalently, all the edge weights are nonnegative and the direction of the optimality is minimization) and we require that the node set should be partitioned into two nonempty subsets, then this variation of the max cut problem is called *min cut* problem. The min cut problem is polynomially solvable; see e.g. Ahuja et al. [1993]. The BQP or equivalently the MCP arises in a variety of contexts. For example, the problem of determining the partitioning function for the *Ising model* of spin glasses having nonzero magnetization arising in Statistical Physics can be formulated as the BQP; see Barahona and Casari [1987].

The max cut problem is also equivalent to the problem of finding a maximum edge weight bipartite subgraph in a graph and has been studied by Barahona et al. [1985] if all the edge weights are non-negative.

2.10 A Classification of Boolean Quadratic Problems

Given a set $V = \{1, \ldots, v\}$ and a vector $\mathbf{x} \in \mathbb{R}^v$ with components $x_1, \ldots, x_v$, the *constrained Boolean quadratic* problem (BQP$_C$) is the quadratic zero-one optimization problem

$$\min \quad \mathbf{cx} + \mathbf{x}^T \mathbf{Q} \mathbf{x}$$
$$\text{subject to} \quad \sum_{i \in S_j} x_i = b_j \quad \text{for } j = 1, \ldots, k \quad (2.15)$$
$$x_i \in \{0, 1\} \quad \text{for } 1 \leq i \leq v, \quad (2.16)$$

where $S_j \subseteq V$ for $1 \leq j \leq k$ are nonempty subsets of V and $\cup_{j=1}^{k} S_j = V$ for some $k \geq 0$. The BQP is formally the special case of the BQP$_C$ if $k = 0$. The EQP has a single constraint (2.15) with $S_1 = V$ and $\mathbf{b} = b_1 = \lfloor |V|/2 \rfloor$.

We will now unify and schematize all problems presented in the first two chapters, by expressing them as Boolean quadratic problems with special ordered sets constraints (BQPS). The BQPS is the special case of the BQP$_C$ where $b_j = 1$ for $1 \leq j \leq k$ in (2.15). The BQPS is evidently $\mathcal{NP}$-hard in general.

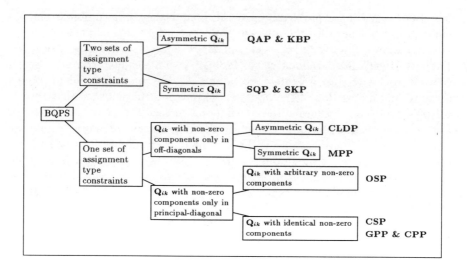

Figure 2.10 A classification of BQPSs

Our classification scheme of BQPSs is based on three characteristics, which we will utilize to derive "locally ideal" linearizations for each one of these problem classes. These three classification parameters are:

(i) number of classes of assignment type constraints, since one set of assignment type constraints leads to a disjoint set of constraints while two sets of assignment type constraints lead to a constraint set with nonempty but well-defined intersections;

(ii) symmetry/asymmetry of the submatrices $\mathbf{Q}_{ik}$ for $1 \leq i < k \leq m$ or $1 \leq i < k \leq n$ in the partitioning (1.15) or (2.3) of $\mathbf{Q}$, as the case may be;

(iii) variability of the diagonal elements of the submatrices $\mathbf{Q}_{ik}$ in case of those problems which have all off-diagonal elements equal to zero.

Figure 2.10 summarizes the membership of all BQPSs, that we have considered so far, according to the various strata in our classification scheme.

3
SOLUTION APPROACHES

The quadratic assignment problem (QAP) has attracted a surpassing *algorithmic* research interest since its introduction in 1957 by Koopmans and Beckmann. A wide variety of algorithms and heuristics have been developed to solve the QAP exactly or approximately. Moreover, since all the problems described in Chapter 2 are closely related to the QAP, one could modify the available exact and approximate techniques for the QAP and utilize them to "solve" every one of these problems. While this is conceptually correct, we do not recommend to solve e.g. traveling salesman problems this way, because the largest size QAP solved to optimality, so far, has $n = 30$; see Clausen [1994], Mans et al. [1992], Pardalos et al. [1994], and Resende et al. [1994]. More to the point, this means that existing algorithms for QAPs are nowhere close to solving practical problems arising from real-life applications to optimality. This state of affairs is unsatisfactory, but not surprising since very little is known about the mathematical properties of QAPs. A straight-forward application of the appropriately modified QAP algorithms to solve its variants can thus not be expected to solve large-scale instances of these problems. While many authors propose (different) mixed zero-one formulations of QAPs, they are hardly exploited in the numerical computations and the *facial structure* of the associated integer polyhedra has not been studied in any detail.

On the other hand, researchers who pursued the polyhedral approach and studied the facial structure of the integer polyhedra associated with combinatorial optimization problems other than the QAP have utilized their results to develop astoundingly successful *polyhedral cutting plane* algorithms. This is the case e.g. for the traveling salesman problem, see Applegate et al. [1994], Grötschel and Padberg [1985], Padberg and Grötschel [1985], Padberg and Rinaldi [1991], the set partitioning problem, see Hoffman and Padberg [1993], Padberg [1973], the linear ordering problem, see Grötschel et al. [1984], the clique partitioning problem, see Grötschel and Wakabayashi [1989], the fixed-charge network problem, see Padberg et al. [1985], Van Roy and Wolsey [1985, 1987], Wolsey [1989], the capacitated network problem, see Araque et al. [1990], etc. In all these cases, the research focused first on developing the mathematical foundations for the respective problems. Computational studies were performed in all cases after the first step was done, i.e. *after* the underlying integer polyhedra were mathematically understood to a sufficient degree. Notable among the computational studies, Padberg and Rinaldi [1991] outline the following key ingredients to a successful application of polyhedral cutting plane algorithms to solve $\mathcal{NP}$-hard problems:

(i) a heuristic procedure to find good feasible solutions,
(ii) efficient *separation* algorithms to find violated inequalities of a partial description of the associated polyhedra,
(iii) a carefully designed interface with the linear programming solver and
(iv) a branching procedure that combines the ideas of branch and bound and polyhedral cutting plane techniques.

This relatively recent approach to combinatorial optimization goes frequently (but not always) by the name of *branch-and-cut*. Using this approach Padberg and Rinaldi [1991] optimize 42 different traveling salesman problems on nodes ranging from 48 to 2,392 cities, which give rise to integer programming problems on up to more than two million variables. A more recent study by Hoffman and Padberg [1993] reports the optimization of 55 pure *set partitioning* problems having up to one million variables and 13 set partitioning problems with base constraints with up to 85,000 variables arising in the real-life context of *airline crew scheduling*. For other successful applications of polyhedral cutting plane methods, see Barahona et al. [1989], Crowder et al. [1983], Grötschel et al. [1992], Van Roy and Wolsey [1987] and others. A substantial body of literature on the facial structure of polytopes associated with some of the problems described in Chapter 2, e.g. the Boolean quadric problem, the max cut problem, the equi-partitioning problem, the graph partitioning problem, already exists and provides leads to the study of the facial structure of BQPSs.

Polyhedral cutting plane methods are robust, versatile and utilize the existing body of knowledge accumulated through research from various perspectives on a given class of problems. In Chapters 4-7 we study the facial structure of several of previously described BQPSs to lay the foundations for a polyhedral cutting plane algorithm to solve reasonably large size practical problem instances of BQPSs. Since it may be possible to utilize some of the key elements of the presently available solution techniques within the framework of a polyhedral cutting plane algorithm for BQPSs, we review some of the current solution approaches to quadratic zero-one problems with assignment type constraints.

A number of both exact and approximate solution techniques to solve QAPs has been reported in the literature. The exact techniques fall into four categories:
(i) enumeration (simple and straight-forward);
(ii) branch-and-bound algorithms; see e.g. Burkard and Derigs [1980], Edwards [1980], Gavett and Plyter [1966], Land [1963], Lawler [1963], Mans et al. [1992, 1993], Nugent et al. [1968], Pardalos and Crouse [1989], Roucairol [1987];
(iii) traditional cutting plane algorithms; see e.g. Balas and Mazzola [1980], Bazaraa and Sherali [1980], Kaufman and Broeckx [1978];
(iv) dynamic programming algorithms; see Christofides and Benavent [1989].

3.1 Mixed zero-one formulations of QAPs

The quadratic assignment problem is a *nonlinear* zero-one optimization problem and as such very little is known about it. While several authors attempt to attack nonlinear integer optimization problems in a nonlinear framework, it is fair to state that these approaches have failed so far to produce any tangible numerical results of significant proportions. Rather the prevailing tendency is to linearize the corresponding nonlinear problem and to cast it as a pure or mixed integer linear optimization problem. Most nonlinear optimization problems in integer variables are tractable and some become treatable this way. In the case of the quadratic assignment problem one introduces new variables

$$y_{ij}^{k\ell} = x_{ij} x_{k\ell} \quad \text{for } i, j, k, \ell \in N.$$

Lawler [1963] proposes the following mixed zero-one formulation of the QAP.

$$\min \sum_{i,j \in N} c_{ij} x_{ij} + \sum_{i,j \in N} \sum_{k,\ell \in N} a_{ij}^{k\ell} y_{ij}^{k\ell}$$

subject to
$$\mathbf{X} \in \mathcal{X}_n \quad (3.1)$$
$$x_{ij} + x_{k\ell} - 2y_{ij}^{k\ell} \geq 0 \quad \text{for } i,j,k,\ell \in N \quad (3.2)$$
$$\sum_{i,j \in N} \sum_{k,\ell \in N} y_{ij}^{k\ell} = n^2 \quad (3.3)$$
$$y_{ij}^{k\ell} \in \{0,1\} \quad \text{for } i,j,k,\ell \in N, \quad (3.4)$$

where $\mathcal{X}_n$ is the set of all $n \times n$ permutation matrices. It is an easy exercise to show that the above formulates the QAP correctly, i.e. if $\mathbf{X} \in \mathcal{X}_n$ then $y_{ij}^{k\ell} = x_{ij} x_{k\ell}$ satisfies the constraints (3.1), ..., (3.4) and vice versa, if $(\mathbf{x}, \mathbf{y})$ satisfies the constraints (3.1), ..., (3.4) then $y_{ij}^{k\ell} = x_{ij} x_{k\ell}$. Since $x_{ij} \in \{0, 1\}$ is part of the constraints of $\mathcal{X}_n$ and $y_{ij}^{ij} = x_{ij}x_{ij} = x_{ij}$ for all $i,j \in N$ one can reduce the number of necessary new variables somewhat. Moreover, $y_{ij}^{k\ell} = x_{ij}x_{k\ell} = x_{k\ell}x_{ij} = y_{k\ell}^{ij}$ and $\mathbf{X} \in \mathcal{X}_n$ implies that $y_{ij}^{kj} = x_{ij}x_{kj} = 0$ and $y_{ji}^{jk} = x_{ji}x_{jk} = 0$ for all $i \neq k \in N$. Consequently, it suffices to introduce $n^2(n-1)^2/2$ new variables in addition to the n^2 variables x_{ij} to formulate the QAP as a mixed zero-one linear programming problem correctly; see also Chapter 1.6 on this point and on how the objective function is affected by the preceding. It follows that $1 + 2n + n^2(n-1)^2/2$ linear constraint in $n^2 + n^2(n-1)^2/2$ zero-one variables suffice to formulate the QAP as a zero-one linear program.

From a *geometric* point of view the formulation (3.1),..., (3.4) is a particulary *bad* formulation: it pays no heed to such things as the linear description of the *affine hull* of the convex hull of the discrete solution set of the QAP nor the proximity of the linear inequalities (3.2) to the facets of the corresponding polytope. Maybe indicative of the common knowledge that (3.1),...,(3.4) is a rather "loose" formulation of the QAP is the fact that we have been unable to track any numerical computation using this formulation. To satisfy our curiosity and to confirm the predictable experimentally, we have generated the corresponding linear program for the five-city plant-location example of Chapter 1.3 (using all n^4 new variables of the original formulation). The lower bound obtained this way is the most trivial bound obtainable, namely zero. Yet the contemporary literature repeats the above formulation and does so without any criticism, see e.g. Burkard [1990], except to note that "...a large additional amount of variables and constraints ..." is needed. If you linearize, a large number of variables is unavoidable and a huge number of constraints may be dictated by the geometry of the problem. Since we have learned how to optimize large scale traveling salesman problems for instance, the sheer number of variables and constraints should hardly impress anybody anymore. It remains to address the underlying mathematics and geometry of the problem.

Rather than attempting to review all formulations of QAPs that have been proposed in the literature – most of them are interrelated anyway – let us consider the following formulation of the QAP in the same set of new variables $y_{ij}^{k\ell}$ introduced above, see Drezner [1995], Frieze and Yadegar [1983], Resende et al. [1994].

$$\min \quad \sum_{i,j \in N} c_{ij} x_{ij} + \sum_{i,j \in N} \sum_{k,\ell \in N} a_{ij}^{k\ell} y_{ij}^{k\ell}$$

$$\text{subject to} \quad \sum_{j \in N} x_{ij} = 1 \quad \text{for } i \in N \quad (3.5)$$

$$\sum_{j \in N} x_{ij} = 1 \quad \text{for } i \in N \quad (3.6)$$

$$\sum_{i=1}^{n} y_{ij}^{k\ell} = x_{k\ell} \quad \text{for } j,k,\ell \in N \quad (3.7)$$

Solution Approaches

$$\sum_{j=1}^{n} y_{ij}^{k\ell} = x_{k\ell} \qquad \text{for } i, k, \ell \in N \qquad (3.8)$$

$$\sum_{k=1}^{n} y_{ij}^{k\ell} = x_{ij} \qquad \text{for } i, j, \ell \in N \qquad (3.9)$$

$$\sum_{\ell=1}^{n} y_{ij}^{k\ell} = x_{ij} \qquad \text{for } i, j, k \in N \qquad (3.10)$$

$$y_{ij}^{ij} = x_{ij} \qquad \text{for } i, j \in N \qquad (3.11)$$

$$y_{ij}^{k\ell} \geq 0 \qquad \text{for } i, j, k, \ell \in N, \qquad (3.12)$$

$$x_{ij} \in \{0, 1\} \qquad \text{for } i, j, k, \ell \in N, \qquad (3.13)$$

To verify the correctness of the formulation for the QAP is left as an exercise for the reader. At first sight we need thus about $n^2 + n^4$ variables and $2n + 4n^3$ equations to formulate the QAP, not counting (3.11), (3.12) and (3.13). Clearly, the new variables $y_{ij}^{k\ell}$ satisfy all of the equations that we have stated and thus instead of Lawler's $2n + 1$ equations we have now considerably more. (To derive Lawler's formulation (3.1), ..., (3.3) as a relaxation of (3.5), ..., (3.13) is left as a recommended exercise for the reader.) Note that the zero-one requirement (3.4) has been replaced by the weaker requirement (3.12); so the resulting linear program has precisely n^2 zero-one variables. Like we did before – see also Chapters 1.1, 1.2 and 1.6 – we can reduce the number of new variables that must be considered to $n^2(n-1)^2/2$ using elementary properties of the feasible solutions to the problem that we have also used above. This shows that some of the equations (3.7), ..., (3.10) are superfluous – utilizing the symmetries $y_{ij}^{k\ell} = y_{k\ell}^{ij}$ some of them are simply "repeats" of others – and it is not difficult to see that $2n + 2n^2(n-1)$ constraints in $n^2 + n^2(n-1)^2/2$ variables suffice to formulate the problem correctly.

Now we have considerably more equations in the same set of variables and it remains to show how many equations are truly required. As we shall see in Chapter 7.1 a proper analysis of the formulation shows that the geometry of the problem requires exactly $2n(n-1)^2 - (n-1)(n-2)$ equations if $n \geq 3$. While we will reduce the necessary number to the bare minimum, this means nevertheless that roughly n^3 equations are required to formulate the problem in *geometric* terms correctly. Traditionally, such geometric considerations have been ignored – we can get away with $2n + 1$ equations, right? – , but mathematically and numerically this kind of thinking has not gotten very far either. Nevertheless, authors continue to propose formulations that have "as few constraints as possible" for the QAP and other difficult combinatorial optimization problems. For instance, probably due to the sheer *size* of their natural formulation, Frieze and Yadegar [1983] propose the following "reduced" formulation for the QAP; see also Assad and Xu [1985], Bazaraa and Sherali [1980] and Carraresi and Malucelli [1992b] for similarly "shortened" formulations that pay no attention to the underlying geometry of the problem.

$$\min \quad \sum_{i,j \in N} c_{ij} x_{ij} + \sum_{i,j \in N} \sum_{k,\ell \in N} a_{ij}^{k\ell} y_{ij}^{k\ell}$$

subject to $(3.5), (3.6), (3.11), (3.12), (3.13)$ and

$$\sum_{i,j \in N} y_{ij}^{k\ell} = n x_{k\ell} \qquad \text{for } k, \ell \in N \quad (3.14)$$

$$\sum_{k,\ell \in N} y_{ij}^{k\ell} = n x_{ij} \qquad \text{for } i, j \in N \quad (3.15)$$

It is not difficult to see that this is a "relaxation" of $(3.5), \ldots, (3.13)$ which formulates the QAP correctly as well. Now we have about $2n + 2n^2$ equations and thus a *substantial* reduction in terms of the number of equations. Or so it seems. Of course, this kind of thinking has nothing to do with the geometry of the problem, for if the reduction of the number of equations is the goal of problem formulation (and by consequence, of numerical problem solving) then we can do vastly better. It has been known since the early 1970s, see e.g. Padberg [1972], that *every* integer program in bounded variables can be formulated using a single equation. More precisely, it follows e.g. from Lemma 1 of Padberg [1972] that the following mixed zero-one program *formulates* the QAP correctly.

$$\min \quad \sum_{i,j \in N} c_{ij} x_{ij} + \sum_{i,j \in N} \sum_{k,\ell \in N} a_{ij}^{k\ell} y_{ij}^{k\ell}$$

subject to $(3.5), (3.6), (3.11), (3.12), (3.13)$ and

$$- \sum_{i,j \in N} n(1 + 2^{n^2}) 2^{n^2 - (i-1)n - j} x_{ij}$$

$$+ \sum_{i,j \in N} \sum_{k,\ell \in N} (2^{n^2 - (k-1)n - \ell} + 2^{2n^2 - (i-1)n - j}) y_{ij}^{k\ell} = 0. \quad (3.16)$$

Indeed, by a full application of Lemma 1 of Padberg [1972], we can reduce the resulting number of equations from $2n + 1$ to 1 and the *digital size* of the coefficients of the resulting constraint matrix is about n^2, i.e. their digital size is polynomially bounded in the parameter n of the QAP. Thus theoretically at least we can reduce the "staggering" number of about $4n^3$ to a single one while ensuring "polynomiality" of the resulting transformation. Evidently, the "chase" for *compact* formulations of the QAP has taken place many years ago – with meager computational and numerical results – and we hasten to state explicitly that (3.16) is **not** recommended for numerical computation. If the method of solution for QAPs is based *exclusively* on some form of *enumeration* – implicit or otherwise – then *compactness* of the formulation, i.e. the formulation of a combinatorial optimization problem with as few linear constraints as possible, may matter. But these considerations do not matter at all if the overall problem is embedded into a *continuum*, such as it is done when we use linear programming, assignment problem-type relaxations and the like, in the numerical solution of such problems. A *minimal* system of equations to represent the linearized formulation of the quadratic assignment problem in the space of $n^2 + n^2(n-1)^2/2$ variables of this section is given in Chapter 7.1.

3.2 Branch-and-bound algorithms for QAPs

Branch-and-bound is an *implicit enumeration* method utilizing, typically, embedded linear programming problems to solve pure-integer or mixed-integer optimization problems. Assuming a finite set of integer values for the integer variables it proceeds by partitioning the integer solutions into – typically – mutually exclusive sets. By refining the partitioning and solving a relaxed problem over the restricted solution set, a sequence of lower bounds is generated that is weakly monotonously increasing when we assume minimization as the sense of the overall optimization problem. If it so happens – and given the finiteness of the solution sets, it must happen eventually – that a solution is found with integer values in the required components, the solution is compared to the best one found so far and, if applicable, it is recorded as the best one with its corresponding objective function value. This gives an upper bound on the objective function and the objective of branch-and-bound is to assure that the worst lower bound coincides with the best upper bound, at which point the algorithm terminates. The algorithm typically proceeds by creating a *binary search tree* which is obtained by *branching* on a single variable that looks somehow "promising" for the creation of two new subproblems. This basic idea for branch-and-bound dates from the 1950s and for many years it was the only integer programming algorithm that was commercially available. This has changed since about 1990 with the introduction of ideas from branch-and-cut into commercial software systems such as CPLEX of CPLEX Optimization, Inc and IBM's OSL optimization package.

Numerous strategic games are possible within the general framework of branch-and-bound and we refer the reader to Nemhauser and Wolsey [1988] for an overview. The questions that are typically addressed are the selection of branching variables, the selection of the next subproblem to be worked on, "look-aheads" to limit the search, etc. Rather than creating two new problems every time the algorithm branches, the exploitation of parallel computers to create $p \geq 2$ branches at a time has been investigated as well, see e.g. Cannon [1988] and Cannon and Hoffman [1990] in the context of the branch-and-cut algorithms for linear zero-one optimization problems. Here p is the number of "processors" that are available at the time when branching takes place. In the context of the quadratic assignment problem, Roucairol [1987] and others have devised special branching schemes to exploit parallel processing. Like in the case of general zero-one problems a linear speed-up can typically be realized as the number of parallel processors is increased. At present this appears to be true when the number of processors is relatively small and we are not aware of pertaining studies for massively parallel computers and their potential for speeding up branch-and-bound algorithms for difficult combinatorial problems. Due to communication problems between the processors a less-than-linear speed-up is predictable.

The application of branch-and-bound to the solution of QAPs relies on the philosophy of generating lower bounds quickly and cheaply. The pertaining work starts apparently with Gilmore [1962] and Lawler [1963] who derived the following *Gilmore-Lawler lower bound* for QAPs. Let us denote by

$$z_{QAP} = min \left\{ \sum_{i,j \in N} c_{ij} x_{ij} + \sum_{i,j \in N} \sum_{k,l \in N} a_{ij}^{kl} x_{ij} x_{kl} : \mathbf{X} \in \mathcal{X}_n \right\}$$

the optimal solution value of QAP and for $i, j \in N$

$$f_{ij} = min \left\{ \sum_{k,l \in N} a_{ij}^{kl} x_{kl} : \mathbf{X} \in \mathcal{X}_n, x_{ij} = 1 \right\}.$$

f_{ij} can be computed by solving a linear assignment problem with the additional restriction that $x_{ij} = 1$ for some $i, j \in N$. By construction we have

$$(c_{ij} + f_{ij}) x_{ij} \leq x_{ij} (c_{ij} + \sum_{k,l \in N} a_{ij}^{kl} x_{kl})$$

for all $i, j \in N$ and $\mathbf{X} \in \mathcal{X}_n$. Consequently,

$$GLB = min \left\{ \sum_{i,j \in N} (c_{ij} + f_{ij}) x_{ij} : \mathbf{X} \in \mathcal{X}_n \right\} \leq z_{QAP}$$

and thus by solving $n^2 + 1$ linear assignment problems a lower bound on z_{QAP} is obtained. Moreover, if $\mathbf{x}^*$ solves the linear assignment problem GLB on the left hand side of the inequality, then we get an upper bound for z_{QAP} by evaluating the objective function value of QAP in terms of $\mathbf{x}^*$. By comparison to the overall problem that we wish to solve the computation of the Gilmore-Lawler bound GLB is relatively cheap, we can partition the set of all permutations

Solution Approaches

by assigning some x_{ij} the value of one and/or zero and iterate. In addition, we can utilize the dual variable information provided for by the calculation of GLB to cleverly select promising subproblems to be chosen in the branching scheme. This gives a basic branch-and-bound algorithm for the QAP which leaves many strategic choices to play with.

Example 1. For the data of our example problem of Chapter 1.3, see Table 1.1, we calculate the Gilmore-Lawler matrix with elements $c_{ij} + f_{ij}$ for $i, j \in N$

$$\mathbf{F} = \begin{pmatrix} 632 & 440 & 228 & 334 & 290 \\ 720 & 466 & 361 & 447 & 339 \\ 564 & 512 & 191 & 265 & 209 \\ 500 & 359 & 168 & 219 & 296 \\ 618 & 375 & 250 & 377 & 218 \end{pmatrix}.$$

This is done by solving the n^2 linear assignment problems. Solving the resulting linear assignment problem GLB we get a lower bound of 1,677 and as it so happens, an upper bound of 2,010 on the optimal value $z_{QAP} = 1,812$ of this particular problem.

The objective function of the QAP consists of a linear and a quadratic part. Using the assignment constraint (3.5) and (3.6) it is possible to "shift" some of the data from the quadratic part to the linear part – like we did in Chapter 1.2 in order to reduce the number of off-diagonal nonzero entries of the flow matrix. The intuitive reason behind such a "reduction" of the quadratic part is the desire to reduce the relative impact of the quadratic part of the objective function and to increase the relative importance of its linear part. As we have seen in Chapter 1.2 this intuitive reasoning has the definite consequence of reducing the number of new variables that are necessary when we linearize the quadratic terms. "Reduction" has attracted a great deal of interest in the literature.

In the context of the Koopmans-Beckmann problem the following rules have been investigated, see also Chapter 1.2:

- Burkard [1973] subtracts from each column of the flow matrix **T** and the distance matrix **D** its minimal off-diagonal element.
- Edwards [1980] reduces **T** and **D** to yield matrices $\overline{\mathbf{T}}$ and $\overline{\mathbf{D}}$, respectively, which have zero principal diagonals and off diagonal elements given by:

$$\overline{t}_{ik} = t_{ik} - \frac{\sum_{i \in N} t_{ik}}{(n-1)} - \frac{\sum_{k \in N} t_{ik}}{(n-1)} + \frac{\sum_{i,k \in N} t_{ik}}{(n-1)(n-2)}$$

$$\overline{d}_{j\ell} = d_{j\ell} - \frac{\sum_{j \in N} d_{j\ell}}{(n-1)} - \frac{\sum_{\ell \in N} d_{j\ell}}{(n-1)} + \frac{\sum_{j,\ell \in N} d_{j\ell}}{(n-1)(n-2)}.$$

- Roucairol [1987] proposes two different reduction schemes. The first consists of subtracting from each row of $\mathbf{T}$ and $\mathbf{D}$ its minimal off-diagonal element and then to subtract from each column of the reduced matrices its minimal off-diagonal element. The second reduction scheme is iterative and goes as follows: for each one of $\mathbf{T}$ and $\mathbf{D}$ pick $2n$ elements sequentially such that the greatest element of the reduced matrix at each iteration is decreased by as much as possible without letting any entry in the reduced matrices become negative.

Evidently, one can play endless games with different reduction schemes and the set of choices is rather unlimited. Before we come back to the question of a rational choice of the reduction parameters let us illustrate reduction by way of an example.

Example 2. In Chapter 1.1 we have stated explicit formulas for a particular reduction and in Table 1.3 we give its application to the five-city example of Chapter 1.3. Calculating the corresponding Gilmore-Lawler matrix like we did in Example 1 we find

$$\mathbf{F} = \begin{pmatrix} 762 & 522 & 322 & 448 & 362 \\ 1154 & 741 & 594 & 745 & 556 \\ 396 & 398 & 122 & 174 & 145 \\ 324 & 243 & 104 & 131 & 222 \\ 412 & 243 & 114 & 211 & 102 \end{pmatrix}.$$

Solving the corresponding linear assignment problem GLB we get a lower bound of 1,619, which is worse than the one obtained without any reduction, and as it so happens, an upper bound of 1,812 which is the optimal value $z_{QAP} = 1,812$ for this particular problem, except that we have no proof of this fact yet.

It follows from the example that reduction *per se* does not guarantee a better lower bound on z_{QAP}. The question of "reducing the data optimally" so as to guarantee e.g. a best possible Gilmore-Lawler bound for the given data ensues and has been dealt with in a very interesting paper by Frieze and Yadegar [1983]. They consider the reduction of the $a_{ij}^{k\ell}$ of the objective function of the QAP in a very general form. Write the reduced coefficients $b_{ij}^{k\ell}$ in the following decomposed form:

$$b_{ij}^{k\ell} = a_{ij}^{k\ell} - \alpha_{jk\ell} - \beta_{ik\ell} - \gamma_{ij\ell} - \delta_{ijk}, \tag{3.17}$$

where $\boldsymbol{\alpha}, \boldsymbol{\beta}, \boldsymbol{\gamma}, \boldsymbol{\delta} \in \mathbb{R}^{n^3}$ are arbitrary real vectors. Substituting (3.17) into the objective function of QAP transforms it into

$$\sum_{i,j \in N} d_{ij} x_{ij} + \sum_{i,j \in N} \sum_{k,\ell \in N} b_{ij}^{k\ell} x_{ij} x_{k\ell}, \tag{3.18}$$

Solution Approaches

where $d_{ij} = c_{ij} + \sum_{\ell \in N} \alpha_{\ell ij} + \sum_{\ell \in N} \beta_{\ell ij} + \sum_{\ell \in N} \gamma_{ij\ell} + \sum_{\ell \in N} \delta_{ij\ell}$ for $i, j \in N$. Now let us denote like we did before in the GLB calculation

$$\overline{f}_{ij} = min \left\{ \sum_{k,\ell \in N} b_{ij}^{k\ell} x_{k\ell} : \mathbf{X} \in \mathcal{X}_n, x_{ij} = 1 \right\}.$$

For given $\mathbf{\alpha}, \mathbf{\beta}, \mathbf{\gamma}, \mathbf{\delta} \in \mathbb{R}^{n^3}$ we can compute all $\overline{f}_{ij}$ as before and it follows that

$$GLB(\mathbf{\alpha}, \mathbf{\beta}, \mathbf{\gamma}, \mathbf{\delta}) = min \left\{ \sum_{i,j \in N} (c_{ij} + \overline{f}_{ij}) x_{ij} : \mathbf{X} \in \mathcal{X}_n \right\} \leq z_{QAP},$$

for all possible choices of $\mathbf{\alpha}, \mathbf{\beta}, \mathbf{\gamma}, \mathbf{\delta} \in \mathbb{R}^{n^3}$. Consequently, to find a best possible (generalized) Gilmore-Lawler bound using a most general form of decomposition of the objective function coefficients of the quadratic part of QAP, we are interested in finding

$$max\{GLB(\mathbf{\alpha}, \mathbf{\beta}, \mathbf{\gamma}, \mathbf{\delta}) : \mathbf{\alpha}, \mathbf{\beta}, \mathbf{\gamma}, \mathbf{\delta} \in \mathbb{R}^{n^3}\}. \quad (3.19)$$

To make matters short, Frieze and Yadegar [1983] show that $\mathbf{\gamma} \in \mathbb{R}^{n^3}$ and $\mathbf{\delta} \in \mathbb{R}^{n^3}$ do not matter at all in the reduction scheme, i.e. we might as well set them equal to zero. Moreover, they show that the maximum (3.19) equals the minimum objective function value of the linear programming relaxation (3.5),...,(3.12) of the QAP. Their result shows that the best lower bound that reduction plus a bounding scheme in the spirit of Gilmore [1962] and Lawler [1963] can provide for is obtainable via the solution of a single linear program. Similar, less complete results of this variety can be found in Assad and Xu [1985] and Carraresi and Malucelli [1992a, 1992b]; see Rijal [1995] for more detail. Frieze and Yadegar investigate the use of *Lagrangian relaxation* to find/approximate the maximum value of $GLB(\mathbf{\alpha}, \mathbf{\beta}, \mathbf{\gamma}, \mathbf{\delta})$. While the avoidance of the solution of a large-scale linear program may have been a reason to explore alternatives in the past, we think that the progress in linear optimization made in the meantime warrants a different thinking, especially in view of the limited size of QAPs actually optimized to date.

A different approach to obtaining lower bounds for QAPs and KBPs utilizes the algebraic properties of the eigen values of symmetric matrices. To facilitate the discussion of these approaches to lower bounds for the KBP, we consider the following nonlinear programming problem:

$$min \quad \sum_{i,j \in N} c_{ij} x_{ij} + \sum_{i,j \in N} \sum_{k,\ell \in N} t_{ik} d_{j\ell} y_{ij} y_{k\ell}$$

subject to $\quad\quad\quad\quad\quad\quad\quad\quad\quad\quad \mathbf{x} \in AP_n$ (3.20)

$$y_{ij} = x_{ij} \quad \text{for } i,j \in N \quad (3.21)$$
$$\sum_{j \in N} y_{ij} y_{ij} = 1 \quad \text{for } i \in N \quad (3.22)$$
$$\sum_{j \in N} y_{ij} y_{kj} = 0 \quad \text{for } i \neq k \in N. \quad (3.23)$$

If $\mathbf{x} = \mathbf{y} \in AP_n$ and $x_{ij} = y_{ij} = 1$ then $y_{ij}y_{ij} = 1$ and $y_{ij}y_{kj} = 0$ for $1 \leq i \neq k \leq n$; thus, the constraints (3.22) and (3.23) are redundant. Hence, the nonlinear programming problem is a formulation of the KBP. Moreover, if we replace the constraint (3.21) by $\mathbf{y} \in AP_n$, we obtain a relaxation of the KBP; consequently, the optimal objective function value of this relaxation problem is a lower bound for the KBP. Since, in this relaxed problem the variables $\mathbf{x}$ and $\mathbf{y}$ are unrelated, the problem decomposes into two subproblems

$$min \left\{ \sum_{i,j \in N} c_{ij} x_{ij} : \mathbf{y} \in AP_n \right\}, \quad (3.24)$$

$$min \left\{ \sum_{i,j \in N} \sum_{k,\ell \in N} a_{ij}^{k\ell} y_{ij} y_{k\ell} : \mathbf{y} \in AP_n, \mathbf{y} \text{ satisfies (3.22) and (3.23)} \right\}. \quad (3.25)$$

The subproblem (3.24) is a linear assignment problem, which can be solved using a variety of network optimization techniques or simply by any linear programming solver. The subproblem (3.25) is a nonlinear programming problem, which is difficult to solve. It has been shown, using Lagrangian multiplier techniques of solving unconstrained nonlinear programming problems if the matrices $\mathbf{T}$ and $\mathbf{D}$ are asymmetric, see Rendl and Wolkowicz [1992], and using the orthogonal diagonalization property of symmetric matrices if the matrices $\mathbf{T}$ and $\mathbf{D}$ are symmetric, see Finke et al. [1987], that the objective function value of this nonlinear programming problem lies between $min \sum_{i=1}^{n} \lambda_i \gamma_{k_i}$ and $max \sum_{i=1}^{n} \lambda_i \gamma_{k_i}$, see Hoffman and Wielandt [1953] and Finke et al. [1987], where λ_i and γ_i for $1 \leq i \leq n$ are respectively the eigen values of the matrices $\mathbf{T}$ and $\mathbf{D}$. Moreover, if the requirement that $\mathbf{y} \in AP_n$ is dropped, then the objective function value of the relaxation problem is, in fact, given by $min \sum_{i=1}^{n} \lambda_i \gamma_i$, see Finke et al. [1987], which is equal to the ranked product of these two sets of eigen values whereby the largest eigen value from one set is paired with the smallest eigen value from the other set.

It has been empirically verified that if the matrices $\mathbf{T}$ and $\mathbf{D}$ are not reduced further, a lower bound for many instances of the KBP obtained using the eigen value decomposition is negative, see e.g. Hadley et al. [1992]; this lower bound is dominated by a trivial lower bound of 0 for the KBP with only nonnegative cost coefficients. Hence, all algorithms that utilize the eigen value approach to calculate a lower bound for the KBP work on matrices obtained by decomposing **T** and **D** in order to augment the influence of the linear assignment subproblem and to reduce the influence of the the nonlinear subproblem in the calculation of the overall lower bound for the KBP. Since a smaller fluctuation of eigen values of the matrices **T** and **D** is likely to lead to a smaller bandwidth within which the ranked products of these two sets of eigen values lie, the matrices **T** and **D** are decomposed so that the *spreads* of these matrices are minimized. The spread of a square matrix **T** is given by, $sp(\mathbf{T}) = max_{i,j}|\lambda_i - \lambda_j|$ for $i \neq j \in N$. Since there is no simple formula to compute the spread, Finke *et al.* [1987] propose to minimize the upper bounds of spreads of these matrices by utilizing Mirsky's approximation [1956]. Mirsky's formula for calculating an upper bound of spread of eigen values of a square matrix **T** is

$$sp(\mathbf{T}) \leq \left[2\sum_{i=1}^{n}\sum_{k=1}^{n} t_{ik}^2 - (2/n)\left(\sum_{i=1}^{n} t_{ii}\right)^2 \right]^{1/2}.$$

Finke *et al.* [1987] decompose the matrices **T** and **D** as follows

$$\bar{t}_{ik} = t_{ik} - f_i - f_k - r_{ik} \quad \text{for all } i, k \in N, i \neq k,$$
$$\bar{d}_{j\ell} = d_{j\ell} - h_j - h_\ell - s_{j\ell} \quad \text{for all } j, \ell \in N, j \neq \ell,$$

where the reduction parameters that minimize an upper bound of $sp(\mathbf{T})$ are

$$f_i = \left(\sum_{k=1}^{n} t_{ik} - t_{ii} - z\right)/(n-2)$$
$$r_{ik} = \begin{cases} t_{ii} - 2f_i & \text{for } i = k \\ 0 & \text{otherwise,} \end{cases}$$
where $z = \left(\sum_{i=1}^{n}\sum_{k=1}^{n} t_{ik} - \sum_{i=1}^{n} t_{ii}\right)/2(n-1)$.

The reduction parameters h_j and $s_{j\ell}$ for $1 \leq j, \ell \leq n$ that minimize $sp(\mathbf{D})$ can be calculated similarly. Rendl and Wolkowicz [1992] show that this reduction scheme is equivalent to minimizing the variance of the corresponding set of eigen values. The reduced matrices $\overline{\mathbf{T}} = (\bar{t}_{ik})$ and $\overline{\mathbf{D}} = (\bar{d}_{j\ell})$ have row and column sums equal to zero and zeroes along the main diagonals. Moreover, this reduction scheme not only reduces the magnitude of quadratic terms in the objective function but it also preserves symmetry of these matrices. Resultant to this reduction scheme, the objective function coefficients c_{ij} in the linear assignment subproblem are replaced by $\bar{c}_{ij} = 2h_j \sum_{i \neq k=1}^{n} t_{ik}$.

Rendl and Wolkowicz [1992] state that this lower bound can be further improved since the matrices **T** and **D** are reduced independently without any consideration of the linear cost matrix **C** in the reduction scheme of Finke, Burkard and Rendl [1987]. Rendl and Wolkowicz [1992] outline an iterative eigen value decomposition approach that also works for the cases when the matrices **T** and **D** are not necessarily symmetric. To improve the overall lower bound for the original problem, Rendl and Wolkowicz [1992] compute the derivative of a suitably perturbed minimal scalar product of the ranked eigen values as well as the subdifferential of the lower bound for linear part in the cost function and move along the steepest ascent direction, which improves the overall lower bound by taking a step size which preserves the optimal basis of the linear assignment problem. The linear assignment problem that is considered there is given by $min\left\{\sum_{i=1}^{n}\sum_{j=1}^{n}\bar{c}_{ij}x_{ij} : \mathbf{x} \in AP_n\right\}$ with

$$\bar{c}_{ij} = c_{ij} + 2h_j \sum_{k=1}^{n} t_{ik} + 2f_i \sum_{\ell=1}^{n} d_{j\ell} - 2h_j \sum_{k=1}^{n} f_k \\ - 2nf_i h_j + (t_{ii} - 2f_i - r_{ii})s_{jj} + r_{ii}(d_{jj} - 2h_j - s_{jj}) + r_{ii}s_{jj},$$

f_i, h_i, r_{ii} and s_{jj} for $i, j \in N$ as defined above. The lower bound derived there is given by the sum of the minimum of the ranked products of the eigen vectors and the objective function value of an optimal solution to the linear assignment problem. The resulting iterative procedure for deriving a lower bound has a complexity of $\mathcal{O}(n^3)$ per iteration.

3.3 Traditional cutting plane algorithms

Several researchers have pursued methods based on Benders' decomposition, see Benders [1962], to solve QAPs or at least, to derive a lower bound for the QAP. The basic idea behind these algorithms is to use an enlarged nonlinear formulation of the QAP by introducing a set of new variables and constraints. The iterative approach works with a master problem and a subproblem. The subproblem is a linear programming problem obtained by fixing some of the variables (usually, the original variables) in this reformulated problem, while the master problem is a reformulation of the original problem with primal and dual solution vectors of the subproblem as its parameters; hence, it is also a linear programming problem. A subproblem in this scheme is usually a simple problem that has a closed-form solution which can be derived using the duality theory of linear programming or can be solved using an efficient algorithm, e.g. a network flow algorithm. Starting with some feasible solution vector, first the master problem is solved. Given this solution to the master problem, the subproblem is solved to yield both primal and dual solution vectors. Assuming the feasibility of the original problem (which is true in all the problems of interest to us), if the solution to the subproblem satisfies all the constraints of the master problem, we have an optimal solution to the original problem. Otherwise, any

Solution Approaches

violated constraint (also called a cutting plane since it cuts off the current solution obtained as a solution to the subproblem) is added to the master problem and the enlarged master problem is solved and the whole procedure is repeated again. This reiterative procedure is continued until a solution to the subproblem that satisfies all the constraints of the master problem is found. Moreover, every solution to the master problem corresponding to a feasible solution to the subproblem at any stage in the iterative procedure furnishes a lower bound for the overall problem. On the other hand, the objective function value of the original problem corresponding to a feasible solution to the subproblem yields an upper bound for the overall problem.

In practice, this method usually turns out to be computationally very expensive due to poor convergence of the lower and upper bounds to a single bound. Kaufman and Broeckx [1978], for instance, use this procedure to derive a lower bound, but they couple it with a suboptimal heuristic solution to the original problem. The whole scheme is accelerated by terminating this iterative algorithm when the difference between lower and upper bounds falls within a certain prespecified range.

Various reformulations of QAPs which lend themselves very well to Benders' decomposition have been proposed in the literature. Kaufman and Broeckx [1978] formulate the QAP using n^2 additional variables and n^2 additional constraints as follows:

$$\min \quad \sum_{i,j \in N} y_{ij}$$
$$\text{subject to} \quad \mathbf{x} \in AP_n$$
$$f_{ij} x_{ij} + \sum_{k=1}^{n} \sum_{\ell=1}^{n} a_{ij}^{k\ell} x_{k\ell} - y_{ij} \leq f_{ij} \quad \text{for } 1 \leq i,j \leq n,$$

where AP_n is defined in (1.14), $y_{ij} = x_{ij} \sum_{k=1}^{n} \sum_{\ell=1}^{n} a_{ij}^{k\ell} x_{k\ell}$ and f_{ij} is the optimal objective function of the linear programming problem given by

$$f_{ij} = \max \left\{ \sum_{k,\ell \in N} a_{ij}^{k\ell} x_{k\ell} : \mathbf{x} \in AP_n \right\},$$

The master problem, they consider is

$$\min\{z : z \geq \sum_{i,j \in N} u_{ij}^p (\sum_{k,\ell \in N} a_{ij}^{k\ell} x_{ij} - f_{ij}) \text{ for } p \in P, z \geq 0, \mathbf{x} \in AP_n\},$$

where $\mathbf{u}^p = (u_{ij}^p, \text{ for } 1 \leq i,j \leq n)$ for $p \in P$ are the finite set of extreme points of the dual of the subproblem given by

$$\min\{y_{ij} : \sum_{k,\ell \in N} a_{ij}^{k\ell} x_{k\ell} - y_{ij} \leq f_{ij}, y_{ij} \geq 0 \text{ for } i,j \in N\}.$$

Bazaraa and Sheriali [1980] introduce another formulation of the QAP using $n^2(n-1)^2/2$ new variables and $2n^2$ new constraints as follows:

$$\min \sum_{i<k\in N} \sum_{j\neq \ell\in N} a_{ij}^{k\ell} y_{ij}^{k\ell}$$

$$\text{subject to} \quad \mathbf{x} \in AP_n \quad (3.26)$$

$$\sum_{k=i+1}^{n} \sum_{j\neq \ell\in N} y_{ij}^{k\ell} - (n-i)x_{ij} = 0 \quad \text{for } 1 \leq i \leq n-1, j \in N \quad (3.27)$$

$$\sum_{i=1}^{k-1} \sum_{\ell\neq j\in N} y_{ij}^{k\ell} - (k-1)x_{k\ell} = 0 \quad \text{for } 2 \leq k \leq n, \ell \in N \quad (3.28)$$

$$0 \leq y_{ij}^{k\ell} \leq 1 \quad \text{for } i < k \in N, j \neq \ell \in N, \quad (3.29)$$

where $y_{ij}^{k\ell} = x_{ij}x_{k\ell}$ and $a_{ij}^{k\ell} = (c_{ij} + c_{k\ell} + a_{ijk\ell})/2$ for $1 \leq i < k \leq n$ and $1 \leq j \neq \ell \leq n$. Bazaraa and Sheriali [1980] also outline an algorithm based on Benders' decomposition. The master problem, they consider is given by

$$\min\{z : z \geq \sum_{i=1}^{n-1}\sum_{j\in N} u_{ij}^p(n-i)x_{ij} + \sum_{k=2}^{n}\sum_{\ell\in N} v_{k\ell}^p(k-1)x_{k\ell} - w^p \text{ for } p \in P, \mathbf{x} \in AP_n\}$$

where $w^p = \sum_{i<k\in N}\sum_{j\neq\ell\in N} w_{ijk\ell}^p$ and $\mathbf{u}^p = (u_{ij}^p, v_{k\ell}^p, w_{ijk\ell}^p,$ for $i < k \in N, j \neq \ell \in N)$ for $p \in P$ are the finite set of extreme points of the dual of the subproblem given by

$$\min\{\sum_{i<k\in N}\sum_{j,\ell\in N} a_{ij}^{k\ell}y_{ij}^{k\ell} : \mathbf{y} \text{ satisfies } (3.27), (3.28) \text{ and } (3.29)\}.$$

Balas and Mazzola [1980] give the following formulation for the QAP with interaction cost terms $a_{ij}^{k\ell} \geq 0$ for all $i, j, k, \ell \in N$:

$$\min \quad z$$
$$\text{subject to} \quad z \geq \sum_{k,\ell\in N}(f_{k\ell}y_{k\ell} + \sum_{i,j\in N} a_{ij}^{k\ell}y_{ij})x_{k\ell} - \sum_{k,\ell\in N} f_{k\ell}y_{k\ell} \quad (3.30)$$
$$\mathbf{x}, \mathbf{y} \in AP_n, \quad (3.31)$$

where $f_{k\ell} = max\{\sum_{k,\ell\in N} a_{ij}^{k\ell}x_{ij} : \mathbf{x} \in AP_n\}$. Balas and Mazzola [1984] outline a cutting plane algorithm to solve nonlinear zero-one programming problems in general. Their algorithm starts by generating some linear inequalities, e.g. generalized cover inequalities, implied by the constraint set of the original problem.

Solution Approaches   75

These inequalities furnish a set of constraints to a linear programming relaxation of the original problem. If an optimal solution to this linear programming relaxation is feasible to the original problem, we have an optimal solution to the original problem. Otherwise, additional linear inequalities implied by the original problem and violated by the solution to the relaxation problem are identified and appended to the linear relaxation problem. This process is reiterated until the linear relaxation yields an optimal solution that does not violate any constraints implied by the original problem. Burkard and Bönniger [1983] utilize the formulation due to Balas and Mazzola [1980] and develop a heuristic cutting plane procedure to find possibly several suboptimal solutions to the QAP.

3.4 Heuristic procedures

There is a plethora of traditional and modern heuristic procedures for the quadratic assignment problem. The major traditional heuristic approaches outlined in the literature either utilize construction methods, see Gilmore [1962], to find one or more suboptimal solutions by enlarging a partial permutation according to some criteria or perform one or more pairwise exchanges, see Heider [1972], until no further improvement can be made. Other so-called *meta-heuristics* like *simulated annealing*, see Burkard and Rendl [1984], Lutton and Bonomi [1986] and Wilhem and Ward [1987], *tabu search*, see Skorin-Kapov [1990] and Taillard [1991] and *genetic algorithms*, see Brown *et al.* [1989] and Mühlenbein [1989], have also been used to find suboptimal solutions to the QAP. Despite their interesting and entertaining names which are borrowed from thermodynamics, psychology and genetics it seems, these *heuristics* are just that – hit-and-run attempts to solve difficult problems with as little mathematics as possible. This is vain, of course, because for the QAP even the problem of finding a feasible solution which is guaranteed to approximate the optimal objective function value by some $\varepsilon > 0$ is $\mathcal{NP}$ hard, see Sahni and Gonzales [1976]. In other words, no polynomial time heuristic can provide any guarantee as to the quality of the solution. Moreover, Dyer *et al.* [1986] show that solving an average case takes exponential time, if the objective function coefficients of QAPs are taken from some simple sample space of random numbers. Heuristics do play a role in the exact solution of QAPs, however, provided they are designed to run fast and provide "reasonable" solutions quickly.

3.5 Polynomially solvable cases

Quadratic assignment problem. Several researchers have identified conditions on input parameters under which the resultant QAP can be solved in polynomial time. As already pointed out in Chapter 2, the linear assignment problem is a polynomially solvable special case of the QAP. Christofides and Gerrard [1976] show that the KBP can be solved in $\mathcal{O}(n^2)$ time if the matrices **T** and **D** are each a weighted adjacency matrix of a *tree* and – by solving a series of linear assignment problems – if the matrix **T** is a weighted adjacency matrix of a *double star*. Furthermore, Rendl [1986] shows that the KBP can be solved in $\mathcal{O}(n^3)$ time if both matrices **T** and **D** are weighted adjacency matrices of series-parallel graphs containing no bipartite subgraph $K_{2,2}$.

Multi-processor assignment problem. Stone [1977] shows that the MPP for $n = 2$ can be modeled as a *min cut* problem. A general *task graph* can be associated with the MPP. A task graph is an ordered pair of nonempty set of nodes and a family of two-element subset of nodes which represent an edge between the corresponding nodes. Nodes in a task graph represent the set of tasks of a modular program while the edges represent the inter-module linkages. The edge weights indicate the amount of data to be transferred between two tasks. To model the MPP as a minimum cut problem, Stone [1977] modifies the task graph as follows: first, two nodes each representing a processor are added and one of them is designated as a source node while the other is a sink node. For each node other than the source and sink nodes, two edges one each to the source and sink are added. The weight of an edge emanating from the source (sink) carries the weight equal to the amount of time required to process the task corresponding to the sink (source) node. The weight of an edge between a pair of tasks is equal to the total communication time between two processors if any reference occurs between two modules. Now any standard *maximum flow* algorithm can be applied to the modified graph and by virtue of the famous *max flow min cut theorem*, see Ford and Fulkerson [1962], every optimal solution to the max flow problem yields a corresponding minimum weight edge cut set. Moreover, the minimum weight edge cut set also defines an optimal solution to the original MPP with the interpretation that if an edge between a task node and source (sink) is in the min cut, then the corresponding task is assigned to the processor corresponding to the sink (source) node. Thus, the MPP for $n = 2$ is equivalent to the min cut problem and hence can be solved in polynomial time. Moreover, the MPP can be solved in $\mathcal{O}(mn^2)$ if the task graph is a tree, see Bokhari [1981], and in time $\mathcal{O}(mn^3)$ if the task graph is series-parallel, see Bokhari [1987].

Solution Approaches   77

Graph partitioning problem. The GPP is polynomially solvable if the associated graph is series-parallel or 4-wheel free, see Chopra [1992] or if the quadratic interaction cost matrix is positive semidefinite, see Carlson and Nemhauser [1966].

Boolean quadric problem and relatives. In a special case in which a_{ij} are nonnegative for $1 \leq i < k \leq m$ and c_i are arbitrary for $i = 1, \ldots, m$, the BQP is solvable in polynomial time, see Balinski [1970], Rhys [1970], Picard and Ratliff [1975], Hansen [1979] and Padberg [1989]. A graph can be associated with the Boolean quadric problem as follows: create a node corresponding to all variables and join a pair of nodes by an edge if they have a nonzero interaction cost in the objective function. Various polynomially solvable cases of the BQP have been characterized on this associated graph. The BQP is polynomially solvable if this graph is series parallel, acyclic or bipartite with $a_{ik} < 0$ for all $1 \leq i < k \leq m$, see Barahona [1986] and Padberg [1989].

The max cut problem is polynomially solvable for planar graphs, see Hadlock [1975], graphs that are not contractible to K_5, see Barahona [1983], *weakly bipartite graphs*, see Grötschel and Pulleyblank [1981], or graphs with no *long odd cycles*, see Grötschel and Nemhauser [1984].

3.6 Computational experience to date

Branch-and-bound type algorithms, some of which utilize a linear programming relaxation of the QAP, have so far been the most successful methods for obtaining optimal solution to the QAP. An instance of the QAP of size $n = 30$ and four instances of the QAP of size $n = 20$ (including one from the Nugent *et al.* test problem collection) available from the test problem file QAPLIB, see Burkard *et al.* [1991], have been reportedly solved to optimality so far, see Mans *et al.* [1992], Clausen [1994], Resende *et al.* [1994]. Mans *et al.* [1992] have solved QAPs of size $n = 20$ in reasonable times by using the branch-and-bound algorithm developed by Mautor and Roucairol [1992] which exploits the parallel computer technology available today. Clausen [1994] solves an instance of the QAP of size $n = 20$ from Nugent *et al.* [1968] and likewise, Resende *et al.* [1994] solve three other instances of the QAP of size $n = 20$. In addition, Christofides and Benavent [1989] report the solution of several instances of the tree QAPs in which the flow matrix is the weighted adjacency matrix of a tree; the largest size of the QAP, they solved using a *dynamic programming* algorithm, has $n = 25$.

Besides exact solution methods, various algorithms have been proposed to obtain the lower and upper bounds of the QAP. Skorin-Kapov (1990) calculates upper bounds of problems of size up to $n = 90$ by applying the *tabu search* technique of obtaining suboptimal solutions. Resende *et al.* [1994] calculate lower bounds of all 63 instances of the problems with $n \leq 30$ from the QAPLIB by solving linear programming relaxations of the associated QAPs. In 54 out of 63 instances, their lower bounds are at least as good as or better than best available lower bounds reported in the literature. The linear programming based lower bounds originally proposed by Frieze and Yadegar [1983] and significantly improved since then are uniformly better than the lower bounds obtained from all other algorithms, except the eigen value based algorithms. Though in some instances including one instance of the test problem of size $n = 30$ from Nugent *et al.* [1968], the eigen value based algorithm reportedly produced the best available lower bounds, the linear programming based lower bounds are better than the former ones in a substantially large majority of the problems from the QAPLIB.

4

LOCALLY IDEAL LP FORMULATIONS I

In this chapter and the next one we discuss linear programming (LP) *formulations* of the scheduling, design and assignment problems described in Chapters 1 and 2 as classes of BQPSs (Boolean quadratic problems with specially structured special ordered set (SOS) constraints). A *formulation* of a combinatorial optimization problem is any system of equations and/or inequalities the integer, mixed-integer, zero-one or mixed zero-one solutions of which are in one-to-one correspondence with the "feasible" configurations or objects over which we wish to optimize. In most cases of practical interest many, seemingly different formulations of a combinatorial optimization problem exist if it can be formulated at all in this sense. The LP formulations of the BQPSs that we derive in this chapter are based on the concept of a "locally ideal" linearization. A locally ideal linearization is a linearization that yields an ideal, i.e., minimal and complete, linear description of the *polytope* corresponding to each pair or certain sets of pairs of variables in the quadratic interaction terms of the objective function; see Padberg [1995] for a complete treatment of polyhedral/polytopal theory and any definitions that we leave unexplained in this monograph. In a way, using the concept of local idealization to formulate BQPSs is analogous to investigating thoroughly a few threads of a cobweb as a starting point for a full-fledged study of the entire cobweb.

An illustrative example of a locally ideal linearization is due to Padberg [1976]. For every pair of variables (x_i, x_k) giving rise to quadratic terms in the unconstrained Boolean quadratic optimization problem (BQP), a new variable $y = x_i x_k$ is introduced; and hence, corresponding to (x_i, x_k, y) there are exactly the four feasible zero-one vectors given by (0,0,0), (1,0,0), (0,1,0) and (1,1,1). The following constraints have been suggested in the literature to linearize each resulting quadratic product term: $x_i + x_k - y \leq 1, -x_i - x_k + 2y \leq 0, x_i, x_k \leq$

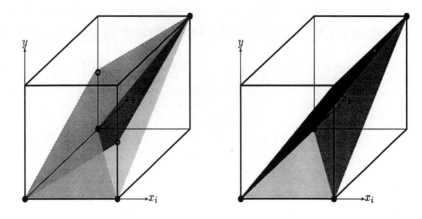

Figure 4.1 Traditional and locally ideal linearizations of the BQP

$1, x_i, x_k, y \geq 0$; see e.g. Fortet [1959], Lawler [1963] and others. The six extreme points corresponding to the polytope in $\mathbb{R}^3$ defined by these seven inequalities include two *fractional*(=non-integer) points $(1,0,1/2)$ and $(0,1,1/2)$ in addition to the four zero-one extreme points; see the left part of Figure 4.1. On the other hand, Padberg [1976, 1989] linearizes the quadratic product term using the constraints: $x_i + x_k - y \leq 1, -x_i + y \leq 0, -x_k + y \leq 0, y \geq 0$. These constraints are an *ideal linear description* of the convex hull of the four feasible solution vectors given above, see the right part of Figure 4.1, because their extreme points are precisely the four zero-one points over which we wish to optimize. With the necessary generalizations this is what we mean by a "locally ideal" linear description of a combinatorial optimization problem.

We denote throughout this chapter $M = \{1, \ldots, m\}$ and $N = \{1, \ldots, N\}$. Linearizing every pair of variables giving rise to a quadratic term in the objective function of the BQP, Padberg [1989] formulates the BQP as the LP problem given by

$$max \left\{ \sum_{i=1}^{m} c_i x_i + \sum_{i<k \in M} q_{ik} y_{ik} : (\mathbf{x}, \mathbf{y}) \in QP_m \right\}, \qquad (\mathcal{O}QP_m)$$

where QP_m is the polytope defined by the convex hull of solutions $(\mathbf{x}, \mathbf{y}) \in \mathbb{R}^{m(m+1)/2}$ to the following system of linear inequalities in zero-one variables:

$$-x_i + y_{ik} \leq 0 \quad \text{for } i < k \in M \qquad (4.1)$$
$$-x_k + y_{ik} \leq 0 \quad \text{for } i < k \in M \qquad (4.2)$$
$$x_i + x_k - y_{ik} \leq 1 \quad \text{for } i < k \in M \qquad (4.3)$$

Locally Ideal LP Formulations I   81

$$y_{ik} \geq 0 \quad \text{for } i < k \in M \tag{4.4}$$
$$x_i \in \{0,1\} \quad \text{for } i \in M. \tag{4.5}$$

For every pair $1 \leq i < k \leq m$ each one of the inequalities (4.1),...,(4.4) describes a *facet* of the polytope QP_m which is another aspect of a locally ideal formulation of a combinatorial optimization problem. This means, in particular, that the system of inequalities (4.1),...,(4.4), like the traditional system mentioned above, is a formulation of the BQP. It is a *better* formulation than the traditional one because the solution set of the linear programming relaxation of (4.1),...,(4.4) is properly contained in the relaxed solution set of the traditional formulation. More precisely, two fractional extreme points per quadratic term of the traditional formulation are eliminated by the locally ideal formulation. In its totality the corresponding locally ideal LP relaxation, however, still has many fractional extreme points that must be "cut off" by facets of QP_m other than those given by (4.1),...,(4.4). There are, of course, plenty of facets of QP_m other than the "trivial" ones given by (4.1),...,(4.4); see Padberg [1989]. Indeed, the BQP is an $\mathcal{NP}$-hard optimization problem which is as difficult as the traveling salesman problem.

The SOS constraints in the BQPSs have a special structure. All of these SOS are of equal cardinality; in addition, they either are disjoint or have well-defined joins. This special structure suggests that we should be able to modify and specialize the linearization of the (unconstrained) BQP to obtain locally ideal linearizations of our problems. As a general rule, it is always advantageous to use all the information that is available from the structure of a given problem to derive its locally ideal linearization and thereby a formulation of optimization problem. In what follows, we derive LP formulations of the major classes of the BQPSs described in Chapters 1 and 2 following this general approach. To do so we proceed as follows: first we derive locally ideal linearizations of the BQPSs introduced in Chapters 1 and 2 by running a computer program for the double description algorithm, see Padberg [1995], to obtain explicit linear descriptions for "small" values of an underlying parameter m or n. In a second step we then generalize our empirical findings to arbitrary values of the parameters in question. In this monograph we give – with minor exceptions – the results of the second step only and hide the laborious *experimental* part of our work from the eyes of the reader. It is clear that for $n = 2$ the problems GPP, OSP, MPP and CLDP can be formulated as a BQP in a smaller set of variables by elimination and substitution using the equations of the form $x_i + x_k = 1$. So we shall assume $n \geq 3$ throughout the chapter.

Rather than reviewing the proof methodology used throughout this chapter, we refer the reader to the survey paper by Grötschel and Padberg [1985], which contains an excellent summary thereof, or to Chapters 7 and 10 of Padberg [1995].

4.1 Graph Partitioning Problems

We define new variables $y_{ik} = \sum_{j=1}^{n} x_{ij} x_{kj}$ for $i < k \in M$ and $1 \leq j \leq n$ to consider the GPP, see Chapter 2.8, and assume throughout that $m \geq n \geq 3$; counting yields that there are $m(m-1)/2$ **y**-variables. Denoting by $DGPP_n^m$ the discrete set

$$DGPP_n^m = \left\{ \begin{array}{l} (\mathbf{x},\mathbf{y}) \in \mathbb{R}^{mn+m(m-1)/2}: \\ \sum_{j=1}^{n} x_{ij} = 1 \quad \text{for } i \in M \\ y_{ik} = \sum_{j=1}^{n} x_{ij} x_{kj} \quad \text{for } i < k \in M \\ x_{ij} \in \{0,1\} \quad \text{for } i \in M, j \in N \end{array} \right\},$$

the GPP can be written as

$$\min \left\{ \sum_{i=1}^{m-1} \sum_{k=i+1}^{m} q_{ik} y_{ik} : (\mathbf{x},\mathbf{y}) \in DGPP_n^m \right\},$$

where $q_{ik} = a_{ik}(I)$ are defined in Chapter 2.8.

To obtain a linear formulation for $DGPP_n^m$ in zero-one variables, we consider the "local" polytope P given by $P = conv(D)$ where $n \geq 3$ and D is defined by

$$D = \left\{ \begin{array}{l} (\mathbf{x},y) \in \mathbb{R}^{2n+1}: \\ \sum_{j=1}^{n} x_{ij} = 1 \quad \text{for } 1 \leq i \leq 2 \\ y = \sum_{j=1}^{n} x_{1j} x_{2j} \\ x_{ij} \in \{0,1\} \quad \text{for } 1 \leq i \leq 2, j \in N \end{array} \right\}.$$

The set of zero-one vectors of the discrete set D is shown in Table 4.1. Let P_L be the polytope given by $(\mathbf{x},y) \in \mathbb{R}^{2n+1}$ satisfying the equations and inequalities:

$$\sum_{j=1}^{n} x_{ij} = 1 \quad \text{for } 1 \leq i \leq 2 \tag{4.6}$$

$$x_{1j} + x_{2j} - y \leq 1 \quad \text{for } j \in N \tag{4.7}$$

$$\sum_{j \in S} x_{1j} - \sum_{j \in S} x_{2j} + y \leq 1 \quad \text{for } \emptyset \neq S \subset N \tag{4.8}$$

$$x_{ij} \geq 0 \quad \text{for } 1 \leq i \leq 2, j \in N \tag{4.9}$$

$$y \geq 0. \tag{4.10}$$

Locally Ideal LP Formulations I

x_{11}	x_{12}	...	x_{1n}	x_{21}	x_{22}	...	x_{2n}	y
1	0	...	0	1	0	...	0	1
1	0	...	0	0	1	...	0	0
:	:	⋱	:	:	:	⋱	:	:
1	0	...	0	0	0	...	1	0
0	1	...	0	1	0	...	0	0
0	1	...	0	0	1	...	0	1
:	:	⋱	:	:	:	⋱	:	:
0	1	...	0	0	0	...	1	0
:	:	⋱	:	:	:	⋱	:	:
0	0	...	1	1	0	...	0	0
0	0	...	1	0	1	...	0	0
:	:	⋱	:	:	:	⋱	:	:
0	0	...	1	0	0	...	1	1

Table 4.1 The feasible 0-1 vectors of the local polytope P of GPP

Remark 4.1 *The system of equations and inequalities* (4.6), ..., (4.10) *is valid for all* $(\mathbf{x}, y) \in P$ *and thus* $P \subseteq P_L$.

Proof. Let $(\mathbf{x}, y) \in D$. Then $(\mathbf{x}, y)$ satisfies (4.6) and (4.9). Since $y = \sum_{j=1}^{n} x_{1j} x_{2j}$ and $x_{ij} \geq 0$ for $1 \leq i \leq 2$, $j \in N$, $(\mathbf{x}, y)$ satisfies (4.10). To prove that (4.7) is satisfied, we calculate $x_{1j} + x_{2j} - y = x_{1j} + x_{2j} - \sum_{\ell=1}^{n} x_{1\ell} x_{2\ell} \leq x_{1j} + x_{2j} - x_{1j} x_{2j} = x_{1j} + x_{2j}(1 - x_{1j}) \in \{0, 1\} \leq 1$ for all $1 \leq j \leq n$. Moreover, since $\sum_{j \in S} x_{1j} - \sum_{j \in S} x_{2j} + y = 1 - \sum_{j \in N-S} x_{1j} - \sum_{j \in S} x_{2j} + \sum_{j \in N} x_{1j} x_{2j} = 1 - \sum_{j \in N-S} x_{1j}(1 - x_{2j}) - \sum_{j \in S} x_{2j}(1 - x_{1j}) \leq 1$, (4.8) is satisfied. Thus it follows that $D \subseteq P_L$ and hence, $P = \text{conv}(D) \subseteq P_L$. □

We order the components of $(\mathbf{x}, y)$ as $(x_{11}, \ldots, x_{1n}, x_{21}, \ldots, x_{2n}, y)$ and denote by $\bar{\mathbf{u}}_{ij} \in \mathbb{R}^{2n}$ with its components indexed in the same order as $\mathbf{x}$ a unit vector with one in its $(i, j)^{th}$ component. Let $\mathbf{u}_{ij} \in \mathbb{R}^{2n+1}$ be obtained from $\bar{\mathbf{u}}_{ij}$ by appending zero at the end and let $\mathbf{v} \in \mathbb{R}^{2n+1}$ be another unit vector with one in its last component.

Proposition 4.1 *The dimension of P equals $2n - 1$ for all $n \geq 3$.*

Proof. Since the two equations in (4.6) are linearly independent, $dim(P) \leq 2n-1$. We establish $dim(P) \geq 2n-1$ by showing that every equation $\alpha \mathbf{x} + \beta y = \gamma$ that is satisfied by all $(\mathbf{x}, y) \in P$ is a linear combination of (4.6).
(i) Since $(\mathbf{u}_{1k} + \mathbf{u}_{2j}) \in P$ for $j \neq k \in N$, $\alpha_{ij} = \alpha_{ik}$ for all $1 \leq i \leq 2, j, k \in N$.

(ii) Since $(\mathbf{u}_{1j} + \mathbf{u}_{2k}), (\mathbf{u}_{1j} + \mathbf{u}_{2j} + \mathbf{v}) \in P$ for $j \neq k \in N$, using (i), $\beta = 0$. Consequently, $\boldsymbol{\alpha}\mathbf{x} + \beta y = \gamma$ becomes $\sum_{i=1}^{2} \alpha_{i1} \sum_{j=1}^{n} x_{ij} = \alpha_{11} + \alpha_{21}$ for all $(\mathbf{x}, y) \in P$; which is a linear combination of the two equations (4.6). □

Proposition 4.2 *Inequality (4.10) defines a facet of P.*

Proof. By Remark (4.1), (4.10) is valid for P. Let $F = \{(\mathbf{x}, y) \in P : y = 0\}$. Since $(\mathbf{u}_{11} + \mathbf{u}_{21} + \mathbf{v}) \in P$ but not in F, F is a *proper face* of P. Suppose there exists a valid inequality $\boldsymbol{\alpha}\mathbf{x} + \beta y \leq \gamma$ for P such that every $(\mathbf{x}, y) \in F$ satisfies $\boldsymbol{\alpha}\mathbf{x} + \beta y = \gamma$.
  (i) Since $(\mathbf{u}_{11} + \mathbf{u}_{2j}) \in F$ for $2 \leq j \leq n$, $\alpha_{2j} = \alpha_{2k}$ for all $2 \leq j, k \leq n$.
  (ii) Since $(\mathbf{u}_{1j} + \mathbf{u}_{21}) \in F$ for $2 \leq j \leq n$, $\alpha_{1j} = \alpha_{1k}$ for all $2 \leq j, k \leq n$.
  (iii) Since $(\mathbf{u}_{1j} + \mathbf{u}_{21}), (\mathbf{u}_{1j} + \mathbf{u}_{2k}) \in F$ for $2 \leq j \neq k \leq n$, from (i) $\alpha_{21} = \alpha_{2k}$ for all $2 \leq k \leq n$. By a similar argument using (ii), $\alpha_{11} = \alpha_{1k}$ for all $2 \leq k \leq n$.
Consequently, $\boldsymbol{\alpha}\mathbf{x} + \beta y = \gamma$ becomes $\sum_{i=1}^{2} \alpha_{i1} \sum_{j=1}^{n} x_{ij} + \beta y = \alpha_{11} + \alpha_{21}$; equivalently, $\beta y = 0$ for all $(\mathbf{x}, y) \in F$ and the proposition follows. □

Proposition 4.3 *Inequality (4.9) defines a facet of P for $1 \leq i \leq 2$, $j \in N$.*

Proof. Inequality (4.9) is trivially valid for P. WROG we prove this proposition for $i = j = 1$. Let $F = \{(\mathbf{x}, y) \in P : x_{11} = 0\}$. Since $(\mathbf{u}_{11} + \mathbf{u}_{21} + \mathbf{v}) \in P$ but not in F, F is a proper face of P. Suppose there exists a valid inequality $\boldsymbol{\alpha}\mathbf{x} + \beta y \leq \gamma$ for P such that every $(\mathbf{x}, y) \in F$ satisfies $\boldsymbol{\alpha}\mathbf{x} + \beta y = \gamma$.
  (i) Since $(\mathbf{u}_{1j} + \mathbf{u}_{21}) \in F$ for $2 \leq j \leq n$, $\alpha_{1j} = \alpha_{1k}$ for all $2 \leq j, k \leq n$.
  (ii) Since $(\mathbf{u}_{1j} + \mathbf{u}_{21}), (\mathbf{u}_{1j} + \mathbf{u}_{22}) \in F$ for $3 \leq j \leq n$, $\alpha_{21} = \alpha_{22}$.
  (iii) Since $(\mathbf{u}_{12} + \mathbf{u}_{22} + \mathbf{v}), (\mathbf{u}_{12} + \mathbf{u}_{2j}) \in F$ for $j \neq 2, j \in N$, $\alpha_{2j} = \alpha_{2k}$ for all $j, k \in N$ and $\beta = 0$.
Consequently, $\boldsymbol{\alpha}\mathbf{x} + \beta y = \gamma$ becomes $(\alpha_{11} - \alpha_{12})x_{11} + \sum_{i=1}^{2} \alpha_{i2} \sum_{j=1}^{n} x_{ij} = \alpha_{12} + \alpha_{22}$; equivalently, $(\alpha_{11} - \alpha_{12})x_{11} = 0$ for all $(\mathbf{x}, y) \in F$. □

Proposition 4.4 *Inequality (4.7) defines a facet of P for $j \in N$.*

Proof. By Remark (4.1), (4.7) is valid for P. WROG we prove this proposition for $j = 1$. Let $F = \{(\mathbf{x}, y) \in P : x_{11} + x_{21} - y = 1\}$. Since $(\mathbf{u}_{12} + \mathbf{u}_{22} + \mathbf{v}) \in P$ but not in F, F is a proper face of P. Suppose there exists a valid inequality $\boldsymbol{\alpha}\mathbf{x} + \beta y \leq \gamma$ for P such that every $(\mathbf{x}, y) \in F$ satisfies $\boldsymbol{\alpha}\mathbf{x} + \beta y = \gamma$.
  (i) Since $(\mathbf{u}_{1j} + \mathbf{u}_{21}) \in F$ for $2 \leq j \leq n$, $\alpha_{1j} = \alpha_{1k}$ for all $2 \leq j, k \leq n$.
  (ii) Since $(\mathbf{u}_{11} + \mathbf{u}_{2j}) \in F$ for $2 \leq j \leq n$, $\alpha_{2j} = \alpha_{2k}$ for all $2 \leq j, k \leq n$.
  (iii) Since $(\mathbf{u}_{11} + \mathbf{u}_{2j}), (\mathbf{u}_{1j} + \mathbf{u}_{21}), (\mathbf{u}_{11} + \mathbf{u}_{21} + \mathbf{v}) \in F$ for $2 \leq j \leq n$, from (i) and (ii) $\alpha_{11} = \alpha_{1j} - \beta$ and $\alpha_{21} = \alpha_{2j} - \beta$ for all $2 \leq j, k \leq n$.

Locally Ideal LP Formulations I

Consequently, $\alpha\mathbf{x} + \beta y = \gamma$ becomes $-\beta(x_{11} + x_{21} - y) + \sum_{i=1}^{2} \alpha_{i2} \sum_{j=1}^{n} x_{ij} = -\beta + \alpha_{12} + \alpha_{22}$; equivalently, $-\beta(x_{11} + x_{21} - y) = -\beta$ for all $(\mathbf{x}, y) \in F$. □

Proposition 4.5 *Inequality (4.8) defines a facet of P for $\emptyset \neq S \subset N$.*

Proof. By Remark (4.1), (4.8) is valid for P. Let $F = \{(\mathbf{x}, y) \in P : \sum_{j \in S} x_{1j} - \sum_{j \in S} x_{2j} + y = 1\}$. Since $(\mathbf{u}_{1g} + \mathbf{u}_{2p}) \in P$ for all $p \in S$ and $g \in N - S$ but not in F, F is a proper face of P. Suppose there exists a valid inequality $\alpha\mathbf{x} + \beta y \leq \gamma$ for P such that every $(\mathbf{x}, y) \in F$ satisfies $\alpha\mathbf{x} + \beta y = \gamma$.

(i) Since $(\mathbf{u}_{1p} + \mathbf{u}_{2g}) \in F$ for $p \in S$ and $g \in N - S$, $\alpha_{1p} = \alpha_{1r}$ and $\alpha_{2g} = \alpha_{2s}$ for all $p, r \in S$ and $g, s \in N - S$.

(ii) Since $(\mathbf{u}_{1p} + \mathbf{u}_{2g}), (\mathbf{u}_{1r} + \mathbf{u}_{2r} + \mathbf{v}) \in F$ for $p \in S, g \in N - S$ and $r \in N$, $\alpha_{1p} - \alpha_{1g} = \beta = \alpha_{2g} - \alpha_{2p}$ for all $p \in S$ and $g \in N - S$. From (i) $\alpha_{1g} = \alpha_{1s}$ and $\alpha_{2p} = \alpha_{2r}$ for all $p, r \in S$ and $g, s \in N - S$.

Thus $\alpha\mathbf{x} + \beta y = \gamma$ becomes $\beta(\sum_{j \in S} x_{1j} - \sum_{j \in S} x_{2j} + y) + \sum_{i=1}^{2} \alpha_{ig} \sum_{j=1}^{n} x_{ij} = \beta + \alpha_{1g} + \alpha_{2g}$ for some $g \in N - S$; equivalently, $\beta(\sum_{j \in S} x_{1j} - \sum_{j \in S} x_{2j} + y) = \beta$ for all $(\mathbf{x}, y) \in F$. □

Remark 4.2 *An optimal solution to $\max\{\mathbf{cx} + qy : (\mathbf{x}, y) \in P\}$ is characterized by two cases:*

(i) *if there exists $p \neq r \in N$ such that $c_{1p} + c_{2r} \geq c_{1i} + c_{2i} + q$ for all $i \in N$ then an optimal solution is $x_{1j} = x_{2\ell} = 1$ and $x_{1i} = x_{2k} = y = 0$ for all $i \neq j \in N$ and $k \neq \ell \in N$ where $j \neq \ell \in N$ and $c_{1j} + c_{2\ell} \geq c_{1p} + c_{2r}$ for all $p \neq r \in N$.*

(ii) *if the condition in (i) does not hold then an optimal solution is $x_{1j} = x_{2j} = y = 1$ and $x_{ik} = 0$ for $1 \leq i \leq 2$, $k \neq j \in N$ where $c_{1j} + c_{2j} \geq c_{1p} + c_{2p}$ for all $p \in N$.*

Proposition 4.6 *The solution of Remark (4.2) is an optimal solution to the LP problem $\max\{\mathbf{cx} + qy : (\mathbf{x}, y) \in P_L\}$ where $(\mathbf{c}, q)$ is an arbitrary cost vector.*

Proof. Let $(\mathbf{x}^*, y^*)$ be the solution vector defined in Remark (4.2). By Remark (4.1), $P \subseteq P_L$ and trivially, $(\mathbf{x}^*, y^*)$ is an extreme point of P_L in both cases of Remark (4.2). We give, in each of these two cases, a polytope $P' \supseteq P_L$ over which $(\mathbf{x}^*, y^*)$ is optimal. Hence $(\mathbf{x}^*, \mathbf{y}^*)$ is, *a forteriori*, optimal over P_L. Suppose we are in case (i) and an optimal solution to P is given by $x_{1j} = x_{2\ell} = 1$ and $x_{1i} = x_{2k} = y = 1$ for all $i \neq j \in N$ and $k \neq \ell \in N$. We consider three subcases. First, assume that $c_{2j} > c_{2\ell}$ and define $P' = \{(\mathbf{x}, y) \in \mathbb{R}^{2n+1} : \sum_{k=1}^{n} x_{ik} = 1$ for $1 \leq i \leq 2, x_{1j} + x_{2j} - y \leq 1, x_{ik} \geq 0$ for $1 \leq i \leq 2, 1 \leq k \leq n, y \geq 0\}$. The dual to this problem is $\min\{u_1 + u_2 + w : u_i \geq c_{ik}$ for $1 \leq i \leq 2, j \neq k \in N, u_1 + w \geq c_{1j}, u_2 + w \geq c_{2j}, -w \geq q, w \geq 0\}$. The vector given by $u_1 = c_{1j} - c_{2j} + c_{2\ell}$, $u_2 = c_{2\ell}$, $w = c_{2j} - c_{2\ell}$, is feasible to the dual problem with objective function value $c_{1j} + c_{2\ell}$. On the other hand, suppose $c_{2j} \leq c_{2\ell}$. WROG assume $c_{21} \leq c_{22} \leq \cdots \leq c_{2n}$ and thus $\ell = n$. Assume $q > 0$ and define $P' = \{(\mathbf{x}, y) \in \mathbb{R}^{2n+1} : \sum_{k=1}^{n} x_{ik} = 1$ for $1 \leq i \leq 2, \sum_{r=k}^{n-1} x_{1r} - \sum_{r=k}^{n-1} x_{2r} + y \leq 1$ for $1 \leq k \leq n-1, x_{ik} \geq 0$ for $1 \leq i \leq 2, k \in N, y \geq 0\}$. The dual to this problem is $min\{u_1 + u_2 + \sum_{k=1}^{n-1} w_k : u_1 + \sum_{r=k}^{n-1} w_r \geq c_{1k}$ for $1 \leq k \leq n-1, u_2 - \sum_{r=k}^{n-1} w_r \geq c_{2k}$ for $1 \leq k \leq n-1, u_1 \geq c_{1n}, u_2 \geq c_{2n}, \sum_{r=1}^{n-1} w_r \geq q, w \geq 0\}$. The vector given by $u_1 = c_{1j} - q$, $u_2 = c_{2n}$, $w_k = 0$ for $1 \leq k \leq p-2$, $w_{p-1} = q - c_{2n} + c_{2p}$, $w_k = c_{2,k+1} - c_{2k}$ for $p \leq k \leq n-1$ where $1 \leq p \leq n-1$ such that $c_{2n} - c_{2p} < q$ and $c_{2n} - c_{2,p-1} \geq q$, is feasible to the dual problem with objective function value $c_{1j} + c_{2n} = c_{1j} + c_{2\ell}$. Next assume $q \leq 0$ and define $P' = \{(\mathbf{x}, y) \in \mathbb{R}^{2n+1} : \sum_{k=1}^{n} x_{ik} = 1$ for $1 \leq i \leq 2, x_{1n} + x_{2n} - y \leq 1, x_{ik} \geq 0$ for $i = 1, 2, k \in N, y \geq 0\}$. The dual to this problem is $min\{u_1 + u_2 + w : u_i \geq c_{ik}$ for $i = 1, 2, k \neq \ell \in N, u_i + w \geq c_{i\ell}$ for $i = 1, 2, w \geq q, w \geq 0\}$. The vector given by $u_1 = c_{1j}, u_2 = c_{2\ell} - w, w = min\{c_{2\ell} - c_{2k} : k \neq \ell \in N\}$ is feasible to the dual problem with objective function value $c_{1j} + c_{2\ell}$. Moreover, in all subcases, the dual objective function value equals $c_{1j} + c_{2\ell}$, which is also equal to that of $(\mathbf{x}^*, y^*)$ and hence, by LP duality $(\mathbf{x}^*, y^*)$ is optimal over P'. Next consider case (ii) of Remark 4.2 and assume that an optimal solution to P is given by $x_{1j} = x_{2j} = y = 1$ and $x_{ik} = 0$ for $1 \leq i \leq 2, k \neq j \in N$. WROG assume $c_{21} \leq c_{22} \leq \cdots \leq c_{2n}$ and define $P' = \{(\mathbf{x}, y) \in \mathbb{R}^{2n+1} : \sum_{k=1}^{n} x_{ik} = 1$ for $1 \leq i \leq 2, \sum_{k=1}^{\ell} x_{1k} - \sum_{k=1}^{\ell} x_{2k} + y \leq 1$ for $1 \leq \ell \leq n-1, x_{1j} + x_{2j} - y \leq 1, x_{ik} \geq 0$ for $1 \leq i \leq 2, k \in N, y \geq 0\}$. The dual to this problem is $min\{u_1 + u_2 + \sum_{k=1}^{n-1} v_k + w : u_1 + \sum_{k=j}^{n-1} v_k \geq c_{1\ell}$ for $j \neq \ell \in N, u_1 + \sum_{k=j}^{n-1} v_k + w \geq c_{1j}, u_2 - \sum_{k=\ell}^{n-1} v_k \geq c_{2\ell}$ for $j \neq \ell \in N, u_2 - \sum_{k=j}^{n-1} v_k + w \geq c_{2j}, \sum_{k=1}^{n-1} v_k - w \geq q, w \geq 0\}$. If $q < 0$ then the vector given by $u_1 = c_{1j} + q, u_2 = c_{2j} + q, w = -q, v_k = 0$ for all $1 \leq k \leq n-1$ is feasible to the dual problem with objective function value $c_{1j} + c_{2j} + q$. On the other hand, if $q \geq 0$ then the vector given by $u_1 = c_{1j} + c_{2j} - c_{2n}, u_2 = c_{2n}, w = 0, v_k = 0$ for all $1 \leq k \leq p-1, v_p = q - c_{2n} + c_{2p}, v_k = c_{2,k+1} - c_{2k}$ for all $p < k \leq n-1$ where $1 \leq p \leq j$ is such that $q \geq c_{2n} - c_{2p}$ and $q < c_{2n} - c_{2,p-1}$, is feasible to the dual problem with objective function value $c_{1j} + c_{2j} + q$. Moreover, the dual objective function value is equal to that of $(\mathbf{x}^*, y^*)$ and hence, by LP duality $(\mathbf{x}^*, y^*)$ is optimal over P'. $\square$

Summarizing we have just proven the following.

Proposition 4.7 *The system of equations and inequalities* (4.6),...,(4.10) *is an ideal linear description of the local polytope* P, *i.e.* $P = P_L$.

Locally Ideal LP Formulations I

Considering all equations and inequalities resulting from the locally ideal linearization of the variables giving rise to quadratic terms in the objective function, we formulate the GPP as the LP problem given by

$$\min\left\{\sum_{i=1}^{m-1}\sum_{k=i+1}^{m} q_{ik}y_{ik} : (\mathbf{x},\mathbf{y}) \in GPP_n^m\right\}, \qquad (\mathcal{O}GPP_n^m)$$

where GPP_n^m is the polytope defined by the convex hull of solutions $(\mathbf{x},\mathbf{y}) \in \mathbb{R}^{mn+m(m-1)/2}$ to the following equations and inequalities in zero-one variables:

$$\sum_{j=1}^{n} x_{ij} = 1 \qquad \text{for } i \in M \qquad (4.11)$$

$$x_{ij} + x_{kj} - y_{ik} \leq 1 \qquad \text{for } i < k \in M, j \in N \qquad (4.12)$$

$$\sum_{j\in S} x_{ij} - \sum_{j\in S} x_{kj} + y_{ik} \leq 1 \qquad \text{for } i < k \in M, j \in N, \emptyset \neq S \subset N \qquad (4.13)$$

$$x_{ij} \geq 0 \qquad \text{for } i \in M, j \in N \qquad (4.14)$$

$$y_{ik} \geq 0 \qquad \text{for } i < k \in M \qquad (4.15)$$

$$x_{ij} \in \{0,1\} \qquad \text{for } i \in M, j \in N. \qquad (4.16)$$

It is not difficult to prove that (4.11),...,(4.16) formulates the GPP correctly. Indeed, a similar, less complete formulation of the GPP has been put forth by Chopra and Rao [1989a, 1993]. Their formulation includes the constraints (4.11), (4.12), (4.14) and (4.15) and the constraints (4.14) for $S = \{j\}$ and $S = N - \{j\}$ only, where $j \in N$. It is shown there that these constraints define facets of the polytope GPP_n^m. We will show here only that the rest of the inequalities in (4.13) not included in their formulation of GPP are also facet defining for GPP_n^m. Chopra and Rao [1989a, 1993] also prove that $dim(GPP_n^m) = m(n-1) + m(m-1)/2$.

Proposition 4.8 *Inequality (4.13) is facet defining for GPP_n^m for $\emptyset \neq S \subset N$.*

Proof. Let the components of $\mathbf{x}$ be ordered as $(x_{11},\ldots,x_{1n},x_{21},\ldots,x_{mn})$ and those of $\mathbf{y}$ be ordered as $(y_{12},y_{13},\ldots,y_{1m},y_{23},\ldots,y_{m-1,m})$. Denote by $\bar{\mathbf{u}}_{ij} \in \mathbb{R}^{mn}$ with its components indexed like those of $\mathbf{x}$, a unit vector with one in its $(i,j)^{th}$ component. Likewise denote by $\bar{\mathbf{v}}_{ik} \in \mathbb{R}^{m(m-1)/2}$ indexed like $\mathbf{y}$ another unit vector with one in its $(i,k)^{th}$ component. Let $\mathbf{u}_{ij} \in \mathbb{R}^{mn+m(m-1)/2}$ be obtained from $\bar{\mathbf{u}}_{ij}$ by appending zeroes in the last $m(m-1)/2$ components and $\mathbf{v}_{ik} \in \mathbb{R}^{mn+m(m-1)/2}$ be obtained from $\bar{\mathbf{v}}_{ik}$ by appending zeroes in the first mn components. Let $\mathbf{z}_I(j) = \sum_{i\in I}\mathbf{u}_{ij} + \sum_{i<k\in I}\mathbf{v}_{ik}$ for $j \in N$ where $I \subseteq M = \{1,\ldots,m\}$. By a similar argument as in Remark (4.1), it follows

that (4.13) is valid for GPP_n^m. WROG, we prove the proposition for $i = 1, k = 2$. Let $F = \{(\mathbf{x},y) \in GPP_n^m : \sum_{j \in S} x_{1j} - \sum_{j \in S} x_{2j} + y_{12} = 1\}$. Since $(\mathbf{u}_{1g} + \mathbf{z}_{M \setminus \{1\}}(p)) \in GPP_m^n$ for all $p \in S$ and $g \in N - S$ but not in F, F is a proper face of P. Suppose there exists a valid inequality $\alpha \mathbf{x} + \beta \mathbf{y} \leq \gamma$ for GPP_n^m such that every $(\mathbf{x},y) \in F$ satisfies $\alpha \mathbf{x} + \beta \mathbf{y} = \gamma$.

(i) Since $(\mathbf{u}_{1j}+\mathbf{u}_{2\ell}+\mathbf{u}_{kp}+\mathbf{z}_{M \setminus \{1,2,k\}}(r)), (\mathbf{u}_{1j}+\mathbf{u}_{2\ell}+\mathbf{u}_{kg}+\mathbf{z}_{M \setminus \{1,2,k\}}(r)) \in F$ for $j \in S, \ell \in N - S, 3 \leq k \leq m, p \neq r \neq g, r \in N$ where $p \neq g \in N$, $\alpha_{kp} = \alpha_{kg}$ for all $3 \leq k \leq m$, $p, g \in N$.

(ii) Since $(\mathbf{u}_{1j} + \mathbf{u}_{2\ell} + \mathbf{u}_{kp} + \mathbf{z}_{M \setminus \{1,2,k\}}(r)), (\mathbf{u}_{1j} + \mathbf{u}_{2\ell} + \mathbf{u}_{ip} + \mathbf{u}_{kp} + v_{ik} + \mathbf{z}_{M \setminus \{1,2,i,k\}}(r)) \in F$ for $j \in S, \ell \in N - S, 3 \leq i < k \leq m$, $p \neq r \in N$, $\beta_{ik} = 0$ for all $3 \leq i < k \leq m$.

(iii) By similar arguments as in (i) and (ii) of Proposition (4.5), $\alpha_{ip} = \alpha_{ir}, \alpha_{ig} = \alpha_{is}, \alpha_{1p} - \alpha_{1g} = \beta_{12} = \alpha_{2g} - \alpha_{2p}$ for all $p, r \in S$ and $g, s \in N - S$.

Consequently, $\alpha \mathbf{x} + \beta \mathbf{y} = \gamma$ becomes $\beta_{12}(\sum_{j \in S} x_{1j} - \sum_{j \in S} x_{2j} + y_{12}) + \sum_{i=1}^{m} \alpha_{ig} \sum_{j=1}^{n} x_{ij} = \beta_{12} + \sum_{i=1}^{m} \alpha_{ig}$ for some $g \in N - S$; equivalently, $\beta_{12}(\sum_{j \in S} x_{1j} - \sum_{j \in S} x_{2j} + y_{12}) = \beta_{12}$ for all $(\mathbf{x},y) \in F$. Hence, the proposition follows. □

The LP relaxation of our formulation of GPP has exponentially many constraints. So the first question to ask is whether or not we can solve the resultant LP problem – practically or theoretically – in polynomial time. This is indeed the case. To this end we must show that the *separation* problem, see e.g. Padberg [1995], for the exponentially many constraints (4.13) can be solved in polynomial time. Let $(\overline{\mathbf{x}}, \overline{\mathbf{y}}) \in \mathbb{R}^{mn+m(m-1)/2}$ satisfy (4.11), (4.12), (4.14) and (4.15) and the inequality $y \leq 1$. These are polynomially many constraints in m and n and they can be checked in polynomial time. To check the constraints (4.13) we need to find for fixed i and k with $1 \leq i < k \leq m$, $z_{ik} = \min\{\sum_{j \in S}(-\overline{x}_{ij} + \overline{x}_{kj}) : \emptyset \neq S \subset N\}$. Using (4.11) and that $y \leq 1$, it follows that $z_{ik} = 0$ for $S = \emptyset$ or $S = N$ and hence we can replace the requirement $\emptyset \neq S \subset N$ by $S \subseteq N$. That is, $z_{ik} = \min\{\sum_{j=1}^{n}(-\overline{x}_{ij} + \overline{x}_{kj})z_j : z_j \in \{0,1\}$ for $1 \leq j \leq n\}$. This zero-one LP problem is trivially solvable; an optimal solution is given by $z_j = 1$ if $\overline{x}_{ij} \geq \overline{x}_{kj}$, 0 otherwise for $j \in N$. Hence if $\overline{y}_{ik} > 1 + z_{ik}$ and only then, the corresponding constraint (4.13) is violated. Consequently we can solve the LP relaxation in polynomial time.

4.2 Operations Scheduling Problems

To consider the OSP, see Chapter 2.7, we define new variables $y_{ikj} = x_{ij}x_{kj}$ for $1 \leq i < k \leq m, 1 \leq j \leq n$ and assume $m \geq n \geq 3$; counting yields that

Locally Ideal LP Formulations I

there are $mn(m-1)/2$ **y**-variables. Denoting by $DQSP_n^m$ the discrete set

$$DQSP_n^m = \left\{ \begin{array}{l} (\mathbf{x}, \mathbf{y}) \in \mathbb{R}^{mn + mn(m-1)/2} : \\ \sum_{j=1}^{n} x_{ij} = 1 \quad \text{for } i \in M \\ y_{ikj} = x_{ij} x_{kj} \quad \text{for } i < k \in M, j \in N \\ x_{ij} \in \{0, 1\} \quad \text{for } i \in M, j \in N \end{array} \right\},$$

the OSP can be written as

$$\min \left\{ \sum_{i=1}^{m} \sum_{j=1}^{n} c_{ij} x_{ij} + \sum_{i=1}^{m-1} \sum_{k=i+1}^{m} \sum_{j=1}^{n} q_{ikj} y_{ikj} : (\mathbf{x}, \mathbf{y}) \in DQSP_n^m \right\},$$

where $q_{ikj} = a_{ikj} + a_{kij}$ in terms of the a_{ikj} of Chapter 2.7. For further use we note that the GPP can be obtained from the OSP by the way of the transformation:

$$y_{ik} = \sum_{j=1}^{n} x_{ij} x_{kj} \quad \text{for all } 1 \le i < k \le m. \tag{4.17}$$

To obtain a linear formulation for $DQSP_n^m$, we consider the local polytope P given by $P = conv(D)$ where $n \ge 3$ and D is defined as follows:

$$D = \left\{ \begin{array}{l} (\mathbf{x}, \mathbf{y}) \in \mathbb{R}^{3n} : \\ \sum_{j=1}^{n} x_{ij} = 1 \quad \text{for } 1 \le i \le 2 \\ y_j = x_{1j} x_{2j} \quad \text{for } 1 \le j \le n \\ x_{ij} \in \{0, 1\} \quad \text{for } 1 \le i \le 2, \ 1 \le j \le n \end{array} \right\}.$$

The set of zero-one vectors of the discrete set D is shown in Table 4.2. Let P_L be the polytope given by $(\mathbf{x}, \mathbf{y}) \in \mathbb{R}^{3n}$ satisfying

$$\sum_{j=1}^{n} x_{ij} = 1 \quad \text{for } 1 \le i \le 2 \tag{4.18}$$

$$-x_{ij} + y_j \le 0 \quad \text{for } 1 \le i \le 2, \ 1 \le j \le n \tag{4.19}$$

$$x_{1j} + x_{2j} - y_j + \sum_{j \ne \ell=1}^{n} y_\ell \le 1 \quad \text{for } 1 \le j \le n \tag{4.20}$$

$$y_j \ge 0 \quad \text{for } 1 \le j \le n. \tag{4.21}$$

Remark 4.3 *The system of equations and inequalities* (4.18), ..., (4.21) *is valid for all* $(\mathbf{x}, \mathbf{y}) \in P$ *and thus* $P \subseteq P_L$.

x_{11}	x_{12}	...	x_{1n}	x_{21}	x_{22}	...	x_{2n}	y_1	y_2	...	y_n
1	0	...	0	1	0	...	0	1	0	...	0
1	0	...	0	0	1	...	0	0	0	...	0
:	:	⋱	:	:	:	⋱	:	:	:	⋱	:
1	0	...	0	0	0	...	1	0	0	...	0
0	1	...	0	1	0	...	0	0	0	...	0
0	1	...	0	0	1	...	0	0	1	...	0
:	:	⋱	:	:	:	⋱	:	:	:	⋱	:
0	1	...	0	0	0	...	1	0	0	...	0
:	:	⋱	:	:	:	⋱	:	:	:	⋱	:
0	0	...	1	1	0	...	0	0	0	...	0
0	0	...	1	0	1	...	0	0	0	...	0
:	:	⋱	:	:	:	⋱	:	:	:	⋱	:
0	0	...	1	0	0	...	1	0	0	...	1

Table 4.2 The feasible 0-1 vectors of the local polytope P of OSP

Proof. Let $(\mathbf{x}, y) \in D$. Then $(\mathbf{x}, \mathbf{y})$ satisfies (4.18). Since $y_j = x_{1j}x_{2j}$, $(\mathbf{x}, \mathbf{y})$ satisfies (4.21). To prove that (4.19) is satisfied, we calculate $-x_{ij} + y_j = -x_{ij}(1 - x_{kj}) \in \{0, -1\} \leq 0$ for all $1 \leq i \neq k \leq 2, 1 \leq j \leq n$. If $\sum_{j \neq \ell = 1}^{n} y_\ell = \sum_{j \neq \ell = 1}^{n} x_{1\ell}x_{2\ell} = 1$ then $x_{1j} = x_{2j} = 0$ and thus, $x_{1j} + x_{2j} - x_{1j}x_{2j} = 0$ for all $(\mathbf{x}, \mathbf{y}) \in D$. If $x_{1j} + x_{2j} - x_{1j}x_{2j} = 1$ then $\sum_{j \neq \ell = 1}^{n} x_{1\ell}x_{2\ell} = 0$ and thus, $x_{1j} + x_{2j} - y_j + \sum_{j \neq \ell = 1}^{n} y_\ell = x_{1j} + x_{2j} - x_{1j}x_{2j} + \sum_{j \neq \ell = 1}^{n} x_{1\ell}x_{2\ell} \leq 1$ for all $(\mathbf{x}, \mathbf{y}) \in D$; consequently, (4.20) is satisfied as well. Thus $D \subseteq P_L$ and $P = conv(D) \subseteq P_L$. □

We order the components of $(\mathbf{x}, \mathbf{y})$ by $(x_{11}, \ldots, x_{1n}, x_{21}, \ldots, x_{2n}, y_1, \ldots, y_n)$ and denote by $\bar{\mathbf{u}}_{ij} \in \mathbb{R}^{2n}$ with its components indexed in the order $(11, \ldots, 1n, 21, \ldots, 2n)$ a unit vector with one in its $(i,j)^{th}$ component. By $\bar{\mathbf{v}}_j \in \mathbb{R}^n$ we denote another unit vector with one in its jth component. Let $\mathbf{u}_{ij} \in \mathbb{R}^{3n}$ be obtained from $\bar{\mathbf{u}}_{ij}$ by appending n zeroes in the last n components and $\mathbf{v}_j \in \mathbb{R}^{3n}$ be obtained from $\bar{\mathbf{v}}_j$ by appending $2n$ zeroes at the beginning.

Proposition 4.9 *The dimension of P given by $dim(P) = 3n - 2$ for all $n \geq 3$.*

Proof. Since the two equations (4.18) are linearly independent, $dim(P) \leq 3n - 2$. We establish $dim(P) \geq 3n - 2$ by showing that every equation $\alpha \mathbf{x} + \beta \mathbf{y} = \gamma$ that is satisfied by all $(\mathbf{x}, \mathbf{y}) \in P$ is a linear combination of (4.18).

(i) Since $(\mathbf{u}_{1k} + \mathbf{u}_{2j}) \in P$ for $1 \leq j \neq k \leq n$, $\alpha_{ij} = \alpha_{ik}$ for all $1 \leq i \leq 2$, $1 \leq j, k \leq n$.

Locally Ideal LP Formulations I 91

(ii) Since $(\mathbf{u}_{1j} + \mathbf{u}_{2k}), (\mathbf{u}_{1j} + \mathbf{u}_{2j} + \mathbf{v}_j) \in P$ for $1 \leq j \neq k \leq n$, $\beta_j = 0$ for all $1 \leq j \leq n$.

Consequently, $\boldsymbol{\alpha}\mathbf{x} + \boldsymbol{\beta}\mathbf{y} = \gamma$ becomes $\sum_{i=1}^{2} \alpha_{i1} \sum_{j=1}^{n} x_{ij} = \sum_{i=1}^{2} \alpha_{i1}$ for all $(\mathbf{x}, \mathbf{y}) \in P$; which is a linear combination of (4.18). □

Proposition 4.10 *Inequality (4.21) defines a facet of P for $1 \leq j \leq n$.*

Proof. By Remark (4.3), (4.21) is valid for P. WROG we prove this proposition for $j = 1$. Let $F = \{(\mathbf{x}, \mathbf{y}) \in P : y_1 = 0\}$. Since $(\mathbf{u}_{11} + \mathbf{u}_{21} + \mathbf{v}_1) \in P$ but not in F, F is a proper face of P. Suppose there exists a valid inequality $\boldsymbol{\alpha}\mathbf{x} + \boldsymbol{\beta}\mathbf{y} \leq \gamma$ for P such that every $(\mathbf{x}, \mathbf{y}) \in F$ satisfies $\boldsymbol{\alpha}\mathbf{x} + \boldsymbol{\beta}\mathbf{y} = \gamma$.

(i) Since $(\mathbf{u}_{11} + \mathbf{u}_{2j}) \in F$ for $2 \leq j \leq n$, $\alpha_{2j} = \alpha_{2k}$ for all $2 \leq j, k \leq n$.
(ii) Since $(\mathbf{u}_{1j} + \mathbf{u}_{21}) \in F$ for $2 \leq j \leq n$, $\alpha_{1j} = \alpha_{1k}$ for all $2 \leq j, k \leq n$.
(iii) Since $(\mathbf{u}_{1j} + \mathbf{u}_{2k}), (\mathbf{u}_{1j} + \mathbf{u}_{2j} + \mathbf{v}_j) \in F$ for $2 \leq j \neq k \leq n$ where $1 \leq k \leq n$, $\alpha_{2k} - \alpha_{2j} = \beta_j$ for all $2 \leq j \neq k \leq n$ and hence from (i), $\beta_j = 0$ and $\alpha_{21} = \alpha_{2j}$ for $2 \leq j \leq n$.
(iv) Since $(\mathbf{u}_{11} + \mathbf{u}_{2j}), (\mathbf{u}_{12} + \mathbf{u}_{2j}) \in F$ for $2 \leq j \leq n$, using (i), $\alpha_{11} = \alpha_{1j}$ for all $2 \leq j \leq n$.

Consequently, $\boldsymbol{\alpha}\mathbf{x} + \boldsymbol{\beta}\mathbf{y} = \gamma$ becomes $\beta_1 y_1 + \sum_{i=1}^{2} \alpha_{i1} \sum_{j=1}^{n} x_{ij} = \alpha_{11} + \alpha_{21}$; equivalently, $\beta_1 y_1 = 0$ for all $(\mathbf{x}, \mathbf{y}) \in F$ and the proposition follows. □

Proposition 4.11 *(4.19) defines a facet of P for $1 \leq i \leq 2$, $1 \leq j \leq n$.*

Proof. By Remark (4.3), (4.19) is valid for P. WROG we prove this proposition for $i = j = 1$. Let $F = \{(\mathbf{x}, \mathbf{y}) \in P : -x_{11} + y_1 = 0\}$. Since $(\mathbf{u}_{11} + \mathbf{u}_{22}) \in P$ but not in F, F is a proper face of P. Suppose there exists a valid inequality $\boldsymbol{\alpha}\mathbf{x} + \boldsymbol{\beta}\mathbf{y} \leq \gamma$ for P such that every $(\mathbf{x}, \mathbf{y}) \in F$ satisfies $\boldsymbol{\alpha}\mathbf{x} + \boldsymbol{\beta}\mathbf{y} = \gamma$.

(i) Since $(\mathbf{u}_{12} + \mathbf{u}_{2j}) \in F$ for $2 \leq j \leq n$, $\alpha_{2j} = \alpha_{2k}$ for all $2 \leq j, k \leq n$.
(ii) Since $(\mathbf{u}_{1j} + \mathbf{u}_{21}) \in F$ for $2 \leq j \leq n$, $\alpha_{1j} = \alpha_{1k}$ for all $2 \leq j, k \leq n$.
(iii) Since $(\mathbf{u}_{1k} + \mathbf{u}_{2j}), (\mathbf{u}_{1k} + \mathbf{u}_{2k} + \mathbf{v}_k) \in F$ for $1 \leq j \neq k \leq n$, $\alpha_{2j} - \alpha_{2k} = \beta_k$ for all $1 \leq j \neq k \leq n$ and hence from (i), $\beta_j = 0$ and $\alpha_{21} = \alpha_{2j}$ for $2 \leq j \leq n$.
(iv) Since $(\mathbf{u}_{1j} + \mathbf{u}_{21}), (\mathbf{u}_{11} + \mathbf{u}_{21} + \mathbf{v}_1) \in F$ for $2 \leq j \leq n$, from (ii), $\alpha_{11} = \alpha_{1j} - \beta_1$ for all $2 \leq j \leq n$.

Consequently, $\boldsymbol{\alpha}\mathbf{x} + \boldsymbol{\beta}\mathbf{y} = \gamma$ becomes $\beta_1(-x_{11} + y_1) + \sum_{i=1}^{2} \alpha_{i2} \sum_{j=1}^{n} x_{ij} = \alpha_{12} + \alpha_{22}$; equivalently, $\beta_1(-x_{11} + y_1) = 0$ for all $(\mathbf{x}, \mathbf{y}) \in F$. □

Proposition 4.12 *Inequality (4.20) defines a facet of P for $1 \leq j \leq n$.*

Proof. By Remark (4.3), (4.20) is valid for P. WROG we prove this proposition for $i = j = 1$. Let $F = \{(\mathbf{x}, \mathbf{y}) \in P : x_{11} + x_{21} - y_1 + \sum_{\ell=2}^{n} y_\ell = 1\}$. Since $(\mathbf{u}_{12} + \mathbf{u}_{23}) \in P$ but not in F, F is a proper face of P. Suppose there exists a valid inequality $\alpha\mathbf{x} + \beta\mathbf{y} \leq \gamma$ for P such that every $(\mathbf{x}, \mathbf{y}) \in F$ satisfies $\alpha\mathbf{x} + \beta\mathbf{y} = \gamma$.

(i) Since $(\mathbf{u}_{11} + \mathbf{u}_{2j}) \in F$ for $2 \leq j \leq n$, $\alpha_{2j} = \alpha_{2k}$ for all $2 \leq j, k \leq n$.
(ii) Since $(\mathbf{u}_{11} + \mathbf{u}_{21} + \mathbf{v}_1), (\mathbf{u}_{11} + \mathbf{u}_{2j}) \in F$ for $2 \leq j \leq n$, $\alpha_{2j} - \alpha_{21} = \beta_1$ for all $2 \leq j \leq n$.
(iii) Since $(\mathbf{u}_{1j} + \mathbf{u}_{2j} + \mathbf{v}_j) \in F$ for $1 \leq j \leq n$, using (ii), $\alpha_{11} - \alpha_{1j} = \beta_j$ for all $2 \leq j \leq n$.
(iv) Since $(\mathbf{u}_{1j} + \mathbf{u}_{21}), (\mathbf{u}_{1j} + \mathbf{u}_{2j} + \mathbf{v}_j) \in F$ for $2 \leq j \leq n$, $\alpha_{21} - \alpha_{2j} = \beta_j$ for all $2 \leq j \leq n$ and hence using (ii), $\beta_1 = -\beta_j$ for $2 \leq j \leq n$.

Thus, $\alpha\mathbf{x} + \beta\mathbf{y} = \gamma$ becomes $\beta_2(x_{11} + x_{21} - y_1 + \sum_{\ell=2}^{n} y_\ell) + \sum_{i=1}^{2} \alpha_{i2} \sum_{j=1}^{n} x_{ij} = \alpha_{12} + \alpha_{22} + \beta_2$; i.e. $\beta_2(x_{11} + x_{21} - y_1 + \sum_{\ell=2}^{n} y_\ell) = \beta_2$ for all $(\mathbf{x}, \mathbf{y}) \in F$. $\square$

Remark 4.4 *An optimal solution to $max\{\mathbf{cx} + \mathbf{qy} : (\mathbf{x}, \mathbf{y}) \in P\}$ is characterized by two cases:*

(i) if there exists $1 \leq p \neq r \leq n$ such that $c_{1p} + c_{2r} \geq c_{1i} + c_{2i} + q_i$ for all $1 \leq i \leq n$ then an optimal solution is $x_{1j} = x_{2\ell} = 1, x_{1i} = x_{2k} = y_t = 0$ for all $1 \leq i \neq j \leq n, 1 \leq k \neq \ell \leq n$ and $1 \leq t \leq n$ where $1 \leq j \neq \ell \leq n$ and $c_{1j} + c_{2\ell} \geq c_{1p} + c_{2r}$ for all $1 \leq p \neq r \leq n$.

(ii) if the condition in (i) does not hold then an optimal solution is $x_{1j} = x_{2j} = y_j = 1$ and $x_{ik} = y_k = 0$ for $1 \leq i \leq 2, 1 \leq k \neq j \leq n$ where $c_{1j} + c_{2j} + q_j \geq c_{1p} + c_{2p} + q_p$ for all $1 \leq p \leq n$.

Proposition 4.13 *The solution of Remark (4.4) is an optimal solution to the LP problem $max\{\mathbf{cx} + \mathbf{qy} : (\mathbf{x}, \mathbf{y}) \in P_L\}$ where $(\mathbf{c}, \mathbf{q})$ is an arbitrary cost vector.*

Proof. Let $(\mathbf{x}^*, \mathbf{y}^*)$ be the solution defined in Remark (4.4). By Remark (4.3), $P \subseteq P_L$ and trivially $(\mathbf{x}^*, \mathbf{y}^*)$ is an extreme point of P_L in both cases of Remark (4.4). We give, in both cases, a polytope $P' \supseteq P_L$ over which $(\mathbf{x}^*, \mathbf{y}^*)$ is optimal. Hence $(\mathbf{x}^*, \mathbf{y}^*)$ is, *a forteriori*, optimal over the polytope P_L.

Suppose we are in case (i) and an optimal solution to P is given by $x_{1j} = x_{2\ell} = 1$ and $x_{1i} = x_{2k} = y_t = 0$ for all $1 \leq i \neq j \leq n, 1 \leq k \neq \ell \leq n$ and $1 \leq t \leq n$. We consider two subcases. First, assume that $c_{2j} > c_{2\ell}$ and define $P' = \{(\mathbf{x}, \mathbf{y}) \in \mathbb{R}^{3n} : \sum_{k=1}^{n} x_{ik} = 1$ for $1 \leq i \leq 2, -x_{ik} + y_k \leq 0, x_{1j} + x_{2j} - y_j + \sum_{j \neq k=1}^{n} y_k \leq 1, x_{ik} \geq 0, y_k \geq 0$ for $1 \leq i \leq 2, 1 \leq k \leq n\}$. The dual to this problem is $min\{u_1 + u_2 + w : u_i - v_{ik} \geq c_{ik}$ for $1 \leq i \leq 2, 1 \leq j \neq k \leq n, u_i - v_{ij} + w \geq c_{ij}$ for $i = 1, 2, v_{1k} + v_{2k} + w \geq q_k$ for $1 \leq j \neq k \leq n, v_{1j} + v_{2j} - w \geq q_j, v_{ik} \geq 0$ for $1 \leq i \leq 2, 1 \leq k \leq n\}$. The vector given by $u_1 = c_{1j} - c_{2j} + c_{2\ell}, u_2 = c_{2\ell}, w = c_{2j} - c_{2\ell}, v_{1k} = max\{c_{1k} +$

Locally Ideal LP Formulations I

$q_k - c_{2k}, 0\}, v_{2k} = c_{2\ell} - c_{2k}$ for all $1 \leq j \neq k \leq n, v_{1j} = v_{2j} = 0$ is feasible to the dual problem with objective function value $c_{1j} + c_{2\ell}$. Next, assume that $c_{2j} \leq c_{2\ell}$ and define $P' = \{(\mathbf{x}, \mathbf{y}) \in \mathbb{R}^{3n} : \sum_{k=1}^{n} x_{ik} = 1$ for $1 \leq i \leq 2, -x_{ik} + y_k \leq 0, x_{ik} \geq 0, y_k \geq 0$ for $1 \leq i \leq 2, 1 \leq k \leq n\}$. The dual to this problem is $min\{u_1 + u_2 : u_i - v_{ik} \geq c_{ik}$ for $1 \leq i \leq 2, 1 \leq k \leq n, v_{1k} + v_{2k} \geq q_k$ for $1 \leq k \leq n, v_{ik} \geq 0$ for $1 \leq i \leq 2, 1 \leq k \leq n\}$. The vector given by $u_1 = c_{1j}, u_2 = c_{2\ell}, v_{1k} = c_{1j} - c_{1k}, v_{2k} = c_{2\ell} - c_{2k}$ for all $1 \leq k \leq n$ is feasible to the dual problem with objective function value $c_{1j} + c_{2\ell}$. In both subcases the dual objective function value is equal to that of $(\mathbf{x}^*, \mathbf{y}^*)$ and hence, by LP duality $(\mathbf{x}^*, \mathbf{y}^*)$ is optimal over P'.

Next consider case (ii) of Remark 4.4 and assume that an optimal solution to P is given by $x_{1j} = x_{2j} = y_j = 1$ and $x_{ik} = y_k = 0$ for $1 \leq i \leq 2, 1 \leq k \neq j \leq n$. Define $P' = \{(\mathbf{x}, \mathbf{y}) \in \mathbb{R}^{3n} : \sum_{k=1}^{n} x_{ik} = 1$ for $1 \leq i \leq 2, -x_{ik} + y_k \leq 0$ for $1 \leq i \leq 2, 1 \leq k \leq n, x_{1j} + x_{2j} - y_j + \sum_{j \neq k=1}^{n} y_k \leq 1, x_{ik} \geq 0, y_k \geq 0$ for $1 \leq i \leq 2, 1 \leq k \leq n\}$. The dual to this problem is $min\{u_1 + u_2 + w : u_i - v_{ik} \geq c_{ik}$ for $1 \leq i \leq 2, 1 \leq j \neq k \leq n, u_i - v_{ij} + w \geq c_{ij}$ for $1 \leq i \leq 2, v_{1j} + v_{2j} - w \geq q_j, v_{ik} \geq 0$ for $1 \leq i \leq 2, 1 \leq k \leq n, w \geq 0\}$. If $q_j < 0$ then the vector given by $u_i = c_{ij} + q_j$ for $1 \leq i \leq 2, v_{ik} = c_{ik} - c_{ij} + q_k$ for $1 \leq i \leq 2, 1 \leq j \neq k \leq n, v_{1j} = v_{2j} = 0, w = -q_j$ is feasible to the dual problem with objective function value $c_{1j} + c_{2j} + q_j$. On the other hand, if $q_j \geq 0$ then the vector given by $u_1 = max_j c_{1j}, u_2 = c_{1j} + c_{2j} + q_j - u_1, v_{1k} = u_1 - c_{1k}, v_{2k} = u_2 - c_{2k}$ for all $1 \leq k \leq n, w = 0$, is feasible to the dual problem with the same objective function value. Moreover, the dual objective function value is equal to that of $(\mathbf{x}^*, \mathbf{y}^*)$ and hence, by LP duality $(\mathbf{x}^*, \mathbf{y}^*)$ is optimal over P'. □

We now state a proposition which summarizes the preceding.

Proposition 4.14 *The system of equations and inequalities* (4.18), ..., (4.21) *is an ideal linear description of the local polytope P, i.e. $P = P_L$.*

Considering all equations and inequalities resulting from the locally ideal linearization of the variables giving rise to quadratic terms in the objective function of the OSP, we formulate the OSP as the LP problem given by:

$$min \left\{ \sum_{i=1}^{m} \sum_{j=1}^{n} c_{ij} x_{ij} + \sum_{i=1}^{m-1} \sum_{k=i+1}^{m} \sum_{j=1}^{n} q_{ikj} y_{ikj} : (\mathbf{x}, \mathbf{y}) \in QSP_n^m \right\}, \quad (\mathcal{O}QSP_n^m)$$

where QSP_n^m is the polytope defined by the convex hull of solutions $(\mathbf{x}, \mathbf{y}) \in \mathbb{R}^{mn+mn(m-1)/2}$ to the following equations and inequalities in zero-one variables:

$$\sum_{j=1}^{n} x_{ij} = 1 \quad \text{for } 1 \leq i \leq m \quad (4.22)$$

$$-x_{ij} + y_{ikj} \leq 0 \quad \text{for } 1 \leq i < k \leq m,\ 1 \leq j \leq n \quad (4.23)$$

$$-x_{kj} + y_{ikj} \leq 0 \quad \text{for } 1 \leq i < k \leq m,\ 1 \leq j \leq n \quad (4.24)$$

$$x_{ij} + x_{kj} - y_{ikj} + \sum_{j \neq \ell = 1}^{n} y_{ik\ell} \leq 1 \quad \text{for } 1 \leq i < k \leq m,\ 1 \leq j \leq n \quad (4.25)$$

$$y_{ikj} \geq 0 \quad \text{for } 1 \leq i < k \leq m,\ 1 \leq j \leq n \quad (4.26)$$

$$x_{ij} \in \{0,1\} \quad \text{for } 1 \leq i \leq m,\ 1 \leq j \leq n. \quad (4.27)$$

Proposition 4.15 $\mathcal{O}QSP_n^m$ *formulates of the Operations Scheduling Problem.*

Proof. By similar arguments as in Remark (4.3) it follows that $DQSP_n^m \subseteq QSP_n^m$. Let $(\mathbf{x},\mathbf{y}) \in QSP_n^m$. We show that $y_{ikj} = x_{ij}x_{kj}$ for all $1 \leq i < k \leq m$ and $1 \leq j \leq n$. Suppose that there exists $1 \leq p < g \leq m$ and $1 \leq r \leq n$ such that $y_{pgr} \neq x_{pr}x_{gr}$. Using (4.23), (4.24) and (4.26), we conclude $y_{pgr} = 0$ whenever $x_{pr} = 0$ or $x_{gr} = 0$. So necessarily $x_{pr} = x_{gr} = 1$. But from (4.26) we get contradiction to (4.25) and hence, $y_{pgr} = 1$. Since all the extreme points of QSP_n^m are zero-one valued and in $DQSP_n^m$, the proposition follows. □

In Chapter 5.2 we give more results about the polytope QSP_n^m. The LP relaxation of our formulation of the OSP has polynomially many variables and polynomially many equations and inequalities and hence, it is polynomially solvable. We also note that the OSP with *machine independent* quadratic interaction costs for all pairs of jobs was shown to be identical to the GPP in Chapter 2. Thus we have the option of working either with the OSP, which formulates the problem in a larger space of variables with polynomially many constraints, or with the GPP, which is defined in a smaller space of variables but with an exponential number of constraints. The choice of formulation in such a situation has to be based on the relative strength of alternative formulations in approximating the associated polyhedra. We will show in Chapter 5.1 that the linear relaxation of the OSP formulation, in this special case, is dominated by the GPP formulation but equivalent to the formulation due to Chopra and Rao [1989a, 1993].

4.3 Multi-Processor Assignment Problems

To consider the MPP, see Chapter 2.5, we define new variables $y_{ij}^{k\ell} = x_{ij}x_{k\ell} + x_{i\ell}x_{kj}$ for $1 \le i < k \le m$ and $1 \le j < \ell \le n$ and assume $m \ge n \ge 3$; counting yields that there are $mn(m-1)(n-2)/4$ **y**-variables. Denoting by $DQPP_n^m$ the discrete set

$$DQPP_n^m = \left\{ \begin{array}{l} (\mathbf{x},\mathbf{y}) \in \mathbb{R}^{mn+mn(m-1)(n-1)/4}: \\ \sum_{j=1}^n x_{ij} = 1 \quad \text{for } i \in M \\ y_{ij}^{k\ell} = x_{ij}x_{k\ell} + x_{i\ell}x_{kj} \quad \text{for } i < k \in M, j < \ell \in N \\ x_{ij} \in \{0,1\} \quad \text{for } i \in M, j \in N \end{array} \right\},$$

the MPP can be written as

$$\min \left\{ \sum_{i,j \in N} c_{ij} x_{ij} + \sum_{i<k\in M} \sum_{j<\ell\in N} q_{ij}^{k\ell} y_{ij}^{k\ell} : (\mathbf{x},\mathbf{y}) \in DQPP_n^m \right\},$$

where $q_{ij}^{k\ell} = a_{ijk\ell} + a_{k\ell ij}$ in terms of the $a_{ijk\ell}$ of Chapter 2.5.

To obtain a linear formulation for $DQPP_n^m$ in zero-one variables, we consider the local polytope P given by $P = conv(D)$ where $n \ge 3$ and D is

$$D = \left\{ \begin{array}{l} (\mathbf{x},\mathbf{y}) \in \mathbb{R}^{n(n+3)/2}: \\ \sum_{j=1}^n x_{ij} = 1 \quad \text{for } 1 \le i \le 2 \\ y_{1j}^{2\ell} = x_{1j}x_{2\ell} + x_{1\ell}x_{2j} \quad \text{for } j < \ell \in N \\ x_{ij} \in \{0,1\} \quad \text{for } 1 \le i \le 2, j \in N \end{array} \right\}.$$

In Table 4.3 we show all zero-one vectors of the discrete set D where we have abbreviated $y_{1j}^{2\ell}$ to y_j^ℓ for $1 \le j < \ell \le n$. Let P_L be the polytope of all $(\mathbf{x},\mathbf{y}) \in \mathbb{R}^{n(n+3)/2}$ satisfying

$$\sum_{j=1}^n x_{ij} = 1 \qquad \text{for } 1 \le i \le 2 \quad (4.28)$$

$$-x_{1j} - x_{2j} + \sum_{\ell=1}^{j-1} y_{1\ell}^{2j} + \sum_{\ell=j+1}^n y_{1j}^{2\ell} \le 0 \qquad \text{for } j \in N \quad (4.29)$$

$$\sum_{j \in S}(x_{1j} - x_{2j}) - \sum_{j>\ell \in N-S} y_{1\ell}^{2j} - \sum_{j<\ell \in N-S} y_{1j}^{2\ell}) \le 0 \qquad \text{for } \emptyset \ne S \subset N \quad (4.30)$$

$$y_{1j}^{2\ell} \ge 0 \qquad \text{for } j < \ell \in N. \quad (4.31)$$

x_{11}	x_{12}	...	x_{1n}	x_{21}	x_{22}	x_{23}	...	$x_{2,n-1}$	x_{2n}	y_1^2	...	y_1^n	y_2^3	...	y_2^n	...	y_{n-1}^n
1	0	...	0	1	0	0	...	0	0	0	...	0	0	...	0	...	0
1	0	...	0	0	1	0	...	0	0	0	...	0	1	...	0	...	0
⋮		⋱	⋮	⋮			⋱	⋮			⋱	⋮		⋱	⋮	⋱	⋮
1	0	...	0	0	0	0	...	0	1	0	...	1	0	...	0	...	0
0	1	...	0	1	0	0	...	0	0	1	...	0	0	...	0	...	0
0	1	...	0	0	1	0	...	0	0	0	...	0	0	...	0	...	0
0	1	...	0	0	0	1	...	0	0	0	...	0	1	...	0	...	0
⋮		⋱	⋮	⋮			⋱	⋮			⋱	⋮		⋱	⋮	⋱	⋮
0	1	...	0	0	0	0	...	0	1	0	...	0	0	...	1	...	0
⋮		⋱	⋮	⋮			⋱	⋮			⋱	⋮		⋱	⋮	⋱	⋮
0	0	...	1	1	0	0	...	0	0	0	...	1	0	...	0	...	0
0	0	...	1	0	1	0	...	0	0	0	...	0	0	...	1	...	0
⋮		⋱	⋮	⋮			⋱	⋮			⋱	⋮		⋱	⋮	⋱	⋮
0	0	...	1	0	0	0	...	1	0	0	...	0	0	...	0	...	1
0	0	...	1	0	0	0	...	0	1	0	...	0	0	...	0	...	0

Table 4.3 The feasible 0-1 vectors of the local polytope P of MPP

Remark 4.5 *The system of equations and inequalities* (4.28),...,(4.31) *is valid for all* $(\mathbf{x},\mathbf{y}) \in P$ *and thus* $P \subseteq P_L$.

Proof. Let $(\mathbf{x},y) \in D$. Then $(\mathbf{x},\mathbf{y})$ satisfies (4.28). Since $y_{1j}^{2\ell} = x_{1j}x_{2\ell} + x_{1\ell}x_{2j} \geq 0$, (4.31) is satisfied. To prove that (4.29) is satisfied, we calculate $-x_{1j} - x_{2j} + \sum_{\ell=1}^{j-1} y_{1\ell}^{2j} + \sum_{\ell=j+1}^{n} y_{1j}^{2\ell} = -x_{1j} - x_{2j} + \sum_{j \neq \ell=1}^{n}(x_{1j}x_{2\ell} + x_{1\ell}x_{2j}) = -x_{1j} - x_{2j} + x_{1j}(1-x_{2j}) + (1-x_{1j})x_{2j} = -2x_{1j}x_{2j} \in \{0,-2\} \leq 0$. Likewise, we calculate $\sum_{j \in S}(x_{1j} - x_{2j} - \sum_{j > \ell \in N-S} y_{1\ell}^{2j} - \sum_{j < \ell \in N-S} y_{1j}^{2\ell}) = \sum_{j \in S}(x_{1j} - x_{2j} - \sum_{\ell \in N-S}(x_{1j}x_{2\ell} + x_{1\ell}x_{2j})) = \sum_{j \in S}(x_{1j} - x_{2j} - x_{1j}(1 - \sum_{\ell \in S} x_{2\ell}) - (1 - \sum_{\ell \in S} x_{1\ell})x_{2j}) = -2\sum_{j \in S}(x_{2j} - \sum_{\ell \in S} x_{1j}x_{2\ell}) = -2\sum_{j \in S} x_{2j}(1 - \sum_{\ell \in S} x_{1j}) \in \{0,-2\} \leq 0$; i.e., (4.30) is satisfied as well. Thus, $D \subseteq P_L$ and hence, $P = conv(D) \subseteq P_L$. □

We order the components of $\mathbf{x}$ by $(x_{11},\ldots,x_{1n},x_{21},\ldots,x_{2n})$ and those of $\mathbf{y}$ by $(y_{11}^{22},\ldots,y_{11}^{2n}, y_{12}^{23}, y_{12}^{24},\ldots y_{12}^{2n},\ldots, y_{1,n-1}^{2n})$. Let $\bar{\mathbf{u}}_{ij} \in \mathbb{R}^{2n}$ with its components ordered like those of $\mathbf{x}$ be a unit vector with one in its $(i,j)^{th}$ component and by $\bar{\mathbf{v}}_{1j}^{2\ell} \in \mathbb{R}^{n(n-1)/2}$ with its components ordered like those of $\mathbf{y}$ be another unit vector with one in its $\binom{2,\ell}{1,j}^{th}$ component. Let $\mathbf{u}_{ij} \in \mathbb{R}^{n(n+3)/2}$ be obtained from $\bar{\mathbf{u}}_{ij}$ by appending $n(n-1)/2$ zeroes in the last $n(n-1)/2$ components and $\mathbf{v}_{1j}^{2\ell} \in \mathbb{R}^{n(n+3)/2}$ be obtained from $\bar{\mathbf{v}}_{1j}^{2\ell}$ by appending $2n$ zeroes at the beginning.

Proposition 4.16 *The dimension of P equals $n(n+3)/2 - 2$ for $n \geq 3$.*

Locally Ideal LP Formulations I

Proof. Since the two equations (4.28) are linearly independent, $dim(P) \leq n(n+3)/2 - 2$. We establish $dim(P) \geq n(n+3)/2 - 2$ by showing that every equation $\boldsymbol{\alpha}\mathbf{x} + \boldsymbol{\beta}\mathbf{y} = \gamma$ that is satisfied by all $(\mathbf{x}, \mathbf{y}) \in P$ is a linear combination of the equations (4.28).

(i) Since $(\mathbf{u}_{1j} + \mathbf{u}_{2j}), (\mathbf{u}_{1j} + \mathbf{u}_{2\ell} + \mathbf{v}_{1j}^{2\ell}) \in P$ for $1 \leq j < \ell \leq n$, $\beta_{1j}^{2\ell} = \alpha_{2j} - \alpha_{2\ell}$ for all $1 \leq j < \ell \leq n$. Since $(\mathbf{u}_{1\ell} + \mathbf{u}_{2\ell}), (\mathbf{u}_{1\ell} + \mathbf{u}_{2j} + \mathbf{v}_{1j}^{2\ell}) \in P$ for $1 \leq j < \ell \leq n$, $\beta_{1j}^{2\ell} = \alpha_{2\ell} - \alpha_{2j}$ for all $1 \leq j < \ell \leq n$. Hence, $\beta_{1j}^{2\ell} = 0$ and $\alpha_{2j} = \alpha_{2\ell}$ for all $1 \leq j < \ell \leq n$.

(ii) Since $(\mathbf{u}_{1j} + \mathbf{u}_{2j}), (\mathbf{u}_{1\ell} + \mathbf{u}_{2j} + \mathbf{v}_{1j}^{2\ell}) \in P$ for $1 \leq j < \ell \leq n$, $\beta_{1j}^{2\ell} = \alpha_{1j} - \alpha_{1\ell}$ for all $1 \leq j < \ell \leq n$. Moreover, by (i), $\alpha_{1j} = \alpha_{1\ell}$ for all $1 \leq j < \ell \leq n$.

Consequently, $\boldsymbol{\alpha}\mathbf{x} + \boldsymbol{\beta}\mathbf{y} = \gamma$ becomes $\sum_{i=1}^{2} \alpha_{i1} \sum_{j=1}^{n} x_{ij} = \sum_{i=1}^{2} \alpha_{i1}$ for all $(\mathbf{x}, \mathbf{y}) \in P$; which is a linear combination of the equations (4.28). □

Proposition 4.17 *Inequality (4.31) defines a facet of P for $1 \leq j < \ell \leq n$.*

Proof. By Remark (4.5), (4.31) is valid for P. Let $F = \{(\mathbf{x}, \mathbf{y}) \in P : y_{1j}^{2\ell} = 0\}$. Since $(\mathbf{u}_{1j} + \mathbf{u}_{2\ell} + \mathbf{v}_{1j}^{2\ell}) \in P$ but not in F, F is a proper face of P. Suppose there exists a valid inequality $\boldsymbol{\alpha}\mathbf{x} + \boldsymbol{\beta}\mathbf{y} \leq \gamma$ for P such that every $(\mathbf{x}, \mathbf{y}) \in F$ satisfies $\boldsymbol{\alpha}\mathbf{x} + \boldsymbol{\beta}\mathbf{y} = \gamma$.

(i) Since $(\mathbf{u}_{1p} + \mathbf{u}_{2p}), (\mathbf{u}_{1p} + \mathbf{v}_{2r} + \mathbf{v}_{1p}^{2r}) \in F$ for $1 \leq j < p \leq n$ except when $p = j$ and $r = \ell$, $\beta_{1p}^{2r} = \alpha_{2p} - \alpha_{2r}$ for $1 \leq p < r \leq n$ except when $p = j$ and $r = \ell$. Since $(\mathbf{u}_{1r} + \mathbf{u}_{2r}), (\mathbf{u}_{1r} + \mathbf{v}_{2p} + \mathbf{v}_{1p}^{2r}) \in F$ for $1 \leq p < r \leq n$ except when $p = j$ and $r = \ell$, $\beta_{1p}^{2r} = \alpha_{2r} - \alpha_{2p}$. Hence, $\alpha_{2p} = \alpha_{2r}$ for $1 \leq p < r \leq n$ and $\beta_{1p}^{2r} = 0$ for all $1 \leq p < r \leq n$ except when $p = j$ and $r = \ell$.

(ii) Since $(\mathbf{u}_{1p} + \mathbf{u}_{2p}), (\mathbf{u}_{1r} + \mathbf{u}_{2r}), (\mathbf{u}_{1p} + \mathbf{v}_{2r} + \mathbf{v}_{1p}^{2r}) \in F$ for $1 \leq p < r \leq n$ except when $p = j$ or $r = \ell$, by a similar argument as in (i), $\alpha_{1p} = \alpha_{1r}$ for all $1 \leq p < r \leq n$.

Consequently, $\boldsymbol{\alpha}\mathbf{x} + \boldsymbol{\beta}\mathbf{y} = \gamma$ becomes $\sum_{i=1}^{2} \alpha_{i1} \sum_{j=1}^{n} x_{ij} + \beta_{1j}^{2\ell} y_{1j}^{2\ell} = \sum_{i=1}^{2} \alpha_{i1}$; equivalently, $\beta_{1j}^{2\ell} y_{1j}^{2\ell} = 0$ for all $(\mathbf{x}, \mathbf{y}) \in F$. □

Proposition 4.18 *Inequality (4.29) defines a facet of P for $1 \leq j < \ell \leq n$.*

Proof. By Remark (4.5), (4.29) is valid for P. Let $F = \{(\mathbf{x}, \mathbf{y}) \in P : -x_{1j} - x_{2j} + \sum_{\ell=1}^{j-1} y_{1\ell}^{2j} + \sum_{\ell=j+1}^{n} y_{1j}^{2\ell} = 0\}$. Since $(\mathbf{u}_{1j} + \mathbf{u}_{2j}) \in P$ but not in F, F is a proper face of P. Suppose there exists a valid inequality $\boldsymbol{\alpha}\mathbf{x} + \boldsymbol{\beta}\mathbf{y} \leq \gamma$ for P such that every $(\mathbf{x}, \mathbf{y}) \in F$ satisfies $\boldsymbol{\alpha}\mathbf{x} + \boldsymbol{\beta}\mathbf{y} = \gamma$.

(i) Since $(\mathbf{u}_{1p}, \mathbf{u}_{2p}), (\mathbf{u}_{1p} + \mathbf{u}_{2r} + \mathbf{v}_{1p}^{2r}) \in F$ for $j \neq p$, $1 \leq p < r \leq n$, $\beta_{1p}^{2r} = \alpha_{2p} - \alpha_{2r}$ for all $j \neq p$, $1 \leq p < r \leq n$. Since $(\mathbf{u}_{1r} + \mathbf{u}_{2r}), (\mathbf{u}_{1r} + \mathbf{u}_{2p} + \mathbf{v}_{1p}^{2r}) \in F$ for $j \neq r$, $1 \leq p < r \leq n$, and thus $\beta_{1p}^{2r} = \alpha_{2r} - \alpha_{2p}$ for all $r \neq j$, $1 \leq p < r \leq$. Hence, $\beta_{1p}^{2r} = 0$ and $\alpha_{2r} = \alpha_{2p}$ for all $p \neq j \neq r$, $1 \leq p < r \leq n$.

(ii) Since $(\mathbf{u}_{1p}, \mathbf{u}_{2p}), (\mathbf{u}_{1r} + \mathbf{u}_{2p} + \mathbf{v}_{1p}^{2r}) \in F$ for $j \neq p, 1 \leq p < r \leq n$, $\alpha_{1p} = \alpha_{1r}$ for all $p \neq j \neq r, 1 \leq p < r \leq n$.

(iii) Since $(\mathbf{u}_{1p}, \mathbf{u}_{2p}), (\mathbf{u}_{1p} + \mathbf{u}_{2j} + \mathbf{v}_{1p}^{2j}), (\mathbf{u}_{1j} + \mathbf{u}_{2p} + \mathbf{v}_{1p}^{2j}) \in F$ for $1 \leq p < j \leq n$, $\beta_{1p}^{2j} = \alpha_{1p} - \alpha_{1j} = \alpha_{2p} - \alpha_{2j}$ for all $1 \leq p < j \leq n$. By a similar argument, $\beta_{1p}^{2j} = \alpha_{1p} - \alpha_{1j} = \alpha_{2p} - \alpha_{2j}$, i.e., $\alpha_{1j} = \alpha_{1p} - \beta_{1p}^{2j}$ and $\alpha_{2j} = \alpha_{2p} - \beta_{1p}^{2j}$ for all $1 \leq p < j \leq n$. Moreover, $\alpha_{1j} = \alpha_{1p} - \beta_{1j}^{2p}$ and $\alpha_{2j} = \alpha_{2p} - \beta_{1j}^{2p}$ for all $1 \leq j < p \leq n$ and $\beta_{1p}^{2j} = \beta_{1j}^{2r}$ for $1 \leq p < j < r \leq n$.

Consequently, $\boldsymbol{\alpha}\mathbf{x} + \boldsymbol{\beta}\mathbf{y} = \gamma$ becomes $\sum_{i=1}^{2} \alpha_{ip} \sum_{\ell=1}^{n} x_{i\ell} - \beta_{1p}^{2j}(x_{1j} + x_{2j} - \sum_{\ell=1}^{j-1} y_{1\ell}^{2j} + \sum_{\ell=j+1}^{n} y_{1j}^{2\ell}) = \sum_{i=1}^{2} \alpha_{ip}$ for some p such that $1 \leq p < j$; equivalently, $-\beta_{1p}^{2j}(x_{1j} + x_{2j} - \sum_{\ell=1}^{j-1} y_{1\ell}^{2j} - \sum_{\ell=j+1}^{n} y_{1j}^{2\ell}) = 0$ for all $(\mathbf{x}, \mathbf{y}) \in F$. $\square$

Proposition 4.19 *Inequality (4.30) defines a facet of P for $\emptyset \neq S \subset N$.*

Proof. By Remark (4.5) it follows that (4.30) is valid for P. WROG, assume $S = \{1, \ldots, s\}$ and let $F = \{(\mathbf{x}, \mathbf{y}) \in P : \sum_{j=1}^{s}(x_{1j} - x_{2j} - \sum_{\ell=s+1}^{n} y_{1j}^{2\ell}) = 0\}$. Since $(\mathbf{u}_{1\ell} + \mathbf{u}_{2j} + \mathbf{u}_{1j}^{2\ell})$ for some $1 \leq j \leq s, s+1 \leq \ell \leq n$ is in P but not in F, F is a proper face of P. Suppose there exists a valid inequality $\boldsymbol{\alpha}\mathbf{x} + \boldsymbol{\beta}\mathbf{y} \leq \gamma$ for P such that every $(\mathbf{x}, \mathbf{y}) \in F$ satisfies $\boldsymbol{\alpha}\mathbf{x} + \boldsymbol{\beta}\mathbf{y} = \gamma$.

(i) Since $(\mathbf{u}_{1p} + \mathbf{u}_{2p}), (\mathbf{u}_{1p} + \mathbf{u}_{2r} + \mathbf{v}_{1p}^{2r}) \in F$ for $1 \leq p < r \leq s$, $\beta_{1p}^{2r} = \alpha_{2p} - \alpha_{2r}$ for all $1 \leq p < r \leq s$. Since $(\mathbf{u}_{1r} + \mathbf{u}_{2r}), (\mathbf{u}_{1r} + \mathbf{u}_{2p} + \mathbf{v}_{1p}^{2r}) \in F$ for $1 \leq p < r \leq s$, $\beta_{1p}^{2r} = \alpha_{2r} - \alpha_{2p}$ for all $1 \leq p < r \leq s$. Thus, $\beta_{1p}^{2r} = 0$ and $\alpha_{2p} = \alpha_{2r}$ for all $1 \leq p < r \leq s$.

(ii) Since $(\mathbf{u}_{1p} + \mathbf{u}_{2p}), (\mathbf{u}_{1r} + \mathbf{u}_{2r}), (\mathbf{u}_{1p} + \mathbf{u}_{2r} + \mathbf{v}_{1p}^{2r}), (\mathbf{u}_{1r} + \mathbf{u}_{2p} + \mathbf{v}_{1p}^{2r}) \in F$ for $s+1 \leq p < r \leq n$, $\beta_{1p}^{2r} = 0$ and $\alpha_{2p} = \alpha_{2r}$ for all $s+1 \leq p < r \leq n$.

(iii) Since $(\mathbf{u}_{1r} + \mathbf{u}_{2r}), (\mathbf{u}_{1p} + \mathbf{u}_{2r} + \mathbf{v}_{1p}^{2r}) \in F$ for $1 \leq p \leq s, s+1 \leq r \leq n$, $\alpha_{1p} = \alpha_{1r} - \beta_{1p}^{2r}$. Since $(\mathbf{u}_{1p} + \mathbf{u}_{2p}), (\mathbf{u}_{1p} + \mathbf{u}_{2r} + \mathbf{v}_{1p}^{2r}) \in F$ for $1 \leq p \leq s, s+1 \leq r \leq n$, $\alpha_{2p} = \alpha_{1r} + \beta_{1p}^{2r}$ for all $1 \leq p \leq s, s+1 \leq r \leq n$.

Consequently, $\boldsymbol{\alpha}\mathbf{x} + \boldsymbol{\beta}\mathbf{y} = \gamma$ becomes $\sum_{i=1}^{2} \alpha_{ir} \sum_{\ell=1}^{n} x_{i\ell} + \beta_{1p}^{2r} \sum_{j \in S}(-x_{1j} + x_{2j} + \sum_{\ell \in N-S} y_{1j}^{2\ell}) = \sum_{i=1}^{2} \alpha_{ir}$; equivalently, $\beta_{1j}^{2p} \sum_{j \in S}(-x_{1j} + x_{2j} + \sum_{\ell \in N-S} y_{1j}^{2\ell}) = 0$ for all $(\mathbf{x}, \mathbf{y}) \in F$ where $1 \leq p \leq s, s+1 \leq r \leq n$. $\square$

Remark 4.6 *An optimal solution to $max\{\mathbf{cx} + \mathbf{qy} : (\mathbf{x}, \mathbf{y}) \in P\}$ is characterized by two cases:*

(i) *if there exists $1 \leq p < r \leq n$ such that $c_{1p} + c_{2r} + q_{1p}^{2r} \geq c_{1i} + c_{2i}$ or $c_{1p} + c_{2r} + q_{1p}^{2r} \geq c_{1i} + c_{2i}$ for all $1 \leq i \leq n$ then an optimal solution is $x_{1p} = x_{2r} = y_{1p}^{2r} = 1$ and $x_{1i} = x_{2k} = y_{1r}^{2t} = 0$ for all $1 \leq i \neq p \leq t$ where $2 \leq t \neq r \leq n$.*

(ii) *if the condition in (i) does not hold then an optimal solution is $x_{1p} = x_{2p} = 1$ and $x_{1i} = x_{2k} = y_{1r}^{2t} = 0$ for $1 \leq i \neq p \leq n, 1 \leq k \neq p \leq n$ and $1 \leq r < t \leq n$ where $c_{1p} + c_{2p} \geq c_{1i} + c_{2i}$ for all $1 \leq i \leq n$.*

Locally Ideal LP Formulations I 99

Proposition 4.20 *The solution of Remark (4.6) is an optimal solution to the LP problem $\max\{\mathbf{cx}+\mathbf{qy} : (\mathbf{x},\mathbf{y}) \in P_L\}$ where $(\mathbf{c},\mathbf{q})$ is an arbitrary cost vector.*

Proof. Let $(\mathbf{x}^*, \mathbf{y}^*)$ be the solution defined in Remark (4.6). By Remark (4.5), $P \subseteq P_L$ and trivially, $(\mathbf{x}^*, \mathbf{y}^*)$ is an extreme point of P_L in both cases of Remark (4.6). We give, in both cases, a polytope $P' \supseteq P_L$ over which $(\mathbf{x}^*, \mathbf{y}^*)$ is optimal. Hence $(\mathbf{x}^*, \mathbf{y}^*)$ is, *a forteriori*, optimal over the polytope P_L.
Suppose we are in case (i) and an optimal solution to P is given by $x_{1,n-1} = x_{2n} = y_{1,n-1}^{2n} = 1$ and $x_{1i} = x_{2k} = y_{1r}^{2t} = 0$ for all $1 \leq i \neq n-1 \leq n, 1 \leq k \leq n-1, 1 \leq r < t \leq n-1$, where we have WROG assumed $p = n-1$ and $r = n$. Define $P' = \{(\mathbf{x},\mathbf{y}) \in \mathbb{R}^{n(n+3)/2} : \sum_{k=1}^{n} x_{ik} = 1$ for $1 \leq i \leq 2, -x_{1k} - x_{2k} + \sum_{\ell=1}^{k-1} y_{1\ell}^{2k} + \sum_{\ell=k+1}^{n} y_{1k}^{2\ell} \leq 0$ for $1 \leq k \leq n, \sum_{\ell=1}^{s}(x_{1\ell} - x_{2\ell} - \sum_{r=s+1}^{n} y_{1\ell}^{2r}) \leq 0$ for $1 \leq s \leq n-1, \sum_{\ell=1}^{s}(-x_{1\ell} + x_{2\ell} - \sum_{r=s+1}^{n} y_{1\ell}^{2r}) \leq 0$ for $1 \leq s \leq n-1\}$.
Using (4.28), the inequality $-\sum_{\ell=1}^{s}(x_{1\ell} + x_{2\ell} - \sum_{r=s+1}^{n} y_{1\ell}^{2r}) \leq 0$ is equivalent to $\sum_{\ell=s+1}^{n}(x_{1\ell} - x_{2\ell} - \sum_{r=1}^{s} y_{1r}^{2\ell}) \leq 0$ for $1 \leq s \leq n-1$; hence, $P' \supseteq P_L$.
The dual to this problem is $\min\{s_1 + s_2 : s_1 - t_j + \sum_{\ell=j}^{n-1} u_\ell - \sum_{\ell=j}^{n-1} v_\ell = c_{1j}$ for $1 \leq j \leq n-1, s_1 - t_n = c_{1n}, s_2 - t_j - \sum_{\ell=j}^{n-1} u_\ell + \sum_{\ell=j}^{n-1} v_\ell = c_{2j}$ for $1 \leq j \leq n-1, s_2 - t_n = c_{2n}, t_r + t_s - \sum_{\ell=r}^{s-1} u_\ell - \sum_{\ell=r}^{s-1} v_\ell \geq q_{1r}^{2s}$ for $1 \leq r \leq n, t_\ell \geq 0$ for $1 \leq \ell \leq n, u_\ell, v_\ell \geq 0$ for $1 \leq \ell \leq n-1\}$. The vector given by $s_1 = (c_{1,n-1} + c_{1n} + q_{1,n-1}^{2n})/2, s_2 = c_{2n} + (c_{1,n-1} - c_{1n} + q_{1,n-1}^{2n})/2, t_\ell = (c_{1,n-1} + c_{2n} + q_{1,n-1}^{2n} - c_{1\ell} - c_{2\ell})$ for $1 \leq \ell \leq n, u_\ell = \max\{(c_{1\ell} - c_{1,\ell+1} - c_{2\ell} + c_{2,\ell+1})/2, 0\}$ for $1 \leq \ell \leq n-1, v_\ell = 0$ if $u_\ell > 0, -(c_{1\ell} - c_{1,\ell+1} - c_{2\ell} + c_{2,\ell+1})/2$ otherwise, for $1 \leq \ell \leq n-1$ is feasible to the dual problem with objective function value $c_{1,n-1} + c_{2n} + q_{1,n-1}^{2n}$. This objective function value is equal to that of $(\mathbf{x}^*, \mathbf{y}^*)$ and hence, by LP duality $(\mathbf{x}^*, \mathbf{y}^*)$ is optimal over P'.
Next consider case (ii) of Remark 4.6 and assume an optimal solution to P is given by $x_{1n} = x_{2n} = 1, x_{1j} = x_{2j} = y_{1r}^{2s} = 0$ for $1 \leq j \leq n-1, 1 \leq r < s \leq n$. Define $P' = \{(\mathbf{x},\mathbf{y}) \in \mathbb{R}^{n(n+3)/2} : \sum_{k=1}^{n} x_{ik} = 1$ for $1 \leq i \leq 2, -x_{1k} - x_{2k} + \sum_{\ell=1}^{k-1} y_{1\ell}^{2k} + \sum_{\ell=k+1}^{n} y_{1k}^{2\ell} \leq 0$ for $1 \leq k \leq n, \sum_{\ell=1}^{s}(x_{1\ell} - x_{2\ell} - \sum_{r=s+1}^{n} y_{1\ell}^{2r}) \leq 0$ for $1 \leq s \leq n-1, \sum_{\ell=1}^{s}(-x_{1\ell} + x_{2\ell} - \sum_{r=s+1}^{n} y_{1\ell}^{2r}) \leq 0$ for $1 \leq s \leq n-1\}$.
Using (4.28), the inequality $-\sum_{\ell=1}^{s}(x_{1\ell} + x_{2\ell} - \sum_{r=s+1}^{n} y_{1\ell}^{2r}) \leq 0$ is equivalent to $\sum_{\ell=s+1}^{n}(x_{1\ell} - x_{2\ell} - \sum_{r=1}^{s} y_{1r}^{2\ell}) \leq 0$ for $1 \leq s \leq n-1$; hence, $P' \supseteq P_L$.
The dual to the corresponding problem is $\min\{s_1 + s_2 : s_1 - t_j + \sum_{\ell=j}^{n-1} u_\ell - \sum_{\ell=j}^{n-1} v_\ell = c_{1j}$ for $1 \leq j \leq n-1, s_1 = c_{1n}, s_2 - t_j - \sum_{\ell=j}^{n-1} u_\ell + \sum_{\ell=j}^{n-1} v_\ell = c_{2j}$ for $1 \leq j \leq n-1, s_2 = c_{2n}, t_r + t_s - \sum_{\ell=r}^{s-1} u_\ell - \sum_{\ell=r}^{s-1} v_\ell \geq q_{1r}^{2s}$ for $1 \leq r < s$ where $2 \leq s \leq n, t_\ell \geq 0$ for $1 \leq \ell \leq n, u_\ell, v_\ell \geq 0$ for $1 \leq \ell \leq n-1\}$. The vector given by $s_1 = c_{1n}, s_2 = c_{2n}, t_\ell = (c_{1n} + c_{2n} - c_{1\ell} - c_{2\ell})/2$ for $1 \leq \ell \leq n, u_\ell = \max\{(c_{1\ell} - c_{1,\ell+1} - c_{2\ell} + c_{2,\ell+1})/2, 0\}$ for $1 \leq \ell \leq n-1, v_\ell = 0$ if $u_\ell > 0, -(c_{1\ell} - c_{1,\ell+1} - c_{2\ell} + c_{2,\ell+1})/2$ otherwise, for $1 \leq \ell \leq n-1$ is feasible to the dual problem with objective function value $c_{1n} + c_{2n}$. This dual objective

objective function value is equal to that of $(\mathbf{x}^*, \mathbf{y}^*)$ and hence, by LP duality $(\mathbf{x}^*, \mathbf{y}^*)$ is optimal over P'. □

The following proposition states that a locally ideal linerization has been obtained.

Proposition 4.21 *The system of equations and inequalities (4.28),...,(4.31) is an ideal linear description of the local polytope P, i.e. $P = P_L$.*

Considering all equations and inequalities resulting from the locally ideal linearization of the variables giving rise to quadratic terms in the objective function of the MPP, we formulate the MPP as the LP problem given by:

$$min \left\{ \sum_{i=1}^{m}\sum_{j=1}^{n} c_{ij}x_{ij} + \sum_{i<k\in M}\sum_{j<\ell\in N} q_{ij}^{k\ell}y_{ij}^{k\ell} : (\mathbf{x}, \mathbf{y}) \in QPP_n^m \right\}, \quad (\mathcal{O}QPP_n^m)$$

where QPP_n^m denotes the convex hull of solutions $(\mathbf{x}, \mathbf{y}) \in \mathbb{R}^{mn+mn(m-1)(n-1)/4}$ to the following equations and inequalities in zero-one variables:

$$\sum_{j=1}^{n} x_{ij} = 1 \quad \text{for } i \in M \qquad (4.32)$$

$$-x_{ij} - x_{kj} + \sum_{\ell=1}^{j-1} y_{i\ell}^{kj} + \sum_{\ell=j+1}^{n} y_{ij}^{k\ell} \leq 0 \quad \text{for } i < k \in M, j \in N \qquad (4.33)$$

$$\sum_{j\in S}(x_{ij} - x_{kj} - \sum_{j>\ell\in N-S} y_{i\ell}^{kj} - \sum_{j<\ell\in N-S} y_{ij}^{k\ell}) \leq 0 \quad \text{for } i < k \in M, \emptyset \neq S \subset N \quad (4.34)$$

$$y_{ij}^{k\ell} \geq 0 \quad \text{for } i < k \in M, j < \ell \in N \qquad (4.35)$$

$$x_{ij} \in \{0,1\} \quad \text{for } i \in M, j \in N. \qquad (4.36)$$

Proposition 4.22 $\mathcal{O}QPP_n^m$ *is a formulation of the Multi Processor Assignment Problem.*

Proof. By similar arguments as in Remark (4.5), $DQPP_n^m \subseteq QPP_n^m$. Let $(\mathbf{x}, \mathbf{y}) \in QMPP_n^m$. We show that $y_{ij}^{k\ell} = x_{ij}x_{k\ell} + x_{i\ell}x_{kj}$ for all $1 \leq i < k \leq m$ and $1 \leq j < \ell \leq n$. Suppose that there exist $1 \leq p < r \leq m, 1 \leq g < s \leq n$ such that $y_{pg}^{rs} \neq x_{pg}x_{rs} + x_{ps}x_{rg}$. If $x_{pg} = x_{rg} = 0$, then using (4.35) it follows from (4.33) that $y_{pg}^{rs} = 0$. On the other hand, if $x_{pg} = x_{rg} = 1$ then from (4.32) and (4.36) $x_{ps} = x_{rs} = 0$; and thus, by a similar argument as above, $y_{pg}^{rs} = 0$. So necessarily $x_{pg} \neq x_{rg} \in \{0,1\}$; WROG we assume $x_{pg} = 1$ and $x_{rh} = 1$

Locally Ideal LP Formulations I 101

for some $1 \leq g < h \leq n$. By a similar argument as above, we have that for $g \neq d \neq h$ and $g \neq t \neq h$, $y_{pd}^{rt} = 0$ for all $1 \leq d < t$ and $y_{pt}^{rd} = 0$ for $t < d \leq n$. Then, using (4.34) for $p = i, r = k$ and $S = \{g\}$, we conclude $1 = x_{pg} \leq \sum_{d=1}^{g-1} y_{pd}^{rg} + \sum_{d=g+1}^{n} y_{pg}^{rd} = y_{pg}^{rh}$. Moreover, using (4.33) we conclude $\sum_{d=1}^{g-1} y_{pd}^{rg} + \sum_{d=g+1}^{n} y_{pg}^{rd} \leq 1$ and thus $y_{pt}^{rd} = 0$ for all $1 \leq t < g$, $y_{pg}^{rt} = 0$ for all $g < t \neq h \leq n$, $y_{pg}^{rh} = 1$. Hence, we get a contradiction to our assumption that $y_{pg}^{rs} \neq x_{pg}x_{rs} + x_{ps}x_{rg}$. Since all extreme points in QPP_n^m are zero-one valued and in $DQPP_n^m$, the proposition follows. □

Though our formulation of the MPP has exponentially many constraints, its LP relaxation can be solved in polynomial time in the parameters m and n because the corresponding *separation* problem is polynomially solvable. Let $(\overline{\mathbf{x}}, \overline{\mathbf{y}}) \in \mathbb{R}^{mn+mn(m-1)(n-1)/4}$ satisfy (4.32), (4.33) and (4.35). These are polynomially many constraints in the parameters m and n and can thus be checked in polynomial time. To check the constraints (4.34) we need to find for fixed i and k with $1 \leq i < k \leq m$, $z_{ik} = max\{\sum_{j \in S}(\overline{x}_{ij} - \overline{x}_{kj}) - \sum_{j > \ell \in N-S} \overline{y}_{i\ell}^{kj} - \sum_{j < \ell \in N-S} \overline{y}_{ij}^{k\ell}) : \emptyset \neq S \subset N\}$. Defining $z_j = 1$ if $j \in S$, 0 otherwise, we can rewrite $z_{ik} = max\{\sum_{j=1}^{n}(\overline{x}_{ij} - \overline{x}_{kj})z_j - \sum_{j=1}^{n}(\sum_{\ell=1}^{j-1} \overline{y}_{i\ell}^{kj}(1-z_\ell) + \sum_{\ell=j+1}^{n} \overline{y}_{ij}^{k\ell}(1-z_\ell))z_j : 1 \leq \sum_{j=1}^{n} z_j \leq n-1, z_j \in \{0,1\}\} = max\{\sum_{j=1}^{n}(\overline{x}_{ij} - \overline{x}_{kj}) - \sum_{\ell=1}^{j-1} \overline{y}_{i\ell}^{kj} - \sum_{\ell=j+1}^{n} \overline{y}_{ij}^{k\ell})z_j + \sum_{j=1}^{n}(\sum_{\ell=1}^{j-1} \overline{y}_{i\ell}^{kj} + \sum_{\ell=j+1}^{n} \overline{y}_{ij}^{k\ell})z_j z_\ell : 1 \leq \sum_{j=1}^{n} z_j \leq n-1, z_j \in \{0,1\}\}$. Using (4.32), it follows that $z_{ik} = 0$ for $\sum_{j=1}^{n} z_j = 0$ or $\sum_{j=1}^{n} z_j = n$; i.e., the inequality (4.34) for fixed i, k is not violated. Hence, we can eliminate the constraint $1 \leq \sum_{j=1}^{n} z_j \leq n-1$ altogether. But then by (4.35), our *separation* problem is an instance of the BQP with nonnegative quadratic cost coefficients, which is polynomially solvable as shown in Picard and Ratliff [1975] and Padberg [1989]; see also Padberg and Wolsey [1983]. Furthermore, if $z_{ik} > 0$ and only then the corresponding constraint (4.34) is violated. Hence we can solve the LP relaxation our formulation of the MPP in polynomial time.

To prove more interesting facts about QPP_n^m, let us order the components of $\mathbf{x} \in \mathbb{R}^{mn}$ as $(x_{11}, \ldots, x_{1n}, \ldots, x_{m1}, \ldots, x_{mn})$ and those of $\mathbf{y} \in \mathbb{R}^{mn(m-1)(n-1)/4}$ as $(y_{11}^{22}, \ldots, y_{11}^{2n}, \ldots, y_{11}^{m2}, \ldots, y_{11}^{mn}, y_{12}^{23}, \ldots, y_{12}^{mn}, \ldots, y_{1,n-1}^{mn}, y_{21}^{32}, \ldots, y_{m-1,n-1}^{mn})$ respectively; that is, $(y_{11}^{22}, y_{11}^{23}, y_{11}^{32}, y_{11}^{33}, y_{12}^{23}, y_{12}^{33}, y_{21}^{32}, y_{21}^{33}, y_{22}^{33})$ explicitly shows the ordering of all components of $\mathbf{y}$ for $m = 3$ and $n = 3$. Let $\overline{\mathbf{u}}_{ij} \in \mathbb{R}^{mn}$ with its components ordered like those of $\mathbf{x}$ be a unit vector with one in its $(i,j)^{th}$ component and $\overline{\mathbf{v}}_{ij}^{k\ell} \in \mathbb{R}^{mn(m-1)(n-1)/4}$ ordered like $\mathbf{y}$ be another unit vector with one in its $\binom{k,\ell}{i,j}^{th}$ component. Let $\mathbf{u}_{ij} \in \mathbb{R}^{mn+mn(m-1)(n-1)/4}$ be obtained from $\overline{\mathbf{u}}_{ij}$ by appending $mn(m-1)(n-1)/4$ zeroes in the last $mn(m-1)(n-1)/4$ components and $\mathbf{v}_{ij}^{k\ell} \in \mathbb{R}^{mn+mn(m-1)(n-1)/4}$ be obtained from $\overline{\mathbf{v}}_{ij}^{k\ell}$ by appending mn zeroes at the beginning. Let $z_S(g,h) = \sum_{i \in S} \mathbf{u}_{ig} + \sum_{i \in M-S} \mathbf{u}_{ih} + $

$\sum_{i \in S} \sum_{i < k \in M-S} v_{ig}^{kh} + \sum_{i \in S} \sum_{i > k \in M-S} v_{kg}^{ih}$ for $1 \leq g < h \leq n$ and $z_S(g,h) = \sum_{i \in S} u_{ig} + \sum_{i \in M-S} u_{ih} + \sum_{i \in S} \sum_{i < k \in M-S} v_{ih}^{kg} + \sum_{i \in S} \sum_{i > k \in M-S} v_{kh}^{ig}$ for $1 \leq h < g \leq n$ where $M = \{1, 2, \ldots, m\}$ and $S \subseteq M$. Likewise, let $z_M(j) = \sum_{i=1}^{m} \mathbf{u}_{ij}$ for $1 \leq j \leq n$.

Proposition 4.23 *The dimension of the MPP polytope $dim(QPP_n^m)$ equals $m(n-1) + mn(m-1)(n-1)/4$ for all $m \geq n \geq 3$.*

Proof. Since the m equations (4.32) are linearly independent, $dim(QPP_n^m) \leq m(n-1) + mn(m-1)(n-1)/4$. We establish $dim(QPP_n^m) \geq m(n-1) + mn(m-1)(n-1)/4$ by showing that every equation $\alpha \mathbf{x} + \beta \mathbf{y} = \gamma$ that is satisfied by all $(\mathbf{x}, \mathbf{y}) \in QPP_n^m$ is a linear combination of the m equations (4.32).

(i) Since $(z_M(j), z_{M-\{k\}}(j, \ell)) \in QPP_n^m$ for $1 \leq k \leq m, 1 \leq j < \ell \leq n$, $\alpha_{kj} - \alpha_{k\ell} = \sum_{i=1}^{k-1} \beta_{ij}^{k\ell} + \sum_{i=k+1}^{m} \beta_{kj}^{i\ell}$ for all $1 \leq j < \ell \leq n$. Since $(z_M(\ell), z_{M-\{k\}}(\ell, j)) \in QPP_n^m$ for $1 \leq j < \ell \leq n$, $\alpha_{k\ell} - \alpha_{kj} = \sum_{i=1}^{k-1} \beta_{ij}^{k\ell} + \sum_{i=k+1}^{m} \beta_{kj}^{i\ell}$. Hence $\alpha_{kj} = \alpha_{k\ell}$ and $\sum_{i=1}^{k-1} \beta_{ij}^{k\ell} + \sum_{i=k+1}^{m} \beta_{kj}^{i\ell} = 0$ for all $1 \leq k \leq m, 1 \leq j < \ell \leq n$.

(ii) Since $(z_{M-\{i\}}(j, \ell), z_{M-\{i,k\}}(j, \ell)) \in QPP_n^m$ for $1 \leq i < k \leq m, 1 \leq j < \ell \leq n$, $\alpha_{kj} + \beta_{ij}^{k\ell} = \alpha_{k\ell} + \sum_{g=1}^{i-1} \beta_{gj}^{i\ell} + \sum_{g=i+1}^{k-1} \beta_{ij}^{g\ell} + \sum_{g=k+1}^{m} \beta_{ij}^{g\ell}$ for all $1 \leq i < k \leq m, 1 \leq j < \ell \leq n$. By (i), $\beta_{ij}^{k\ell} = -\beta_{ij}^{k\ell}$ and hence, $\beta_{ij}^{k\ell} = 0$ for all $1 \leq i < k \leq m$ and $1 \leq j < \ell \leq n$.

Consequently, $\alpha \mathbf{x} + \beta \mathbf{y} = \gamma$ becomes $\sum_{i=1}^{m} \alpha_{i1} \sum_{j=1}^{n} x_{ij} = \sum_{i=1}^{m} \alpha_{i1}$ for all $(\mathbf{x}, \mathbf{y}) \in QPP_n^m$; which is a linear combination of the m equations (4.32). □

Proposition 4.24 *(4.35) defines a facet of QPP_n^m for $1 \leq i < k \leq m, 1 \leq j < \ell \leq n$.*

Proof. By Proposition (4.22), (4.35) is valid for QPP_n^m. Let $F = \{(\mathbf{x}, \mathbf{y}) \in QPP_m^n : y_{ij}^{k\ell} = 0\}$. Since $z_{M-\{i\}}(\ell, j) \in QPP_m^n$ but not in F, F is a proper face of QPP_n^m. Suppose there exists a valid inequality $\alpha \mathbf{x} + \beta \mathbf{y} \leq \gamma$ for P such that every $(\mathbf{x}, \mathbf{y}) \in F$ satisfies $\alpha \mathbf{x} + \beta \mathbf{y} = \gamma$.

(i) Since $(z_M(g), z_{M-\{p\}}(g, h)) \in F$ for $p \in M - \{i, k\}, 1 \leq g < h \leq n$, $\alpha_{pg} - \alpha_{ph} = \sum_{s=1}^{p-1} \beta_{sg}^{ph} + \sum_{s=p+1}^{m} \beta_{pg}^{sh}$ for all $p \in M - \{i, k\}, 1 \leq g < h \leq n$. Since $(z_M(h), z_{M-\{p\}}(h, g)) \in F$ for $p \in M - \{i, k\}, 1 \leq g < h \leq n$, $\alpha_{ph} - \alpha_{pg} = \sum_{s=1}^{p-1} \beta_{sg}^{ph} + \sum_{s=p+1}^{m} \beta_{pg}^{sh}$ and hence, $\alpha_{pg} = \alpha_{ph}$ and $\sum_{s=1}^{p-1} \beta_{sg}^{ph} + \sum_{s=p+1}^{m} \beta_{pg}^{sh} = 0$ for all $p \in M - \{i, k\}, 1 \leq g < h \leq n$.

(ii) Since $(z_{M-\{r\}}(g, h), z_{M-\{p,r\}}(g, h)) \in F$ for $p, r \in M - \{i, k\}, p < r$ and $1 \leq g < h \leq n$, $\alpha_{pg} + \beta_{pg}^{rh} = \alpha_{ph} + \sum_{s=1}^{p-1} \beta_{sg}^{ph} + \sum_{s=p+1}^{r-1} \beta_{pg}^{sh} + \sum_{g=r+1}^{m} \beta_{pg}^{sh}$ for all $p, r \in M - \{i, k\}, p < r$ and $1 \leq g < h \leq n$. By (i), $\beta_{pg}^{rh} = -\beta_{pg}^{rh}$ and hence, $\beta_{pg}^{rh} = 0$ for all $p, r \in M - \{i, k\}, p < r$ and $1 \leq g < h \leq n$.

Locally Ideal LP Formulations I

(iii) Since $(z_M(g), z_M(h), z_{M-\{p\}}(g,h), z_{M-\{p\}}(h,g)) \in F$ for $p \in \{i,k\}, 1 \leq g < h \leq n$ except when $g = j$ and $h = \ell$, $\alpha_{pg} = \alpha_{ph}$ and $\sum_{s=1}^{p-1} \beta_{sg}^{ph} + \sum_{s=p+1}^{m} \beta_{pg}^{sh} = 0$ for all $p \in \{i,k\}, 1 \leq g < h \leq n$ except when $g = j$ and $h = \ell$. But, then from (i), $\beta_{ig}^{kh} = 0$ for all $1 \leq g < h \leq n$ except when $g = j$ and $h = \ell$.

Consequently, $\alpha \mathbf{x} + \beta \mathbf{y} = \gamma$ becomes $\sum_{i=1}^{m} \alpha_{ig} \sum_{\ell=1}^{n} x_{i\ell} + \beta_{1j}^{2\ell} y_{1j}^{2\ell} = \sum_{i=1}^{m} \alpha_{ig}$; equivalently, $\beta_{1j}^{2\ell} y_{1j}^{2\ell} = 0$ for all $(\mathbf{x}, \mathbf{y}) \in F$ where $1 \leq g \leq n$. □

Proposition 4.25 *Inequality (4.33) defines a facet of QPP_n^m for $1 \leq i < k \leq m, 1 \leq j \leq n$.*

Proof. By Proposition (4.22), (4.33) is valid for QPP_n^m. Let $F = \{(\mathbf{x}, \mathbf{y}) \in QPP_n^m : -x_{ij} - x_{kj} + \sum_{\ell=1}^{j-1} y_{i\ell}^{kj} + \sum_{\ell=j+1}^{n} y_{ij}^{k\ell} = 0\}$. Since $z_M(j) \in QPP_m^n$ but not in F, F is a proper face of QPP_n^m. Suppose there exists a valid inequality $\alpha \mathbf{x} + \beta \mathbf{y} \leq \gamma$ for P such that every $(\mathbf{x}, \mathbf{y}) \in F$ satisfies $\alpha \mathbf{x} + \beta \mathbf{y} = \gamma$.

(i) Since $(z_M(g), z_M(h), z_{M-\{p\}}(g,h), z_{M-\{p\}}(h,g)) \in F$ for $p \in M, 1 \leq g < h \leq n$ and $g \neq j \neq h$, it follows like in (i) and (ii) of Proposition (4.24) that $\beta_{pg}^{rh} = \beta_{sg}^{ph} = 0$ for all $1 \leq s < p < r \leq m, 1 \leq g < h \leq n$ and $g \neq j \neq h$.

(ii) Since $(z_M(g), z_{M-\{p\}}(g,j)) \in F$ for $p \in M, 1 \leq g \neq j \leq n$, $\alpha_{pg} = \alpha_{pj} + \sum_{s=1}^{p-1} \beta_{sg}^{pj} + \sum_{s=p+1}^{m} \beta_{pg}^{sj}$ for $1 \leq p \leq m, 1 \leq g < j \leq n$ and $\alpha_{pg} = \alpha_{pj} + \sum_{s=1}^{p-1} \beta_{sj}^{pg} + \sum_{s=p+1}^{m} \beta_{pj}^{sg}$ for $1 \leq p \leq m, 1 \leq j < g \leq n$.

(iii) Since $(z_{M-\{r\}}(g,j), z_{M-\{p,r\}}(g,j)) \in F$ for $1 \leq p < r \leq m, 1 \leq g \neq j \leq n$ except when $p = i$ and $r = k$, WROG assuming $g < j$, we have $\alpha_{pg} + \beta_{pg}^{rj} = \alpha_{pj} + \sum_{s=1}^{p-1} \beta_{sg}^{pj} + \sum_{s=p+1}^{r-1} \beta_{pg}^{sj} + \sum_{s=r+1}^{m} \beta_{pg}^{sj}$. From (ii) $\beta_{pg}^{rj} = -\beta_{pg}^{rj}$ and hence, $\beta_{pg}^{rj} = 0$ for all $1 \leq p < r \leq m, 1 \leq g < j \leq n$ except when $p = i$ and $r = k$. By a similar argument, we get $\beta_{pj}^{pg} = 0$ for all $1 \leq p < r \leq m, 1 \leq j < g \leq n$ except when $p = i$ and $r = k$. Thus from (ii), $\alpha_{pj} = \alpha_{pg}$ for $p \in M - \{i,k\}, 1 \leq j \neq g \leq n$, $\alpha_{pg} = \alpha_{pj} + \beta_{ij}^{kg}$ for $p \in \{i,k\}, 1 \leq j < g \leq n$, $\alpha_{pg} = \alpha_{pj} + \beta_{ij}^{kg}$ for $p \in \{i,k\}, 1 \leq g < j \leq n$ and $\beta_{ig}^{kj} = \beta_{ij}^{kh}$ for $1 \leq g < j < h \leq n$.

Consequently, $\alpha \mathbf{x} + \beta \mathbf{y} = \gamma$ reads $\sum_{i=1}^{m} \alpha_{ig} \sum_{\ell=1}^{n} x_{i\ell} + \beta_{ij}^{kg}(-x_{ij} - x_{kj} + \sum_{\ell=1}^{j-1} y_{i\ell}^{kj} + \sum_{\ell=j+1}^{n} y_{ij}^{k\ell}) = \sum_{i=1}^{m} \alpha_{ip}$; equivalently, $\beta_{ij}^{kg}(-x_{ij} - x_{kj} + \sum_{\ell=1}^{j-1} y_{i\ell}^{kj} + \sum_{\ell=j+1}^{n} y_{ij}^{k\ell}) = 0$ for all $(\mathbf{x}, \mathbf{y}) \in F$ where $1 \leq g \neq j \leq n$. □

Proposition 4.26 *Inequality (4.34) defines a facet of QPP_n^m for $1 \leq i < k \leq m, \emptyset \neq S \subset N$.*

Proof. By Proposition (4.22), (4.34) is valid for QPP_n^m. WROG, let $S = \{1,\ldots,s\}$ and $F = \{(\mathbf{x},\mathbf{y}) \in QPP_m^n : \sum_{j=1}^{s}(x_{ij} - x_{kj} - \sum_{\ell=s+1}^{n} y_{ij}^{k\ell}) = 0\}$. Since $z_{M-\{i\}}(j,\ell) \in QPP_m^n$ but not in F for $j \in S, \ell \notin S$, F is a proper face of QPP_n^m. Suppose there exists a valid inequality $\boldsymbol{\alpha}\mathbf{x} + \boldsymbol{\beta}\mathbf{y} \leq \gamma$ for P such that every $(\mathbf{x},\mathbf{y}) \in F$ satisfies $\boldsymbol{\alpha}\mathbf{x} + \boldsymbol{\beta}\mathbf{y} = \gamma$.

(i) Since $(z_M(g), z_M(h), z_{M-\{p\}}(g,h), z_{M-\{p\}}(h,g)) \in F$ for $p \in M, 1 \leq g < h \leq s$, we get like in (i) and (ii) of Proposition (4.24) $\alpha_{pg} = \alpha_{ph}, \beta_{pg}^{rh} = 0$ for all $p \in M, 1 \leq g < h \leq s$.

(ii) Since $(z_M(g), z_M(h), z_{M-\{p\}}(g,h), z_{M-\{p\}}(h,g)) \in F$ for $p \in M, s+1 \leq g < h \leq n$, by a similar argument as in (i), $\alpha_{pg} = \alpha_{ph}, \beta_{pg}^{rh} = 0$ for all $p \in M, s+1 \leq g < h \leq n$.

(iii) Since $(z_M(g), z_M(h), z_{M-\{p\}}(g,h), z_{M-\{p\}}(h,g)) \in F$ for $1 \leq g \leq s, s+1 \leq h \leq n, p \in M - \{i,k\}$; using similar arguments as in (i), $\alpha_{pg} = \alpha_{ph}$ and $\beta_{pg}^{rh} = 0$ for all $p \in M - \{i,k\}, 1 \leq g \leq s, s+1 \leq h \leq n$.

(iv) Since $(z_M(g), z_{M-\{k\}}(g,h)) \in F$ for $1 \leq g \leq s, s+1 \leq h \leq n$, using (iii) we have $\alpha_{kg} = \alpha_{kh} + \beta_{ig}^{kh}$. Moreover, since $(z_M(h), z_{M-\{i\}}(h,g)) \in F$ for $1 \leq g \leq s, s+1 \leq h \leq n, \alpha_{ih} = \alpha_{ig} + \beta_{ig}^{kh}$ for all $1 \leq g \leq s, s+1 \leq h \leq n$.

Consequently, $\boldsymbol{\alpha}\mathbf{x} + \boldsymbol{\beta}\mathbf{y} = \gamma$ becomes $\sum_{i=1}^{m} \alpha_{ih} \sum_{\ell=1}^{n} x_{i\ell} + \beta_{ig}^{kh} \sum_{j \in S}(-x_{ij} + x_{kj} + \sum_{\ell \in N-S} y_{ij}^{k\ell}) = \sum_{i=1}^{m} \alpha_{ih}$; i.e. $\beta_{ig}^{kh} \sum_{j \in S}(-x_{ij} + x_{kj} + \sum_{\ell \in N-S} y_{ij}^{k\ell}) = 0$ for all $(\mathbf{x},\mathbf{y}) \in F$ where $1 \leq g \leq s, s+1 \leq h \leq n$. □

5
LOCALLY IDEAL LP FORMULATIONS II

In this chapter we continue our investigations into the locally ideal linearization of the major problem classes from Chapters 1 and 2. In particular, we study here the VLSI circuit layout design problem, a general model that comprises all BQPSs considered so far, the quadratic assignment problem and its symmetric relative. Except for the symmetric quadratic assignment problem, complete characterizations of the associated local polytopes are obtained. Like in the case of our results of Chapter 4, these local polytopes are of interest on their own whenever the substructures that we study occur in a quadratic zero-one optimization problem. In all cases we obtain from the locally ideal linearization *formulations* of the respective problems that in most cases improve on existing formulations for these problems.

5.1 VLSI Circuit Layout Design Problems

To consider the CLDP, see Chapter 2.4, we define new variables $y_{ij}^{k\ell} = x_{ij}x_{k\ell}$ for $1 \leq i < k \leq m$ and $1 \leq j \neq \ell \leq n$ and assume $m \geq n \geq 3$; counting yields that there are $mn(m-1)(n-1)/2$ **y**-variables. Denoting by $DQDP_n^m$ the discrete set

$$DQDP_n^m = \left\{ \begin{array}{l} (\mathbf{x},\mathbf{y}) \in \mathbb{R}^{mn+mn(m-1)(n-1)/2}: \\ \quad \sum_{j=1}^n x_{ij} = 1 \quad \text{for } i \in M \\ \quad y_{ij}^{k\ell} = x_{ij}x_{k\ell} \quad \text{for } i < k \in M, j \neq \ell \in N \\ \quad x_{ij} \in \{0,1\} \quad \text{for } i \in M, j \in N \end{array} \right\},$$

the CLDP can be written as

$$\min\left\{\sum_{i=1}^{m-1}\sum_{k=i+1}^{m}\sum_{j=1}^{n}\sum_{j\neq\ell=1}^{n} q_{ij}^{k\ell}y_{ij}^{k\ell} : (\mathbf{x},\mathbf{y}) \in DQDP_n^m\right\},$$

where $q_{ij}^{k\ell} = a_{ijk\ell} + a_{k\ell ij}$ in terms of the $a_{ijk\ell}$ of Chapter 2.4. We note that the $\mathbf{y}$-variables in the MPP can be obtained from the CLDP by the transformation:

$$y_{ij}^{k\ell} = x_{ij}x_{k\ell} + x_{i\ell}x_{kj} \qquad \text{for all } i < k \in M, j < \ell \in N. \qquad (5.1)$$

To obtain a linear formulation for $DQDP_n^m$ in zero-one variables, we consider the local polytope P given by $P = conv(D)$ where $n \geq 3$ and D is defined as follows; see Figure 5.1:

$$D = \left\{\begin{array}{ll} (\mathbf{x},\mathbf{y}) \in \mathbb{R}^{n(n+1)} : & \\ \sum_{j=1}^{n} x_{ij} = 1 & \text{for } 1 \leq i \leq 2 \\ y_{1j}^{2\ell} = x_{1j}x_{2\ell} & \text{for } 1 \leq j \neq \ell \leq n \\ x_{ij} \in \{0,1\} & \text{for } 1 \leq j \leq n \end{array}\right\}.$$

Let P_L be the polytope given by $(\mathbf{x},\mathbf{y}) \in \mathbb{R}^{n(n+1)}$ satisfying

$$\sum_{j=1}^{n} x_{ij} = 1 \qquad \text{for } 1 \leq i \leq 2 \qquad (5.2)$$

$$x_{1j} - x_{2j} - \sum_{j\neq\ell=1}^{n} y_{1j}^{2\ell} + \sum_{j\neq\ell=1}^{n} y_{1\ell}^{2j} = 0 \qquad \text{for } 1 \leq j \leq n-1 \qquad (5.3)$$

$$-x_{1j} + \sum_{j\neq\ell=1}^{n} y_{1j}^{2\ell} \leq 0 \qquad \text{for } 1 \leq j \leq n \qquad (5.4)$$

$$y_{1j}^{2\ell} \geq 0 \qquad \text{for } 1 \leq j \neq \ell \leq n. \qquad (5.5)$$

Remark 5.1 *The system of equations and inequalities (5.2),...,(5.5) is valid for all $(\mathbf{x},\mathbf{y}) \in P$ and thus $P \subseteq P_L$. There are $n+1$ equations in (5.2) and (5.3) and $x_{1n} - x_{2n} - \sum_{\ell=1}^{n-1} y_{1n}^{2\ell} + \sum_{\ell=1}^{n-1} y_{1\ell}^{2n} = 0$ is redundant for P_L.*

Proof. Let $(\mathbf{x},\mathbf{y}) \in D$. Then $(\mathbf{x},\mathbf{y})$ satisfies (5.2). We calculate $x_{1j} - x_{2j} - \sum_{j\neq\ell=1}^{n}(y_{1j}^{2\ell} - y_{1\ell}^{2j}) = x_{1j} - x_{2j} - \sum_{j\neq\ell=1}^{n}(x_{1j}x_{2\ell} - x_{1\ell}x_{2j}) = x_{1j} - x_{2j} - x_{1j}(1 - x_{2j}) + x_{2j}(1 - x_{1j}) = x_{1j}x_{2j} - x_{1j}x_{2j} = 0$ and hence (5.3) is satisfied as well. By calculating $-x_{1j} + \sum_{j\neq\ell=1}^{n} y_{1j}^{2\ell} = -x_{1j} + x_{1j}\sum_{j\neq\ell=1}^{n} x_{2\ell} = -x_{1j}x_{2j} \in \{0,-1\} \leq 0$ $1 \leq j \leq n$, it follows that (5.4) is satisfied. Thus, $D \subseteq P_L$ and

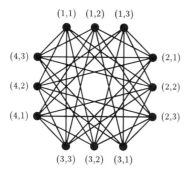
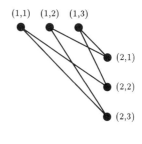

The cobweb of all node and edge variables of QDP_3^4

Node and edge variables used in the locally ideal linearization of QDP_3^4

Figure 5.1 The locally ideal linearization of CLDPs

$P = conv(D) \subseteq P_L$. There are $n+1$ equations (5.2) and (5.3). To show the stated redundancy, we sum all equations (5.3) for $1 \leq j \leq n-1$ and use (5.2) to obtain $x_{1n} - x_{2n} - \sum_{\ell=1}^{n-1} y_{1n}^{2\ell} + \sum_{\ell=1}^{n-1} y_{1\ell}^{2n} = 0$. Hence, this equation is redundant for all $(\mathbf{x}, \mathbf{y}) \in P_L$. □

We order the components of $\mathbf{x}$ as $(x_{11}, \ldots, x_{1n}, x_{21}, \ldots, x_{2n})$ and those of $\mathbf{y}$ as $(y_{11}^{22}, \ldots, y_{11}^{2n}, y_{12}^{21}, y_{12}^{23}, \ldots, y_{12}^{2n}, \ldots, y_{1n}^{21}, \ldots, y_{1n}^{2,n-1})$. Let $\overline{\mathbf{u}}_{ij} \in \mathbb{R}^{2n}$ with its components ordered like those of $\mathbf{x}$ be a unit vector with one in its $(i,j)^{th}$ component and $\overline{\mathbf{v}}_{1j}^{2\ell} \in \mathbb{R}^{n(n-1)}$ ordered like $\mathbf{y}$ be another unit vector with one in its $\binom{2,\ell}{1,j}^{th}$ component. Let $\mathbf{u}_{ij} \in \mathbb{R}^{n(n+1)}$ be obtained from $\overline{\mathbf{u}}_{ij}$ by appending $n(n-1)$ zeroes in the last $n(n-1)$ components and $\mathbf{v}_{1j}^{2\ell} \in \mathbb{R}^{n(n+1)}$ be obtained from $\overline{\mathbf{v}}_{1j}^{2\ell}$ by appending $2n$ zeroes at the beginning.

Proposition 5.1 *The dimension of P equals $n^2 - 1$ for all $n \geq 3$.*

Proof. We write the equations (5.2) and (5.3) in matrix form as $\mathbf{A}_1 \mathbf{x} + \mathbf{A}_2 \mathbf{y} = \mathbf{b}$ where $\mathbf{A} = (\mathbf{A}_1, \mathbf{A}_2)$. Partitioning $\mathbf{A} = (\mathbf{A}', \mathbf{A}'')$ columnwise so that $\mathbf{A}'$ corresponds to $x_{11}, x_{21}, y_{11}^{2n}, \ldots, y_{1,n-1}^{2n}$, we have $\mathbf{A}' = \mathbf{L}_p$ where $\mathbf{L}_p \in \mathbb{R}^{p \times p}$ is a lower triangular matrix and $p = n+1$. Thus, $dim(P) \leq (n+1)(n-1)$. We establish $dim(P) \geq (n+1)(n-1)$ by exhibiting $(n+1)(n-1)+1$ linearly independent zero-one vectors belonging to P. Consider the matrix $\mathbf{Z}$ whose rows are formed by the following vectors:

(i) the vector $\mathbf{u}_{1n} + \mathbf{u}_{2n} \in P$,
(ii) $2(n-1)$ vectors $\mathbf{u}_{1j} + \mathbf{u}_{2\ell} \in P$ for $\ell \in \{j, n\}$ where $1 \leq j \leq n-1$,
(iii) $(n-1)(n-2)$ vectors $\mathbf{u}_{1j} + \mathbf{u}_{2\ell} \in P$ for $1 \leq j \neq \ell \leq n-1$,
(iv) $n-1$ vectors $\mathbf{u}_{1n} + \mathbf{u}_{2j} \in P$ for $1 \leq j \leq n-1$.

Partitioning $\mathbf{Z} = (\mathbf{Z}', \mathbf{Z}'')$ such that $\mathbf{Z}''$ corresponds to $x_{2n}, y_{11}^{2n}, \ldots, y_{1,n-1}^{2n}$, we have that *modulo* row permutations $\mathbf{Z}' = \mathbf{L}_p$ where $\mathbf{L}_p \in \mathbb{R}^{p \times p}$ is a lower triangular matrix and $p = (n+1)(n-1) + 1$. Hence, these $(n+1)(n-1) + 1$ vectors are linearly independent. □

Proposition 5.2 *Inequality (5.5) defines a facet of P for $1 \leq j \neq \ell \leq n$.*

Proof. By Remark (5.1), (5.5) is valid for P. Let $F = \{(\mathbf{x}, \mathbf{y}) \in P : y_{1j}^{2\ell} = 0\}$. Since $(\mathbf{u}_{1j} + \mathbf{u}_{2\ell} + \mathbf{v}_{1j}^{2\ell}) \in P$ but not in F, F is a proper face of P. Since all vectors used in the proof of Proposition 5.1 except $\mathbf{u}_{1g} + \mathbf{u}_{2h} + \mathbf{v}_{1g}^{2h}$ for a pair of indices $1 \leq g \neq h \leq n$ satisfy $y_{1g}^{2h} \geq 0$ at equality, the inequality (5.5) defines a facet of P for $1 \leq j \neq \ell \leq n$. □

Proposition 5.3 *Inequality (5.4) defines a facet of P for $1 \leq j \leq n$.*

Proof. By Remark (5.1), (5.4) is valid for P. Let $F = \{(\mathbf{x}, \mathbf{y}) \in P : -x_{1j} + \sum_{j \neq \ell = 1}^{n} y_{1j}^{2\ell} = 0\}$. Since $(\mathbf{u}_{1j} + \mathbf{u}_{2j}) \in P$ but not in F, F is a proper face of P. Since all vectors used in the proof of Proposition 5.1 except $\mathbf{u}_{1g} + \mathbf{u}_{2g}$ for g satisfy $-x_{1g} + \sum_{g \neq h = 1}^{n} y_{1g}^{2h} \leq 0$ at equality, the inequality (5.4) defines a facet of P for $1 \leq j \leq n$. □

Remark 5.2 *An optimal solution to $max\{\mathbf{cx} + \mathbf{qy} : (\mathbf{x}, \mathbf{y}) \in P\}$ is characterized by two cases:*

(i) *if there exists $1 \leq p \neq r \leq n$ such that $c_{1p} + c_{2r} + q_{1p}^{2r} \geq c_{1i} + c_{2i}$ for all $1 \leq i \leq n$ then an optimal solution is $x_{1p} = x_{2r} = y_{1p}^{2r} = 1$ and $x_{1j} = x_{2\ell} = y_{1j}^{2\ell} = 0$ for all $1 \leq j \neq p \leq n$ and $1 \leq \ell \neq r \leq n$.*

(ii) *if the condition in (i) does not hold then an optimal solution is $x_{1p} = x_{2p} = 1$ and $x_{1j} = x_{2\ell} = y_{1j}^{2\ell} = 0$ for $1 \leq j \neq p \leq n, 1 \leq \ell \neq p \leq n$ where $c_{1p} + c_{2p} \geq c_{1j} + c_{2j}$ for all $1 \leq j \leq n$.*

Proposition 5.4 *The solution of Remark (5.2) is an optimal solution to the LP problem $max\{\mathbf{cx} + \mathbf{qy} : (\mathbf{x}, \mathbf{y}) \in P_L\}$ where $(\mathbf{c}, \mathbf{q})$ is an arbitrary cost vector.*

Proof. Let $(\mathbf{x}^*, \mathbf{y}^*)$ be the solution vector defined in Remark (5.2). By Remark (5.1), $P \subseteq P_L$ and $(\mathbf{x}^*, \mathbf{y}^*)$ is an extreme point of P_L in either case of Remark (5.2). The dual to $max\{\mathbf{cx} + \mathbf{qy} : (\mathbf{x}, \mathbf{y}) \in P_L\}$ is $min\{u_1 + u_2 : u_1 + v_j - w_j = c_{1j}$ for $1 \leq j \leq n-1, u_1 - w_n = c_{1n}, u_2 - v_j = c_{2j}$ for $1 \leq j \leq n-1, u_2 = c_{2n}, -v_j + v_\ell + w_j \geq q_{1j}^{2\ell}$ for all $1 \leq j \neq \ell \leq n\}$. The vector given by $u_1 = z - c_{2n}, u_2 = c_{2n}, v_j = c_{2n} - c_{2j}$ for all $1 \leq j \leq n, w_j = z - c_{1j} - c_{2j}$ for all $1 \leq j \leq n$ where $z = c_{1p} + c_{2r} + q_{1p}^{2r}$ in case (i) of Remark (5.2), $c_{1p} + c_{2p}$ otherwise, is feasible to the dual problem with the same objective function value as that of $(\mathbf{x}^*, \mathbf{y}^*)$. Hence by LP duality, $(\mathbf{x}^*, \mathbf{y}^*)$ is optimal over P_L in both cases. □

Proposition 5.5 summarizes what we have proven in this section.

Proposition 5.5 *The system of equations and inequalities* (5.2), ..., (5.5) *is an ideal linear description of the local polytope* P, *i.e.* $P = P_L$.

Considering all equations and inequalities resulting from the locally ideal linearization of the variables giving rise to quadratic terms in the objective function of the CLDP, we formulate the CLDP as the LP problem given by

$$\min \left\{ \sum_{i=1}^{m-1} \sum_{k=i+1}^{m} \sum_{j=1}^{n} \sum_{j \neq \ell = 1}^{n} q_{ij}^{k\ell} y_{ij}^{k\ell} : (\mathbf{x}, \mathbf{y}) \in QDP_n^m \right\}, \qquad (\mathcal{O}QDP_n^m)$$

where QDP_n^m denotes the convex hull of solutions $(\mathbf{x}, \mathbf{y}) \in \mathbb{R}^{mn+mn(m-1)(n-1)/2}$ to the following system of equations and inequalities in zero-one variables:

$$\sum_{j=1}^{n} x_{ij} = 1 \qquad \text{for } i \in M \qquad (5.6)$$

$$-x_{ij} + x_{kj} + \sum_{j \neq \ell = 1}^{n} y_{ij}^{k\ell} - \sum_{j \neq \ell = 1}^{n} y_{i\ell}^{kj} = 0 \qquad \text{for } i < k \in M, 1 \leq j \leq n-1 \quad (5.7)$$

$$-x_{ij} + \sum_{j \neq \ell = 1}^{n} y_{ij}^{k\ell} \leq 0 \qquad \text{for } i < k \in M, j \in N \qquad (5.8)$$

$$y_{ij}^{k\ell} \geq 0 \qquad \text{for } i < k \in M, j \neq \ell \in N \qquad (5.9)$$

$$x_{ij} \in \{0, 1\} \qquad \text{for } i < k \in M, j \neq \ell \in N. \qquad (5.10)$$

Proposition 5.6 $\mathcal{O}QDP_n^m$ *is a formulation of the VLSI Circuit Layout Design Problem with* $m + m(m-1)(n-1)/2$ *equations, where* $m \geq n \geq 3$.

Proof. By a similar argument as in Remark (5.1), $DQDP_n^m \subseteq QDP_n^m$. Let $(\mathbf{x}, \mathbf{y}) \in QDP_n^m$. We show that $y_{ij}^{k\ell} = x_{ij} x_{k\ell}$ for all $1 \leq i < k \leq m$ and $1 \leq j \neq \ell \leq n$. Suppose that there exist $1 \leq p < g \leq m$ and $1 \leq r \neq s \leq n$ such that $y_{pr}^{gs} \neq x_{pr} x_{gs}$. Using (5.6), ..., (5.9), we conclude $y_{pr}^{gs} = 0$ whenever $x_{pr} = 0$ or $x_{gs} = 0$. So necessarily $x_{pr} = x_{gs} = 1$. But, then using (5.6) and (5.9) and an identical argument as above, we conclude from (5.7) where $i = p, k = g$ and $j = r$ that $1 = x_{pr} = \sum_{p \neq \ell=1}^{n} y_{pr}^{g\ell} = y_{pr}^{gs}$, which contradicts the assumption that $y_{pr}^{gs} \neq x_{pr} x_{gs}$. Since, all extreme points in QDP_n^m are zero-one valued and in $DQDP_n^m$, the first part of the proposition follows. The rest follows by a simple counting argument. $\square$

The LP relaxation of our formulation of the CLDP has polynomially many variables and polynomially many equations and inequalities and hence, it is polynomially solvable.

To say more about the formulation of the CLDP that we have just obtained, let us order the components of $\mathbf{x} \in \mathbb{R}^{mn}$ as $(x_{11}, \ldots, x_{1n}, \ldots, x_{m1}, \ldots, x_{mn})$ and those of $\mathbf{y} \in \mathbb{R}^{mn(m-1)(n-1)/2}$ as $(y_{11}^{22}, \ldots, y_{11}^{2n}, \ldots, y_{11}^{m2}, \ldots, y_{11}^{mn}, y_{12}^{21}, \ldots, y_{12}^{mn}, \ldots, y_{1n}^{m,n-1}, y_{21}^{32}, \ldots, y_{m-1,n}^{m,n-1})$; i.e. $(y_{11}^{22}, y_{11}^{23}, y_{11}^{32}, y_{11}^{33}, y_{12}^{21}, y_{12}^{23}, y_{12}^{31}, y_{12}^{33}, y_{13}^{21}, y_{13}^{22}, y_{13}^{31}, y_{13}^{32}, y_{21}^{32}, y_{21}^{33}, y_{22}^{31}, y_{22}^{33}, y_{23}^{31}, y_{23}^{32})$ explicitly shows the ordering of all components of $\mathbf{y}$ for $m = 3$ and $n = 3$. Let $\bar{\mathbf{u}}_{ij} \in \mathbb{R}^{mn}$ with its components ordered like those of $\mathbf{x}$ be a unit vector with one in its $(i,j)^{th}$ component and $\bar{\mathbf{v}}_{ij}^{k\ell} \in \mathbb{R}^{mn(m-1)(n-1)/2}$ ordered like $\mathbf{y}$ be another unit vector with one in its $\binom{k,\ell}{i,j}^{th}$ component. Let $\mathbf{u}_{ij} \in \mathbb{R}^{mn+mn(m-1)(n-1)/2}$ be obtained from $\bar{\mathbf{u}}_{ij}$ by appending $mn(m-1)(n-1)/2$ zeroes in the last $mn(m-1)(n-1)/2$ components and $\mathbf{v}_{ij}^{k\ell} \in \mathbb{R}^{mn+mn(m-1)(n-1)/2}$ be obtained from $\bar{\mathbf{v}}_{ij}^{k\ell}$ by appending mn zeroes at the beginning. Let $z_{(S_1, S_2, \ldots, S_g)}(t_1, t_2, \ldots, t_g) = \sum_{i=1}^{g} \sum_{i \in S_g} \mathbf{u}_{it_g} + \sum_{i=1}^{g-1} \sum_{k=i+1}^{g} \sum_{j \in S_i} (\sum_{j < \ell \in S_k} \mathbf{v}_{jt_j}^{\ell t_\ell} + \sum_{j > \ell \in S_k} \mathbf{v}_{\ell t_\ell}^{j t_j})$ and $z_M(j) = \sum_{i=1}^{m} \mathbf{u}_{ij}$ where $M = \{1, 2, \ldots, m\}$.

Proposition 5.7 *The dimension of the CLDP polytope $dim(QDP_n^m)$ equals $m(n-1) + m(m-1)(n-1)^2/2$ for all $m \geq n \geq 3$.*

Proof. We write equations (5.6) and (5.7) in matrix form as $\mathbf{A}_1 \mathbf{x} + \mathbf{A}_2 \mathbf{y} = \mathbf{b}$ where $\mathbf{A} = (\mathbf{A}_1, \mathbf{A}_2)$. Partitioning $\mathbf{A} = (\mathbf{A}', \mathbf{A}'')$ so that $\mathbf{A}'$ corresponds to $x_{11}, \ldots, x_{m1}, y_{11}^{2n}, \ldots, y_{11}^{mn}, y_{12}^{2n}, \ldots, y_{12}^{mn}, \ldots, y_{1,n-1}^{mn}, y_{21}^{3n}, \ldots, y_{2,n-1}^{mn}, \ldots, y_{m-1,n-1}^{mn}$, we have that *modulo* row permutations $\mathbf{A}' = \mathbf{L}_p$ where $\mathbf{L}_p \in \mathbb{R}^{p \times p}$ is a lower triangular matrix and $p = m + m(m-1)(n-1)/2$. Thus, $dim(QDP_n^m) \leq m(n-1) + m(m-1)(n-1)^2/2$. We establish $dim(QDP_n^m) \geq m(n-1) + m(m-1)(n-1)^2/2$ by exhibiting $m + m(m-1)(n-1)^2/2 + 1$ linearly independent zero-one vectors that belong to QDP_n^m. Consider the matrix $\mathbf{Z}$ whose rows are formed by
(i) the vector $z_M(n) \in QDP_n^m$,
(ii) $m(n-1)$ vectors $z_{(S_1, S_2)}(j, n) \in QDP_n^m$ for $1 \leq j \leq n-1, S_1 = \{1, \ldots, i\}$, $S_2 = \{i+1, \ldots, m\}$ where $1 \leq i \leq m$,
(iii) $m(m-1)(n-1)(n-2)/2$ vectors $z_{(S_1, S_2, S_3)}(j, \ell, n) \in QDP_n^m$ for $1 \leq j \neq \ell \leq n-1, S_1 = \{1, \ldots, i\}, S_2 = \{i+1, \ldots, k\}, S_3 = \{k+1, \ldots, m\}$ where $1 \leq i < k \leq m$, and
(iv) $m(m-1)(n-1)$ vectors $z_{(S_1, S_2)}(n, j) \in QDP_n^m$ for $1 \leq j \leq n-1, S_1 = \{1, \ldots, i, k+1, \ldots, m\}, S_2 = \{i+1, \ldots, k\}$ where $1 \leq i < k \leq m$.

Locally Ideal LP Formulations II

Partitioning $\mathbf{Z} = (\mathbf{Z}', \mathbf{Z}'')$ such that $\mathbf{Z}''$ corresponds to $x_{2n}, \ldots, x_{nn}, y_{11}^{2n}, \ldots,$
$y_{11}^{mn}, y_{12}^{2n}, \ldots, y_{12}^{mn}, \ldots, y_{1,n-1}^{2n}, \ldots, y_{1,n-1}^{mn}, y_{21}^{3n}, \ldots, y_{2,n-1}^{mn}, \ldots, y_{m-1,n-1}^{mn}$, we have that *modulo* row permutations $\mathbf{Z}' = \mathbf{L}_p$ where $\mathbf{L}_p \in \mathbb{R}^{p \times p}$ is a lower triangular matrix and $p = m(n-1) + m(m-1)(n-1)^2/2 + 1$. Hence, these $m(n-1) + m(m-1)(n-1)^2/2 + 1$ vectors are linearly independent. □

Proposition 5.8 *(5.9) defines a facet of QDP_n^m for $1 \leq i < k \leq m, 1 \leq j \neq \ell \leq n$.*

Proof. By Proposition (5.6), (5.9) is valid for QDP_n^m. Since all vectors of the proof of Proposition 5.7 except $z_{(\{m-1\},\{m\},M-\{m-1,m\})}(g,h,n)$ for all $1 \leq g \neq h \leq n-1$ satisfy $y_{m-1,g}^{mh} \geq 0$ at equality, (5.9) defines a facet of QDP_n^m for $i = m-1, k = m$ and $1 \leq j \neq \ell \leq n-1$. By appropriately permuting the indices of these vectors and using similar arguments as above, it can be shown that all inequalities (5.9) define facets of QDP_n^m. □

Proposition 5.9 *Inequality (5.8) defines a facet of QDP_n^m for $1 \leq i < k \leq m, 1 \leq j \leq n$.*

Proof. By Proposition (5.6), (5.8) is valid for QDP_n^m. Since all vectors of the proof of Proposition 5.7 except $z_M(g) = z_{(M,\{\emptyset\})}(g,n)$ for all $1 \leq g \leq n-1$ satisfy $-x_{m-1,g} + \sum_{g \neq h=1}^{n} y_{m-1,g}^{mh} \leq 0$ at equality, (5.8) defines a facet of QDP_n^m for $i = m-1, k = m$ and $1 \leq j \leq n-1$. By appropriately permuting the indices of these vectors and using similar arguments as above, it can be shown that all inequalities (5.8) define facets of QDP_n^m. □

5.2 A General Model

We now consider a model that generalizes all BQPSs considered so far in this chapter. Define mn zero-one variables x_{ij} for $1 \leq i \leq m$ and $1 \leq j \leq n$ and $n^2 m(m-1)/2$ variables $y_{ij}^{k\ell} = x_{ij} x_{k\ell}$ for $1 \leq i < k \leq m$ and $1 \leq j, \ell \leq n$ with $m \geq n \geq 3$. Denoting by $DQGP_n^m$ the discrete set

$$DQGP_n^m = \left\{ \begin{array}{l} (\mathbf{x},\mathbf{y}) \in \mathbb{R}^{mn + n^2 m(m-1)/2} : \\ \sum_{j=1}^{n} x_{ij} = 1 \quad \text{for } i \in M \\ y_{ij}^{k\ell} = x_{ij} x_{k\ell} \quad \text{for } i < k \in N, j, \ell \in N \\ x_{ij} \in \{0,1\} \quad \text{for } i \in M, j \in N \end{array} \right\},$$

we define this general linear optimization model as

$$\min\left\{\sum_{i=1}^{m}\sum_{j=1}^{n}c_{ij}x_{ij} + \sum_{i=1}^{m-1}\sum_{k=i+1}^{m}\sum_{j=1}^{n}\sum_{\ell=1}^{n}q_{ij}^{k\ell}y_{ij}^{k\ell} : (\mathbf{x},\mathbf{y}) \in DQGP_n^m\right\},$$

where for $1 \leq j \leq n$, $q_{ij}^{kj} = a_{ikj} + a_{kij}$ in terms of the a_{ikj} of Chapter 2.6 and for $1 \leq j \neq \ell \leq n$, $q_{ij}^{k\ell} = a_{ijk\ell} + a_{k\ell ij}$ in terms of the $a_{ijk\ell}$ of Chapter 2.4. Projecting out all y_{ij}^{kj} for $1 \leq j \leq n$ from the general model yields the CLDP, while projecting out all $y_{ij}^{k\ell}$ for $1 \leq j \neq \ell \leq n$ yields the OSP. Since the MPP can be obtained from the CLDP by symmetrization (see 5.1), the GPP can be obtained from the OSP by aggregation (see 4.17) and all problems considered so far in this chapter can be obtained as special cases of this general model.

To obtain a linear formulation for $DQGP_n^m$ in zero-one variables, we consider the local polytope P given by $P = conv(D)$ where $n \geq 3$ and D is defined as follows; see Figure 5.2:

$$D = \left\{\begin{array}{ll}(\mathbf{x},\mathbf{y}) \in \mathbb{R}^{n^2+2n} : & \\ \sum_{j=1}^{n} x_{ij} = 1 & \text{for } 1 \leq i \leq 2 \\ y_{1j}^{2\ell} = x_{1j}x_{2\ell} & \text{for } 1 \leq j, \ell \leq n \\ x_{ij} \in \{0,1\} & \text{for } 1 \leq j \leq n\end{array}\right\}.$$

Let P_L be the polytope given by $(\mathbf{x},\mathbf{y}) \in \mathbb{R}^{n^2+2n}$ satisfying

$$\sum_{j=1}^{n} x_{ij} = 1 \qquad \text{for } 1 \leq i \leq 2 \qquad (5.11)$$

$$-x_{1j} + \sum_{\ell=1}^{n} y_{1j}^{2\ell} = 0 \qquad \text{for } 1 \leq j \leq n \qquad (5.12)$$

$$-x_{2j} + \sum_{\ell=1}^{n} y_{1\ell}^{2j} = 0 \qquad \text{for } 1 \leq j \leq n-1 \qquad (5.13)$$

$$y_{1j}^{2\ell} \geq 0 \qquad \text{for } 1 \leq j, \ell \leq n. \qquad (5.14)$$

Remark 5.3 *The system of equations and inequalities (5.11),...,(5.14) is valid for all* $(\mathbf{x},\mathbf{y}) \in P$ *and thus* $P \subseteq P_L$. *There are* $2n+1$ *equations in (5.11), (5.12) and (5.13); the equation* $-x_{2n} + \sum_{\ell=1}^{n} y_{1\ell}^{2n} = 0$ *is redundant for all* $(\mathbf{x},\mathbf{y}) \in P_L$.

Proof. Let $(\mathbf{x},\mathbf{y}) \in D$. Then $(\mathbf{x},\mathbf{y})$ satisfies (5.11). From (5.11) we calculate $-x_{1j} + \sum_{\ell=1}^{n} y_{1j}^{2\ell} = -x_{1j} + \sum_{\ell=1}^{n} x_{1j}x_{2\ell} = -x_{1j} + x_{1j}\sum_{\ell=1}^{n} x_{2\ell} = -x_{1j} + x_{1j} = 0$

Locally Ideal LP Formulations II

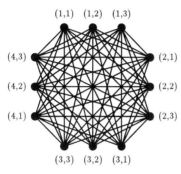
The cobweb of all node and edge variables of QGP_3^4

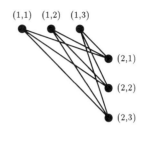
Node and edge variables used in the locally ideal linearization of QGP_3^4

Figure 5.2 The locally ideal linearization of the general model

and hence (5.12) is satisfied as well. By a similar argument, (5.13) is satisfied. Thus, $D \subseteq P_L$ and $P = conv(D) \subseteq P_L$. There are $2n+1$ equations (5.11), (5.12) and (5.13). To prove the stated redundancy, we take the linear combination of (5.11), (5.12) and (5.13) given by $\sum_{j=1}^{n} x_{1j} - \sum_{j=1}^{n} x_{2j} + \sum_{j=1}^{n}(-x_{1j} + \sum_{\ell=1}^{n} y_{1j}^{2\ell}) - \sum_{j=1}^{n-1}(-x_{2j} + \sum_{\ell=1}^{n} y_{1\ell}^{2j}) = -x_{2n} + \sum_{\ell=1}^{n} y_{1\ell}^{2n}$ which equals 0 for all feasible $(\mathbf{x}, \mathbf{y}) \in P_L$; hence, the remark follows. □

We order the components of $\mathbf{x}$ as $(x_{11}, \ldots, x_{1n}, x_{21}, \ldots, x_{2n})$ and those of $\mathbf{y}$ as $(y_{11}^{21}, \ldots, y_{11}^{2n}, y_{12}^{21}, y_{12}^{22}, \ldots, y_{12}^{2n}, \ldots, y_{1n}^{21}, \ldots, y_{1n}^{2,n})$. Let $\overline{\mathbf{u}}_{ij} \in \mathbb{R}^{2n}$ with its components ordered like those of $\mathbf{x}$ be a unit vector with one in its $(i,j)^{th}$ component and $\overline{\mathbf{v}}_{1j}^{2\ell} \in \mathbb{R}^{n^2}$ ordered like $\mathbf{y}$ be another unit vector with one in its $\binom{2,\ell}{1,j}^{th}$ component. Let $\mathbf{u}_{ij} \in \mathbb{R}^{n^2+2n}$ be obtained from $\overline{\mathbf{u}}_{ij}$ by appending n^2 zeroes in the last n^2 components and $\mathbf{v}_{1j}^{2\ell} \in \mathbb{R}^{n^2+2n}$ be obtained from $\overline{\mathbf{v}}_{1j}^{2\ell}$ by appending $2n$ zeroes at the beginning.

Proposition 5.10 *The dimension of P equals $n^2 - 1$ for all $n \geq 3$.*

Proof. We write the equations (5.11) in ascending order of i and those of (5.12) followed by the ones of (5.13) arranged in ascending order of j in matrix form as $\mathbf{A}_1 \mathbf{x} + \mathbf{A}_2 \mathbf{y} = \mathbf{b}$. Let $\mathbf{A} = (\mathbf{A}_1, \mathbf{A}_2)$. Partitioning $\mathbf{A} = (\mathbf{A}', \mathbf{A}'')$ such that $\mathbf{A}'$ corresponds to $x_{1n}, x_{2n}, y_{11}^{2n}, y_{12}^{2n}, \ldots, y_{1n}^{2n}, y_{1n}^{21}, y_{1n}^{22}, \ldots, y_{1n}^{2,n-1}$, we have that $\mathbf{A}'$ is a lower triangular matrix of dimension $2n+1$. Thus, $dim(P) \leq n^2 - 1$. We establish $dim(P) \geq n^2 - 1$ by exhibiting n^2 linearly independent zero-one vectors belonging to P. Consider the matrix $\mathbf{Z}$ whose rows are formed by the vectors $\mathbf{u}_{1j} + \mathbf{u}_{2\ell} + \mathbf{v}_{1j}^{2\ell} \in P$ for $1 \leq j, \ell \leq n$. Partitioning $\mathbf{Z} = (\mathbf{Z}', \mathbf{Z}'')$

columnwise so that $\mathbf{Z}'$ corresponds to the variables $y_{1j}^{2\ell}$ for $1 \leq j, \ell \leq n$, we have that $\mathbf{Z}' = \mathbf{I}_p$ where $\mathbf{I}_p \in \mathbb{R}^{p \times p}$ is an identity matrix and $p = n^2$. Hence, these n^2 vectors are linearly independent. □

Proposition 5.11 *(5.14) defines a facet of P for $1 \leq i \leq m, 1 \leq j, \ell \leq n$.*

Proof. By Remark (5.3), (5.14) is valid for P. Since all vectors used in the proof of Proposition 5.10 except $\mathbf{u}_{1g} + \mathbf{u}_{2h} + \mathbf{v}_{1g}^{2h}$ for a pair $1 \leq g, h \leq n$ satisfy $y_{1g}^{2h} \geq 0$ at equality, (5.14) for $1 \leq j, \ell \leq n$ defines a facet of P. □

Remark 5.4 *An optimal solution to $max\{\mathbf{cx} + \mathbf{qy} : (\mathbf{x}, \mathbf{y}) \in P\}$ is $x_{1p} = x_{2r} = y_{1p}^{2r} = 1$ where $1 \leq p, r \leq n$ and $c_{1p} + c_{2r} + q_{1p}^{2r} \geq c_{1j} + c_{2\ell} + q_{1j}^{2\ell}$ for all $1 \leq j, \ell \leq n$.*

Proposition 5.12 *The solution of Remark (5.4) is an optimal solution to the LP problem $max\{\mathbf{cx} + \mathbf{qy} : (\mathbf{x}, \mathbf{y}) \in P_L\}$ where $(\mathbf{c}, \mathbf{q})$ is arbitrary.*

Proof. Let $(\mathbf{x}^*, \mathbf{y}^*)$ be the solution vector defined in Remark (5.4). By Remark (5.3), $P \subseteq P_L$ and trivially, $(\mathbf{x}^*, \mathbf{y}^*)$ is an extreme point of P_L. Let $P' = P_L \cup \{(\mathbf{x}, \mathbf{y}) \in \mathbb{R}^{n^2+1} : -x_{2n} + \sum_{\ell=1}^n y_{1\ell}^{2n} = 0\}$. By Remark (5.3), $P_L = P'$. We show that $(\mathbf{x}^*, \mathbf{y}^*)$ is optimal over P' and hence optimal over P_L. The dual to the maximization problem over P' is $min\{s_1 + s_2 : s_i - u_{ij} = c_{ij}$ for $1 \leq i \leq 2, 1 \leq j \leq n, u_{1j} + u_{2\ell} \geq q_{1j}^{2\ell}$ for $1 \leq j, \ell \leq n\}$. The vector given by $s_1 = s_2 = (c_{1p} + c_{2r} + q_{1p}^{2r})/2, u_{ij} = s_i - c_{ij}$ for $1 \leq i \leq 2, 1 \leq j \leq n$ is feasible to the dual problem and its objective function value $c_{1p} + c_{2r} + q_{1p}^{2r}$ equals that of $(\mathbf{x}^*, \mathbf{y}^*)$. Thus by LP duality $(\mathbf{x}^*, \mathbf{y}^*)$ is optimal over P'. □

We now state a proposition which summarizes the preceding.

Proposition 5.13 *The system of equations and inequalities (5.11), ..., (5.14) is an ideal linear description of the local polytope P, i.e. $P = P_L$.*

Considering all equations and inequalities resulting from the locally ideal linearization of the variables giving rise to quadratic terms in the objective function of the general model, we formulate the general model as the LP problem

$$min\left\{\sum_{i,j \in N} c_{ij} x_{ij} + \sum_{i<k \in M} \sum_{j<\ell \in N} q_{ij}^{k\ell} y_{ij}^{k\ell} : (\mathbf{x}, \mathbf{y}) \in QGP_n^m\right\}, \quad (OQGP_n^m)$$

Locally Ideal LP Formulations II

where QGP_n^m is the polytope defined by the convex hull of solutions $(\mathbf{x},\mathbf{y}) \in \mathbb{R}^{mn+n^2m(m-1)/2}$ to the following equations and inequalities in zero-one variables:

$$\sum_{j=1}^{n} x_{ij} = 1 \quad \text{for } i \in M \tag{5.15}$$

$$-x_{ij} + \sum_{\ell=1}^{n} y_{ij}^{k\ell} = 0 \quad \text{for } i < k \in M, j \in N \tag{5.16}$$

$$-x_{kj} + \sum_{\ell=1}^{n} y_{i\ell}^{kj} = 0 \quad \text{for } i < k \in M, 1 \le j \le n-1 \tag{5.17}$$

$$y_{ij}^{k\ell} \ge 0 \quad \text{for } i < k \in M, j, \ell \in N \tag{5.18}$$

$$x_{ij} \in \{0,1\} \quad \text{for } i \in M, j \in N \tag{5.19}$$

Proposition 5.14 $\mathcal{O}QGP_n^m$ *is a formulation of the general model with* $m + m(m-1)(2n-1)/2$ *equations where* $m \ge n \ge 3$.

Proof. By a similar argument as in Remark (5.3), $DQGP_n^m \subseteq QGP_n^m$. Let $(\mathbf{x},\mathbf{y}) \in QGP_n^m$. For any pair $1 \le i < k \le m$, consider the linear combination of equations (5.15), (5.16) and (5.17) given by $\sum_{j=1}^n x_{ij} - \sum_{j=1}^n x_{kj} + \sum_{j=1}^n (-x_{ij} + \sum_{\ell=1}^n y_{ij}^{k\ell}) - \sum_{j=1}^{n-1}(-x_{kj} + \sum_{\ell=1}^n y_{i\ell}^{kj}) = -x_{kn} + \sum_{\ell=1}^n y_{i\ell}^{kn}$ which equals 0 for all $(\mathbf{x},\mathbf{y}) \in QGP_n^m$; hence, $-x_{kn} + \sum_{j=1}^n y_{ij}^{kn} = 0$ for all $1 \le i < k \le m$ are redundant. We show that $y_{ij}^{k\ell} = x_{ij}x_{k\ell}$ for all $1 \le i < k \le m$ and $1 \le j, \ell \le n$. Suppose that there exist $1 \le p < g \le m$ and $1 \le r,s \le n$ such that $y_{pr}^{gs} \ne x_{pr}x_{gs}$. Using (5.16), (5.17) and (5.18) and the equations shown to be redundant, we conclude $y_{pr}^{gs} = 0$ whenever $x_{pr} = 0$ or $x_{gs} = 0$. So necessarily $x_{pr} = x_{gs} = 1$. But, then using (5.15) and (5.18) and an identical argument as above, we conclude from (5.16) and (5.17) and the redundant equations that $1 = x_{pr} = \sum_{\ell=1}^n y_{pr}^{g\ell} = y_{pr}^{gs}$, which contradicts the assumption that $y_{pr}^{gs} \ne x_{pr}x_{gs}$. Since all extreme points in QGP_n^m are zero-one valued and in $D\mathcal{Q}GP_n^m$, the first part follows. The rest follows by counting. □

The LP relaxation of our formulation of the general model has polynomially many variables and polynomially many equations and inequalities and hence, it is polynomially solvable.

To get more insight into this model, we order the components of $\mathbf{x} \in \mathbb{R}^{mn}$ as $(x_{11}, \ldots, x_{1n}, \ldots, x_{m1}, \ldots, x_{mn})$ and those of $\mathbf{y} \in \mathbb{R}^{n^2m(m-1)/2}$ as $(y_{11}^{21}, \ldots, y_{11}^{2n}, \ldots, y_{11}^{m1}, \ldots, y_{11}^{mn}, y_{12}^{21}, \ldots, y_{12}^{2n}, \ldots, y_{12}^{mn}, \ldots, y_{1n}^{mn}, y_{21}^{31}, \ldots, y_{m-1,n}^{mn})$. Let $\overline{\mathbf{u}}_{ij} \in \mathbb{R}^{mn}$ with its components ordered like those of $\mathbf{x}$ be a unit vector with one in its $(i,j)^{th}$ component and $\bar{\mathbf{v}}_{ij}^{k\ell} \in \mathbb{R}^{n^2m(m-1)/2}$ ordered like $\mathbf{y}$ be another unit vector with one in its $\binom{k,\ell}{i,j}^{th}$ component. Let $\mathbf{u}_{ij} \in \mathbb{R}^{mn+n^2m(m-1)/2}$ be obtained from $\bar{\mathbf{u}}_{ij}$ by appending $n^2m(m-1)/2$ zeroes in the last $n^2m(m-1)/2$ components and $\mathbf{v}_{ij}^{k\ell} \in \mathbb{R}^{mn+n^2m(m-1)/2}$ be obtained from $\bar{\mathbf{v}}_{ij}^{k\ell}$ by appending mn zeroes at the beginning. Let $z_{(S_1,S_2,\ldots,S_g)}(t_1,t_2,\ldots,t_g) = \sum_{i=1}^{g} \sum_{i \in S_g} \mathbf{u}_{it_g} + \sum_{i=1}^{g-1} \sum_{k=i+1}^{g} \sum_{j \in S_i} \sum_{\ell \in S_k} \mathbf{v}_{jt_j}^{\ell t_\ell}$ where $M = \{1,2,\ldots,m\}$.

Proposition 5.15 *The dimension of the polytope associated with the general model is given by $dim(QGP_n^m) = m(n-1)+m(m-1)(n-1)^2/2$ for $m \geq n \geq 3$.*

Proof. We write all equations (5.15) and (5.16) in matrix form as $\mathbf{A}_1\mathbf{x}+\mathbf{A}_2\mathbf{y} = \mathbf{b}$ and let $\mathbf{A} = (\mathbf{A}_1, \mathbf{A}_2)$. Partitioning $\mathbf{A} = (\mathbf{A}', \mathbf{A}'')$ such that $\mathbf{A}'$ corresponds to $x_{11},\ldots,x_{m1}, y_{11}^{2n},\ldots,y_{11}^{mn}, y_{12}^{2n},\ldots,y_{12}^{mn},\ldots,y_{1n}^{mn}, y_{21}^{3n},\ldots,y_{2n}^{mn},\ldots y_{m-1,n}^{mn}$, we have that *modulo* row permutations $\mathbf{A}' = \mathbf{L}_p$ where $\mathbf{L}_p \in \mathbb{R}^{p \times p}$ is a lower triangular matrix and $p = m(n-1) + m(m-1)(n-1)^2/2$. Thus, $dim(P) \leq m(n-1)+m(m-1)(n-1)^2/2$. We establish $dim(P) \geq m(n-1)+m(m-1)(n-1)^2/2$ by exhibiting $m(n-1) + m(m-1)(n-1)^2/2 + 1$ linearly independent zero-one vectors that belong to QGP_n^m. Consider the matrix $\mathbf{Z}$ whose rows are formed by the following vectors:

(i) the vector $z_M(n) \in QGP_n^m$,
(ii) $m(n-1)$ vectors $z_{(\{i\},M-\{i\})}(j,n) \in QGP_n^m$ for $1 \leq i \leq m, 1 \leq j \leq n-1$, and
(iii) $m(m-1)(n-1)^2/2$ vectors $z_{(\{i\},\{k\},M-\{i,k\})}(j,\ell,n) \in QGP_n^m$ for $1 \leq i < k \leq m, 1 \leq j,\ell \leq n-1$.

Partitioning $\mathbf{Z} = (\mathbf{Z}',\mathbf{Z}'')$ such that $\mathbf{Z}''$ corresponds to $x_{2n},\ldots,x_{mn}, y_{11}^{2n},\ldots, y_{11}^{mn}, y_{12}^{2n},\ldots,y_{12}^{mn},\ldots,y_{1n}^{mn}, y_{21}^{3n},\ldots,y_{2n}^{mn},\ldots y_{m-1,n}^{mn}$, we have that *modulo* row permutations $\mathbf{Z}' = \mathbf{L}_p$ where $\mathbf{L}_p \in \mathbb{R}^{p \times p}$ is a lower triangular matrix and $p = m(n-1)+m(m-1)(n-1)^2/2+1$. Hence, these $m(n-1)+m(m-1)(n-1)^2/2+1$ vectors are linearly independent. □

Proposition 5.16 *Inequality (5.18) defines a facet of QGP_n^m for $1 \leq i < k \leq m, 1 \leq j, \ell \leq n$.*

Proof. By Proposition (5.14), (5.18) is valid for QGP_n^m. Since all vectors of the proof of Proposition 5.15 except $z_{(\{i\},\{k\},M-\{i,k\})}(g,h,n)$ for all $1 \leq g \neq h \leq n-1$ satisfy $y_{ig}^{kh} \geq 0$ at equality, (5.10) defines a facet of QGP_n^m for $1 \leq i < k \leq m$ and $1 \leq g,h \leq n-1$. By appropriately permuting the indices of these vectors and by similar arguments, even when $j = n$ or $\ell = n$ or both, it can be shown that all inequalities (5.18) define facets of QGP_n^m. □

Remark 5.5 *From (5.16) we have $y_{ij}^{kj} = x_{ij} - \sum_{j \neq \ell=1}^{n} y_{ij}^{k\ell}$ for all $1 \leq i < k \leq m$ and $1 \leq j \leq n$. We can thus eliminate the variables y_{ij}^{kj} for $1 \leq i < k \leq m$ and $1 \leq j \leq n$ from the general model and formulate this model also in the same variables as the CLDP by appropriately modifying the objective function coefficients. Moreover, eliminating y_{ij}^{kj} from (5.17) for $1 \leq i < k \leq m$ and $1 \leq j \leq n-1$, we obtain (5.7) for $1 \leq i < k \leq m$ and $1 \leq j \leq n-1$. A similar elimination of y_{ij}^{kj} from (5.18) for $1 \leq i < k \leq m$ and $1 \leq j = \ell \leq n$ yields (5.8) for $1 \leq i < k \leq m$ and $1 \leq j \leq n$. The remaining equations and inequalities in the general model are the same as those of the CLDP; hence, the two formulations are equivalent.*

5.3 Quadratic Assignment Problems

To consider the QAP, see Chapter 1.6, we define new variables $y_{ij}^{k\ell} = x_{ij} x_{k\ell}$ for $1 \leq i < k \leq n$, $1 \leq j \neq \ell \leq n$. Counting yields that there are $n^2(n-1)^2/2$ y-variables. Denoting by $DQAP_n$ the discrete set

$$DQAP_n = \left\{ \begin{array}{l} (\mathbf{x}, \mathbf{y}) \in \mathbb{R}^{n^2 + n^2(n-1)^2/2} : \\ \sum_{i=1}^{n} x_{ij} = 1 \quad \text{for } j \in N \\ \sum_{j=1}^{n} x_{ij} = 1 \quad \text{for } i \in N \\ y_{ij}^{k\ell} = x_{ij} x_{k\ell} \quad \text{for } i < k \in n, j \neq \ell \in N \\ x_{ij} \in \{0, 1\} \quad \text{for } i, j \in N \end{array} \right\},$$

the QAP can be written as

$$\min \left\{ \sum_{i=1}^{n} \sum_{j=1}^{n} c_{ij} x_{ij} + \sum_{i=1}^{n-1} \sum_{k=i+1}^{n} \sum_{j=1}^{n} \sum_{j \neq \ell=1}^{n} q_{ij}^{k\ell} y_{ij}^{k\ell} : (\mathbf{x}, \mathbf{y}) \in DQAP_n \right\},$$

where $q_{ij}^{k\ell} = a_{ijk\ell} + a_{k\ell ij}$ in terms of the $a_{ijk\ell}$ of Chapter 1.6.

To obtain a linear formulation for $DQAP_n$ in zero-one variables, we consider the local polytope P given by $P = \text{conv}(D)$ where D is defined as follows; see Figure 5.3:

$$D = \left\{ \begin{array}{l} (\mathbf{x}, \mathbf{y}) \in \mathbb{R}^{n^2} : \\ \sum_{j=1}^{n} x_{1j} = 1 \\ \sum_{i=1}^{n} x_{i1} = 1 \\ y_{1j}^{i1} = x_{1j} x_{i1} \quad \text{for } 2 \leq i \leq n, 2 \leq j \leq n \\ x_{1j}, x_{i1} \in \{0, 1\} \quad \text{for } 1 \leq j \leq n, 2 \leq i \leq n \end{array} \right\}.$$

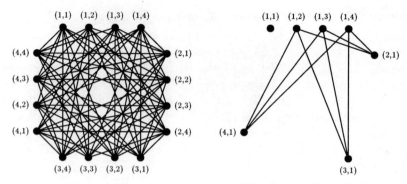

Figure 5.3 The locally ideal linearization of QAPs

Let P_L be the polytope given by $(\mathbf{x}, \mathbf{y}) \in \mathbb{R}^{n^2}$ satisfying

$$\sum_{j=1}^{n} x_{1j} = 1 \tag{5.20}$$

$$\sum_{i=1}^{n} x_{i1} = 1 \tag{5.21}$$

$$-x_{1j} + \sum_{i=2}^{n} y_{1j}^{i1} = 0 \quad \text{for } 2 \leq j \leq n \tag{5.22}$$

$$-x_{i1} + \sum_{j=2}^{n} y_{1j}^{i1} = 0 \quad \text{for } 2 \leq i \leq n-1 \tag{5.23}$$

$$x_{11} \geq 0 \tag{5.24}$$

$$y_{1j}^{i1} \geq 0 \quad \text{for } 2 \leq i \leq n, 2 \leq j \leq n. \tag{5.25}$$

Remark 5.6 *The system of equations and inequalities* (5.20),...,(5.25) *is valid for all* $(\mathbf{x}, \mathbf{y}) \in P$ *and thus* $P \subseteq P_L$. *There are* $2n-1$ *equations in* (5.20), ..., (5.23). *Moreover, the equation* $-x_{n1} + \sum_{j=2}^{n} y_{1j}^{n1} = 0$ *is redundant for all* $(\mathbf{x}, \mathbf{y}) \in P_L$.

Locally Ideal LP Formulations II

Proof. Let $(\mathbf{x}, \mathbf{y}) \in D$. Then $(\mathbf{x}, \mathbf{y})$ satisfies (5.20), (5.21), (5.24) and (5.25). To prove that (5.22) is satisfied as well we calculate

$$-x_{1j} + \sum_{i=2}^{n} y_{1j}^{i1} = -x_{1j} + \sum_{i=2}^{n} x_{i1} x_{1j} = -x_{1j} + (1 - x_{11}) x_{1j} = -x_{1j} x_{11}.$$

But $x_{1j} x_{11} = 0$ for all $(\mathbf{x}, \mathbf{y}) \in D$ and thus, (5.22) is satisfied. The proof that (5.23) are satisfied goes likewise. Thus, $D \subseteq P_L$ and hence, $P = conv(D) \subseteq P_L$. There are $2n-1$ equations in (5.20),..., (5.23). The linear combination of (5.22) and (5.23) given by $\sum_{p=2}^{n}(-x_{1p} + \sum_{i=2}^{n} y_{1p}^{i1}) - \sum_{p=2}^{n-1}(-x_{p1} + \sum_{j=2}^{n} y_{1j}^{p1}) = -\sum_{p=2}^{n} x_{1p} + \sum_{j=2}^{n} y_{1j}^{n1} + \sum_{p=2}^{n-1} x_{p1} = -x_{n1} + \sum_{j=1}^{n} y_{1j}^{p1} = 0$ where we have used (5.20) and (5.21). Hence, $-x_{n1} + \sum_{j=2}^{n} y_{1j}^{n1} = 0$ is redundant. □

We order components of $\mathbf{x}$ as $(x_{11}, \ldots, x_{1n}, x_{21}, \ldots, x_{n1})$ and those of $\mathbf{y}$ as $(y_{12}^{21}, \ldots, y_{12}^{n1}, \ldots, y_{1n}^{21}, \ldots, y_{1n}^{n1})$. Let $\bar{\mathbf{u}}_{ij} \in \mathbb{R}^{2n-1}$ with its components ordered like those of $\mathbf{x}$ be a unit vector with one in its $(i,j)^{th}$ component and by $\bar{\mathbf{v}}_{1j}^{k1} \in \mathbb{R}^{(n-1)(n-1)}$ with its components ordered like those of $\mathbf{y}$ be another unit vector with one in its $\binom{k,1}{1,j}^{th}$ component. Let $\mathbf{u}_{ij} \in \mathbb{R}^{(n^2)}$ be obtained from $\bar{\mathbf{u}}_{ij}$ by appending $(n-1)^2$ zeroes in the last $(n-1)^2$ components and $\mathbf{v}_{1j}^{k1} \in \mathbb{R}^{n^2}$ be obtained from $\bar{\mathbf{v}}_{1j}^{k1}$ by appending $2n-1$ zeroes at the beginning.

Proposition 5.17 *The dimension of P equals $(n-1)^2$ for all $n \geq 2$.*

Proof. We write all equations (5.22) and (5.23) in ascending order of j and i respectively followed by the equation (5.20) and (5.21) as $\mathbf{A}_1 \mathbf{x} + \mathbf{A}_2 \mathbf{y} = \mathbf{b}$ and let $\mathbf{A} = (\mathbf{A}_1, \mathbf{A}_2)$. Partitioning $\mathbf{A} = (\mathbf{A}', \mathbf{A}'')$ columnwise so that $\mathbf{A}'$ corresponds to $x_{12}, \ldots, x_{1n}, x_{21}, \ldots, x_{n-1,1}, x_{11}, x_{n1}$, we have that $\mathbf{A}'$ is a lower triangular and of dimension $2n-1$. Thus $dim(P) \leq n^2 - (2n-1) = (n-1)^2$. We establish $dim(P) \geq (n-1)^2$ by exhibiting $(n-1)^2 + 1$ linearly independent zero-one vectors that belong to P. Consider the matrix $\mathbf{Z}$ whose rows are formed by the vectors $\mathbf{u}_{11}, \mathbf{u}_{1j} + \mathbf{u}_{i1} + \mathbf{v}_{1j}^{i1}$ for $2 \leq i, j \leq n$ which are all in P. Partitioning $\mathbf{Z} = (\mathbf{Z}', \mathbf{Z}'')$ columnwise so that $\mathbf{Z}'$ corresponds to x_{11} and y_{1j}^{i1} for $2 \leq i, j \leq n$, we have $\mathbf{Z}' = \mathbf{I}_p$ where $\mathbf{I}_p \in \mathbb{R}^{p \times p}$ and $p = (n-1)^2 + 1$. Hence, these vectors are linearly independent. □

Proposition 5.18 *Inequality (5.24) defines a facet of P.*

Proof. By Remark (5.6), (5.24) is valid for P. Moreover, all vectors except $\mathbf{u}_{11}$ used in the proof of Proposition 5.17 satisfy (5.24) at equality. □

Proposition 5.19 *Inequality (5.25) defines a facet of P for $2 \leq i, j \leq n$.*

Proof. By Remark (5.6), (5.25) is valid for P. Moreover, all vectors except $\mathbf{u}_{1j}+\mathbf{u}_{21}+\mathbf{v}_{1j}^{i1}$ used in the proof of Proposition 5.17 satisfy (5.25) for $2 \leq i, j \leq n$ at equality. □

Remark 5.7 *An optimal solution to $max\{\mathbf{cx}+\mathbf{qy} : (\mathbf{x},\mathbf{y}) \in P\}$ is characterized by two cases:*

(i) if there exists $2 \leq p, r \leq n$ such that $c_{1p} + c_{r1} + q_{1p}^{r1} \geq c_{11}$ then an optimal solution is $x_{1p} = x_{r1} = y_{1p}^{r1} = 1$ and $x_{1i} = x_{2k} = y_{1j}^{i1} = 0$ for all $2 \leq i, j \leq n$ where $i \neq r$ and $j \neq p$.

(ii) if the condition in (i) does not hold then an optimal solution is $x_{11} = 1$ and $x_{1j} = x_{i1} = y_{1j}^{i1} = 0$ where $2 \leq i, j \leq n$.

Proposition 5.20 *The solution of Remark (5.7) is an optimal solution to the LP problem $max\{\mathbf{cx}+\mathbf{qy} : (\mathbf{x},\mathbf{y}) \in P_L\}$ where $(\mathbf{c}, \mathbf{q})$ is arbitrary.*

Proof. Let $(\mathbf{x}^*, \mathbf{y}^*)$ be the solution vector defined in Remark (5.7). By Remark (5.6), $P \subseteq P_L$ and trivially, $(\mathbf{x}^*, \mathbf{y}^*)$ is an extreme point of P_L in either case of Remark 5.7. The dual to $max\{\mathbf{cx}+\mathbf{qy} : (\mathbf{x},\mathbf{y}) \in P_L\}$ is $min\{s+t : s - u_j = c_{1j} \text{ for } 2 \leq j \leq n, t - v_k = c_{i1} \text{ for } 2 \leq i \leq n, u_j + v_k \geq q_{1j}^{i1} \text{ for } 2 \leq i \leq n-1, 2 \leq j \leq n, u_j \geq q_{1j}^{n1} \text{ for } 2 \leq j \leq n\}$. Let $z = c_{1p}+c_{r1}+q_{1p}^{r1}$ if we are in case (i) of Remark (5.7) and $z = c_{11}$ if we are in case (ii). The vector $s = z - c_{n1}, t = c_{n1}, u_j = z - c_{n1} - c_{1j}$ for $2 \leq j \leq n, v_i = c_{n1} - c_{i1}$ for $2 \leq i \leq n-1$ is feasible to the dual problem. Its objective function value is equal to that of $(\mathbf{x}^*, \mathbf{y}^*)$ and hence, by LP duality $(\mathbf{x}^*, \mathbf{y}^*)$ is optimal over P'. □

We now summarize what we have proven in this section.

Proposition 5.21 *The system of equations and inequalities (5.20), ..., (5.25) is an ideal linear description of the local polytope P, i.e. $P = P_L$.*

Considering all equations and inequalities resulting from the locally ideal linearization of the variables giving rise to quadratic terms in the objective function of the QAP except for $x_{11} \geq 0$ which is redundant for the QAP, we formulate QAP as the LP problem given by:

$$min\left\{\sum_{i,j \in N} c_{ij}x_{ij} + \sum_{i<k \in N}\sum_{j \neq \ell \in N} q_{ij}^{k\ell} y_{ij}^{k\ell} : (\mathbf{x},\mathbf{y}) \in QAP_n\right\}, \qquad (OQAP_n)$$

where QAP_n is the polytope defined by the convex hull of solutions $(\mathbf{x}, \mathbf{y}) \in \mathbb{R}^{n^2+n^2(n-1)^2/2}$ to the following equations and inequalities in zero-one variables:

$$\sum_{j=1}^{n} x_{ij} = 1 \quad \text{for } i \in N \tag{5.26}$$

$$\sum_{i=1}^{n} x_{ij} = 1 \quad \text{for } j \in N \tag{5.27}$$

$$-x_{ij} + \sum_{k=1}^{i-1} y_{k\ell}^{ij} + \sum_{k=i+1}^{n} y_{ij}^{k\ell} = 0 \quad \text{for } i \in N, j \neq \ell \in N \tag{5.28}$$

$$-x_{ij} + \sum_{\ell=1}^{j-1} y_{k\ell}^{ij} + \sum_{\ell=j+1}^{n} y_{k\ell}^{ij} = 0 \quad \text{for } j \in N, 1 \leq k < i \leq n-1 \tag{5.29}$$

$$-x_{ij} + \sum_{\ell=1}^{j-1} y_{ij}^{k\ell} + \sum_{\ell=j+1}^{n} y_{ij}^{k\ell} = 0 \quad \text{for } j \in N, i < k \in N \tag{5.30}$$

$$y_{ij}^{k\ell} \geq 0 \quad \text{for } i < k \in N, j \neq \ell \in N \tag{5.31}$$

$$x_{ij} \in \{0,1\} \quad \text{for } i,j \in N. \tag{5.32}$$

Proposition 5.22 $\mathcal{O}QAP_n$ *is a formulation of Quadratic Assignment Problem with $2n + n(n-1)(2n-1)$ equations.*

Proof. By a similar argument as in Remark (5.6), $DQAP_n \subseteq QAP_n$ and $-x_{nj} + \sum_{\ell=1}^{j-1} y_{k\ell}^{nj} + \sum_{\ell=j+1}^{n} y_{k\ell}^{nj} = 0$ for all $1 \leq k \leq n-1, 1 \leq j \leq n$ are redundant for QAP_n. Let $(\mathbf{x}, \mathbf{y}) \in QAP_n$. We have to show that $y_{ij}^{k\ell} = x_{ij}x_{k\ell}$ for all $1 \leq i < k \leq n$ and $1 \leq j \neq \ell \leq n$. Suppose there exist $1 \leq p < r \leq n$ and $1 \leq g \neq s \leq n$ such that $y_{pg}^{rs} \neq x_{pg}x_{rs}$. WROG we can assume that $g < s$. If $x_{pg} = 0$ then from (5.28) for $i = p, j = g$ and $\ell = s$ we conclude using (5.31) that $y_{pg}^{rs} = 0$ and we conclude likewise when $x_{rs} = 0$. So necessarily $x_{pg} = x_{rs} = 1$. Then by (5.30), $y_{pg}^{rs} + \sum_{\ell \in N \setminus \{s,g\}} y_{pg}^{r\ell} = 1$. Since $x_{rs} = 1$ implies $x_{r\ell} = 0$ for all $1 \leq \ell \neq s \leq n$, by a similar argument, we conclude $\sum_{\ell \in N \setminus \{s,g\}} y_{pg}^{r\ell} = 0$ and thus, $y_{pg}^{rs} = 1$. Since, all extreme points in QAP_n are zero-one valued and in $DQAP_n$, the first part follows. The rest follows by counting. □

The QAP_n formulation takes into account the symmetry $x_{ij}x_{k\ell} = x_{k\ell}x_{ij}$ for $1 \leq i,j,k,\ell \leq n$ as well as the duplication of the equations (2.6) and (2.7) in equations (2.8) and (2.9) of Frieze and Yadegar's [1983] formulation that we have discussed in Chapter 3. Resende et al. [1994] propose a formulation of the QAP similar to our $\mathcal{O}QAP_n$ formulation, but their formulation has $n(n-1)$ more equations than our formulation. Moreover, the above formulation $\mathcal{O}QAP_n$

does not give complete consideration to the *minimality* of the equations describing the *affine hull* of QAP_n. We return to this issue in Chapter 7.1.

5.4 Symmetric Quadratic Assignment Problems

To consider the Symmetric Quadratic Assignment Problem (SQP), see Chapter 1.6, we define new variables $y_{ij}^{k\ell} = x_{ij}x_{k\ell} + x_{i\ell}x_{kj}$ for $1 \leq i < k \leq n$, $1 \leq j < \ell \leq n$. Counting yields that there are $n^2(n-1)^2/4$ y-variables. Denoting by $DSQP_n$ the discrete set

$$DSQP_n = \left\{ \begin{array}{l} (\mathbf{x},\mathbf{y}) \in \mathbb{R}^{n^2+n^2(n-1)^2/4} : \\ \sum_{i=1}^n x_{ij} = 1 \quad \text{for } j \in N \\ \sum_{j=1}^n x_{ij} = 1 \quad \text{for } i \in N \\ y_{ij}^{k\ell} = x_{ij}x_{k\ell} + x_{i\ell}x_{kj} \quad \text{for } i < k \in N, j < \ell \in N \\ x_{ij} \in \{0,1\} \quad \text{for } i,j \in N \end{array} \right\},$$

the SQP can be written as

$$\min \left\{ \sum_{i=1}^n \sum_{j=1}^n c_{ij} x_{ij} + \sum_{i=1}^{n-1} \sum_{k=i+1}^n \sum_{j=1}^{n-1} \sum_{\ell=j+1}^n q_{ij}^{k\ell} y_{ij}^{k\ell} : (\mathbf{x},\mathbf{y}) \in DSQP_n \right\},$$

where $q_{ij}^{k\ell} = a_{ijk\ell} + a_{k\ell ij}$ in terms of the $a_{ijk\ell}$ of Chapter 1.6. We to note that the SQP can be obtained from the QAP by the transformation:

$$y_{ij}^{k\ell} = x_{ij}x_{k\ell} + x_{i\ell}x_{kj} \quad \text{for all } 1 \leq i < k \leq n, \ 1 \leq j < \ell \leq n, \quad (5.33)$$

since in the SQP we assume symmetry of the cost coefficients, i.e. $q_{ij}^{k\ell} = q_{i\ell}^{kj}$ for $1 \leq i < k \leq n$ and $1 \leq j < \ell \leq n$.

To obtain a linear formulation for $DSQP_n$ in zero-one variables, we consider the local polytope P given by $P = conv(D)$ where D is defined as follows:

$$D = \left\{ \begin{array}{ll} (\mathbf{x},\mathbf{y}) \in \mathbb{R}^{3n-1} : \\ \sum_{j=1}^n x_{ij} = 1 & \text{for } 1 \leq i \leq 2 \\ \sum_{i=1}^2 x_{ij} \leq 1 & \text{for } 1 \leq j \leq n \\ y_{11}^{2j} = x_{11}x_{2j} + x_{1j}x_{21} & \text{for } 2 \leq j \leq n \\ x_{ij} \in \{0,1\} & \text{for } 1 \leq i \leq 2, 1 \leq j \leq n \end{array} \right\}.$$

In Table 5.1 we show all zero-one vectors of the discrete set D where we

x_{11}	x_{12}	x_{13}	...	x_{1n}	x_{21}	x_{22}	x_{23}	...	x_{2n}	y^2	y^3	...	y^n
1	0	0	...	0	0	1	0	...	0	1	0	...	0
1	0	0	...	0	0	0	1	...	0	0	1	...	0
:	:	:	⋱	:	:	:	:	⋱	:	:	:	⋱	:
1	0	0	...	0	0	0	0	...	1	0	0	...	1
0	1	0	...	0	1	0	0	...	0	1	0	...	0
0	1	0	...	0	0	0	1	...	0	0	0	...	0
:	:	:	⋱	:	:	:	:	⋱	:	:	:	⋱	:
0	1	0	...	0	0	0	0	...	1	0	0	...	0
0	0	1	...	0	1	0	0	...	0	1	0	...	0
0	0	1	...	0	0	0	1	...	0	0	0	...	0
:	:	:	⋱	:	:	:	:	⋱	:	:	:	⋱	:
0	0	1	...	0	0	0	0	...	1	0	0	...	0
:	:	:	⋱	:	:	:	:	⋱	:	:	:	⋱	:
0	0	0	...	1	1	0	0	...	0	0	0	...	1
0	0	0	...	1	0	1	0	...	0	0	0	...	0
:	:	:	⋱	:	:	:	:	⋱	:	:	:	⋱	:
0	0	0	...	1	0	0	0	...	0	0	0	...	0

Table 5.1 The feasible 0-1 vectors of the local polytope P of SQP

have abbreviated $y_{11}^{2\ell}$ to y^ℓ for $2 \leq \ell \leq n$. Let P_L be the polytope given by $(\mathbf{x}, \mathbf{y}) \in \mathbb{R}^{3n-1}$ satisfying

$$\sum_{j=1}^{n} x_{ij} = 1 \quad \text{for } 1 \leq i \leq 2 \tag{5.34}$$

$$-x_{11} - x_{21} + \sum_{j=2}^{n} y_{11}^{2j} = 0 \tag{5.35}$$

$$-x_{1j} - x_{2j} + y_{11}^{2j} \leq 0 \quad \text{for } 2 \leq j \leq n \tag{5.36}$$

$$x_{11} + x_{1j} + x_{21} + x_{2j} - y_{11}^{2j} \leq 1 \quad \text{for } 2 \leq j \leq n \tag{5.37}$$

$$-x_{i1} - \sum_{j \in S}(x_{ij} - y_{11}^{2j}) \leq 0 \quad \begin{array}{l}\text{for } 1 \leq i \leq 2, \emptyset \neq S \subset N - \{1\}, \\ |S| \leq n-3\end{array} \tag{5.38}$$

$$x_{ij} \geq 0 \quad \text{for } 1 \leq i \leq 2, 1 \leq j \leq n \tag{5.39}$$

$$y_{11}^{2j} \geq 0 \quad \text{for } 2 \leq j \leq n. \tag{5.40}$$

Remark 5.8 *The system of equations and inequalities* (5.34),..., (5.40) *is valid for all* $(\mathbf{x}, \mathbf{y}) \in P$ *and thus* $P \subseteq P_L$.

Proof. Let $(\mathbf{x}, \mathbf{y}) \in D$. Then $(\mathbf{x}, \mathbf{y})$ satisfies (5.34), (5.39) and (5.40). Using (5.33), $-x_{11} - x_{21} + \sum_{j=2}^n y_{11}^{2j} = -x_{11} - x_{21} + \sum_{j=2}^n (x_{11} x_{2j} + x_{1j} x_{21}) = -x_{11}(1 - \sum_{j=2}^n x_{2j}) - x_{21}(1 - \sum_{j=2}^n x_{1j}) = -2x_{11}x_{21} = 0$; hence, (5.35) is satisfied as well. Using (5.33) and (5.39), $-x_{1j} - x_{2j} + y_{11}^{2j} = -x_{1j} - x_{2j} + x_{11}x_{2j} + x_{1j}x_{21} = -x_{1j}(1 - x_{21}) - x_{2j}(1 - x_{11}) \leq 0$ for $2 \leq j \leq n$; hence, (5.36) is satisfied as well. We write $x_{11} + x_{1j} + x_{21} + x_{2j} - y_{11}^{2j} = x_{11} + x_{1j} + x_{21} + x_{2j} - x_{11}x_{2j} - x_{1j}x_{21} = x_{11}(1-x_{2j}) + x_{1j}(1-x_{21}) + x_{21} + x_{2j}$ for $2 \leq j \leq n$. There are two possible cases: (i) if $x_{11} = 1$ or $x_{1j} = 1$, then $x_{11} + x_{1j} + x_{21} + x_{2j} - y_{11}^{2j} = 1$; (ii) if $x_{11} = x_{1j} = 0$ then $x_{11} + x_{1j} + x_{21} + x_{2j} - y_{11}^{2j} = x_{21} + x_{2j} \leq 1$ where the last inequality follows from (5.34) and (5.39). Finally, using (5.33) and (5.39), $-x_{i1} - \sum_{j \in S}(x_{ij} - y_{11}^{2j}) = -x_{i1} - \sum_{j \in S}(x_{ij} - x_{11}x_{2j} - x_{1j}x_{21}) = -x_{i1}(1 - \sum_{j \in S} x_{kj}) - \sum_{j \in S} x_{ij}(1 - x_{k1}) \leq 0$ for $1 \leq i \leq 2, \emptyset \neq S \subset N - \{1\}, |S| \leq n-3$ where $1 \leq i \neq k \leq 2$ and hence, (5.38) is satisfied as well. Thus, $D \subseteq P_L$ and hence, $P = \text{conv}(D) \subseteq P_L$. □

We order the components of $\mathbf{x}$ as $(x_{11}, \ldots, x_{1n}, x_{21}, \ldots, x_{2n})$ and those of $\mathbf{y}$ as $(y_{11}^{22}, \ldots, y_{11}^{2n})$. Let $\bar{\mathbf{u}}_{ij} \in \mathbb{R}^{2n}$ with its components ordered like those of $\mathbf{x}$ be a unit vector with one in its $(i,j)^{th}$ component and $\bar{\mathbf{v}}_{11}^{2\ell} \in \mathbb{R}^{n-1}$ ordered like $\mathbf{y}$ be another unit vector with one in its $\binom{2,\ell}{1,1}^{th}$ component. Let $\mathbf{u}_{ij} \in \mathbb{R}^{3n-1}$ be obtained from $\bar{\mathbf{u}}_{ij}$ by appending $n-1$ zeroes in the last $n-1$ components and $\mathbf{v}_{11}^{2\ell} \in \mathbb{R}^{3n-1}$ be obtained from $\bar{\mathbf{v}}_{11}^{2\ell}$ by appending $2n$ zeroes at the beginning.

Proposition 5.23 *The dimension of P is given by* $\dim(P) = 3n - 4$ *for* $n \geq 4$.

Proof. Since the three equations in (5.34) and (5.35) are linearly independent and $P \subseteq P_L$, $\dim(P) \leq 3n - 4$. We establish $\dim(P) \geq 3n - 4$ by showing that every equation $\boldsymbol{\alpha}\mathbf{x} + \boldsymbol{\beta}\mathbf{y} = \gamma$ that is satisfied by all $(\mathbf{x}, \mathbf{y}) \in P$ is a linear combination of (5.34) and (5.35).

(i) Since $(\mathbf{u}_{1p} + \mathbf{u}_{2r}) \in P$ for $2 \leq p, r \leq n$, $\alpha_{ip} = \alpha_{ir}$ for all $1 \leq i \leq 2, 2 \leq p, r \leq n$.

(ii) Since $(\mathbf{u}_{1p} + \mathbf{u}_{2r}), (\mathbf{u}_{11} + \mathbf{u}_{2r} + \mathbf{v}_{11}^{2r}) \in P$ for $2 \leq p, r \leq n$, $\alpha_{1p} = \alpha_{11} - \beta_{11}^{2r}$ for all $2 \leq p, r \leq n$. Moreover, by (i), $\alpha_{1r} = \alpha_{11} + \beta_{11}^{2r}$ for all $2 \leq r \leq n$.

(iii) Since $(\mathbf{u}_{11} + \mathbf{u}_{2r} + \mathbf{v}_{11}^{2r}), (\mathbf{u}_{1r} + \mathbf{u}_{21} + \mathbf{v}_{11}^{2r}) \in P$ for $2 \leq r \leq n$, $\alpha_{11} + \alpha_{2r} = \alpha_{1r} + \alpha_{21}$ for all $2 \leq r \leq n$. Moreover, by (i) and (ii), $\beta_{11}^{2p} = \beta_{11}^{2r}$ for all $2 \leq p, r \leq n$.

So $\boldsymbol{\alpha}\mathbf{x} + \boldsymbol{\beta}\mathbf{y} = \gamma$ becomes $\sum_{i=1}^2 \alpha_{ir} \sum_{p=1}^n x_{ip} + \beta_{11}^{2r}(-x_{11} - x_{21} + \sum_{p=2}^n y_{11}^{2p}) = \sum_{i=1}^2 \alpha_{ir}$ for $(\mathbf{x}, \mathbf{y}) \in P$ where $2 \leq r \leq n$, i.e. a linear combination of the equations (5.34) and (5.35). □

Locally Ideal LP Formulations II

Proposition 5.24 *Inequality (5.40) defines a facet of P for all $2 \leq j \leq n$.*

Proof. By Remark (5.8), (5.40) is valid for P. Let $F = \{(\mathbf{x}, \mathbf{y}) \in P : y_{11}^{2j} = 0\}$. Since $(\mathbf{u}_{11} + \mathbf{u}_{2j} + \mathbf{v}_{11}^{2j}) \in P$ but not in F, F is a proper face of P. Suppose there exists a valid inequality $\boldsymbol{\alpha}\mathbf{x} + \boldsymbol{\beta}\mathbf{y} \leq \gamma$ for P such that every $(\mathbf{x}, \mathbf{y}) \in F$ satisfies $\boldsymbol{\alpha}\mathbf{x} + \boldsymbol{\beta}\mathbf{y} = \gamma$.

(i) Since $(\mathbf{u}_{1p} + \mathbf{u}_{2r}) \in F$ for $1 \leq p, r \leq n$, $\alpha_{ip} = \alpha_{ir}$ for $1 \leq i \leq 2, 2 \leq p, r \leq n$.

(ii) Since $(\mathbf{u}_{1p} + \mathbf{u}_{2r}), (\mathbf{u}_{11} + \mathbf{u}_{2r} + \mathbf{v}_{11}^{2r}) \in F$ for $2 \leq r \neq j \leq n$, $\alpha_{11} + \beta_{11}^{2r} = \alpha_{1p}$ for all $2 \leq r \neq j \leq n$. Moreover, by (i), $\alpha_{11} = \alpha_{1r} - \beta_{11}^{2r}$ for all $2 \leq r \neq j \leq n$.

(iii) Since $(\mathbf{u}_{11} + \mathbf{u}_{2p} + \mathbf{v}_{11}^{2p}), (\mathbf{u}_{1p} + \mathbf{u}_{21} + \mathbf{v}_{11}^{2p}) \in F$ for $2 \leq p \neq j \leq n$, $\alpha_{11} + \alpha_{2p} = \alpha_{1p} + \alpha_{21}$ for all $2 \leq p \neq j \leq n$. Moreover, by (ii), $\alpha_{11} - \alpha_{1p} = \alpha_{21} - \alpha_{2p} = -\beta_{11}^{2p}$ for all $2 \leq p \neq j \leq n$.

So $\boldsymbol{\alpha}\mathbf{x} + \boldsymbol{\beta}\mathbf{y} = \gamma$ becomes $\sum_{i=1}^{2} \alpha_{ir} \sum_{p=1}^{n} x_{ip} + \beta_{11}^{2r}(-x_{11} - x_{21} + \sum_{p=2}^{n} y_{11}^{2p}) + (\beta_{11}^{2j} - \beta_{11}^{2r}) y_{11}^{2j} = \sum_{i=1}^{2} \alpha_{ir}$; equivalently, $(\beta_{11}^{2j} - \beta_{11}^{2r}) y_{11}^{2j} = 0$ for all $(\mathbf{x}, \mathbf{y}) \in F$ where $2 \leq r \leq n$. Since F is a proper face of P, the proposition follows. □

Proposition 5.25 *Inequality (5.39) defines a facet of P for all $1 \leq i \leq 2, 2 \leq j \leq n$.*

Proof. By Remark (5.8), (5.39) is valid for P. WROG assume $i = 1$ and let $F = \{(\mathbf{x}, \mathbf{y}) \in P : x_{1j} = 0\}$. We consider the two cases: (i) $j = 1$ and (ii) $j \neq 1$. First, consider case (i). Since $(\mathbf{u}_{11} + \mathbf{u}_{2j} + \mathbf{v}_{11}^{2j}) \in P$ but not in F, F is a proper face of P. Suppose there exists a valid inequality $\boldsymbol{\alpha}\mathbf{x} + \boldsymbol{\beta}\mathbf{y} \leq \gamma$ for P such that every $(\mathbf{x}, \mathbf{y}) \in F$ satisfies $\boldsymbol{\alpha}\mathbf{x} + \boldsymbol{\beta}\mathbf{y} = \gamma$.

(i) Since $(\mathbf{u}_{1p} + \mathbf{u}_{2r}) \in F$ for $1 \leq p, r \leq n$, $\alpha_{ip} = \alpha_{ir}$ for $1 \leq i \leq 2, 2 \leq p, r \leq n$.

(ii) Since $(\mathbf{u}_{1p} + \mathbf{u}_{2r}), (\mathbf{u}_{1p} + \mathbf{u}_{21} + \mathbf{v}_{11}^{2p}) \in F$ for $2 \leq p, r \leq n$, $\alpha_{2r} = \alpha_{21} + \beta_{11}^{2p}$ for all $2 \leq p, r \leq n$. Moreover, by (i), $\alpha_{2p} = \alpha_{21} - \beta_{11}^{2p}$ for all $2 \leq p \leq n$.

(iii) Since $(\mathbf{u}_{1p} + \mathbf{u}_{21} + \mathbf{v}_{11}^{2p}), (\mathbf{u}_{1r} + \mathbf{u}_{21} + \mathbf{v}_{11}^{2r}) \in F$ for $2 \leq p, r \leq n$, $\alpha_{1p} + \beta_{11}^{2p} = \alpha_{1r} + \beta_{11}^{2r}$ for all $2 \leq p, r \leq n$. Moreover, by (i), $\beta_{11}^{2p} = \beta_{11}^{2r}$ for all $2 \leq p, r \leq n$.

Consequently, $\boldsymbol{\alpha}\mathbf{x} + \boldsymbol{\beta}\mathbf{y} = \gamma$ becomes $(\alpha_{11} - \alpha_{1r} + \beta_{11}^{2r}) x_{11} + \sum_{i=1}^{2} \alpha_{ir} \sum_{p=1}^{n} x_{ip} + \beta_{11}^{2r}(-x_{11} - x_{21} + \sum_{p=2}^{n} y_{11}^{2p}) = \sum_{i=1}^{2} \alpha_{ir}$; equivalently, $(\alpha_{11} - \alpha_{1r} + \beta_{11}^{2r}) x_{11} = 0$ for all $(\mathbf{x}, \mathbf{y}) \in F$ where $2 \leq r \leq n$. Consider case (ii). Since $(\mathbf{u}_{1j} + \mathbf{u}_{21} + \mathbf{v}_{11}^{2j}) \in P$ but not in F, F is a proper face of P. Suppose there exists a valid inequality $\boldsymbol{\alpha}\mathbf{x} + \boldsymbol{\beta}\mathbf{y} \leq \gamma$ for P such that every $(\mathbf{x}, \mathbf{y}) \in F$ satisfies $\boldsymbol{\alpha}\mathbf{x} + \boldsymbol{\beta}\mathbf{y} = \gamma$.

(i) Since $(\mathbf{u}_{1p} + \mathbf{u}_{2r}) \in F$ for $2 \leq p, r \leq n, p \neq j$, $\alpha_{1p} = \alpha_{1r}$ for $2 \leq p, r \leq n, p \neq j \neq r$ and $\alpha_{2p} = \alpha_{2r}$ for $2 \leq p, r \leq n$.

(ii) Since $(\mathbf{u}_{1p} + \mathbf{u}_{2r}), (\mathbf{u}_{11} + \mathbf{u}_{2r} + \mathbf{v}_{11}^{2r}) \in F$ for $2 \leq p \neq j \leq n, 2 \leq r \leq n$, $\alpha_{1p} = \alpha_{11} + \beta_{11}^{2r}$ for all $2 \leq p \neq j \leq n, 2 \leq r \leq n$. Moreover, by (i), $\alpha_{1r} = \alpha_{11} + \beta_{11}^{2r}$ for all $2 \leq r \neq j \leq n$.

(iii) Since $(\mathbf{u}_{1p} + \mathbf{u}_{2r}), (\mathbf{u}_{1p} + \mathbf{u}_{21} + \mathbf{v}_{11}^{2p}) \in F$ for $2 \leq p \neq j \leq n, 2 \leq r \leq n$, $\alpha_{2r} = \alpha_{21} + \beta_{11}^{2p}$ for all $2 \leq p \neq j \leq n, 2 \leq r \leq n$. Moreover, using (i) and (ii), $\alpha_{2p} = \alpha_{21} + \beta_{11}^{2p}$ and $\beta_{11}^{2p} = \beta_{11}^{2r}$ for all $2 \leq p, r \leq n, p \neq j \neq r$.

Thus $\boldsymbol{\alpha}\mathbf{x} + \boldsymbol{\beta}\mathbf{y} = \gamma$ becomes $\sum_{i=1}^{2} \alpha_{ir} \sum_{p=1}^{n} x_{ip} + (\alpha_{1j} - \alpha_{1r})x_{1j} + \beta_{11}^{2r}(-x_{11} - x_{21} + \sum_{p=2}^{n} y_{11}^{2p}) = \sum_{i=1}^{2} \alpha_{ir}$; equivalently, $(\alpha_{1j} - \alpha_{1r})x_{1j} = 0$ for all $(\mathbf{x}, \mathbf{y}) \in F$ where $2 \leq r \leq n$. □

Proposition 5.26 *Inequality (5.36) defines a facet of P for all $2 \leq j \leq n$.*

Proof. By Remark (5.8), (5.36) is valid for P. Let $F = \{(\mathbf{x}, \mathbf{y}) \in P : -x_{1j} - x_{2j} + y_{11}^{2j} = 0\}$. Since $(\mathbf{u}_{1j} + \mathbf{u}_{2p}) \in P$, for some $2 \leq p, j \leq n$, but not in F, F is a proper face of P. Suppose there exists a valid inequality $\boldsymbol{\alpha}\mathbf{x} + \boldsymbol{\beta}\mathbf{y} \leq \gamma$ for P such that every $(\mathbf{x}, \mathbf{y}) \in F$ satisfies $\boldsymbol{\alpha}\mathbf{x} + \boldsymbol{\beta}\mathbf{y} = \gamma$.

(i) Since $(\mathbf{u}_{1p} + \mathbf{u}_{2r}) \in F$ for $2 \leq p, r \leq n, p \neq j \neq r$, $\alpha_{ip} = \alpha_{ir}$ for $1 \leq i \leq 2, 2 \leq p, r \leq n, p \neq j \neq r$.

(ii) Since $(\mathbf{u}_{1p} + \mathbf{u}_{2r}), (\mathbf{u}_{11} + \mathbf{u}_{2r} + \mathbf{v}_{11}^{2r}) \in F$ for $2 \leq p, r \leq n, p \neq j \neq r$, $\alpha_{1p} = \alpha_{11} + \beta_{11}^{2r}$ for all $2 \leq p, r \leq n, p \neq j \neq r$. Moreover, by (i), $\alpha_{1r} = \alpha_{11} + \beta_{11}^{2r}$ for all $2 \leq r \neq j \leq n$.

(iii) Since $(\mathbf{u}_{1p} + \mathbf{u}_{2r}), (\mathbf{u}_{1p} + \mathbf{u}_{21} + \mathbf{v}_{11}^{2p}) \in F$ for $2 \leq p \leq n, p \neq j \neq r$, $\alpha_{2r} = \alpha_{21} + \beta_{11}^{2p}$ for all $2 \leq p \leq n, p \neq j \neq r$. Moreover, by (i), $\alpha_{2p} = \alpha_{21} + \beta_{11}^{2p}$ and $\beta_{11}^{2p} = \beta_{11}^{2r}$ for all $2 \leq p, r \leq n, p \neq j \neq r$.

(iv) Since $(\mathbf{u}_{11} + \mathbf{u}_{2p} + \mathbf{v}_{11}^{2p}) \in F$ for $2 \leq p \leq n$, $\alpha_{2p} + \beta_{11}^{2p} = \alpha_{2j} + \beta_{11}^{2j}$, i.e., $\alpha_{2p} - \alpha_{2j} = \beta_{11}^{2j} - \beta_{11}^{2p}$ for all $2 \leq p \leq n$.

(v) Since $(\mathbf{u}_{11} + \mathbf{u}_{2p} + \mathbf{v}_{11}^{2p}), (\mathbf{u}_{1p} + \mathbf{u}_{21} + \mathbf{v}_{11}^{2p}) \in F$ for $2 \leq p \leq n$, $\alpha_{11} + \alpha_{2p} = \alpha_{1p} + \alpha_{21}$, i.e., $\alpha_{11} - \alpha_{1p} = \alpha_{21} - \alpha_{2p}$ for all $2 \leq p \leq n$; in particular, $\alpha_{11} - \alpha_{1j} = \alpha_{21} - \alpha_{2j}$. Moreover, from (ii) and (ii), $\alpha_{1j} = \alpha_{1p} - \alpha_{2p} + \alpha_{2j}$ for $2 \leq p \leq n$.

Consequently, $\boldsymbol{\alpha}\mathbf{x} + \boldsymbol{\beta}\mathbf{y} = \gamma$ becomes $\sum_{i=1}^{2} \alpha_{ir} \sum_{p=1}^{n} x_{ip} + \beta_{11}^{2r}(-x_{11} - x_{21} + \sum_{p=2}^{n} y_{11}^{2p}) + (\alpha_{2r} - \alpha_{2j})(-x_{1j} - x_{2j} + y_{11}^{2j}) = \sum_{i=1}^{2} \alpha_{ir}$; equivalently, $(\alpha_{2r} - \alpha_{2j})(-x_{1j} - x_{2j} + y_{11}^{2j}) = 0$ for all $(\mathbf{x}, \mathbf{y}) \in F$ where $2 \leq r \leq n$. Hence, the proposition follows. □

Proposition 5.27 *Inequality (5.37) defines a facet of P for all $2 \leq j \leq n$.*

Proof. By Remark (5.8), (5.37) is valid for P. Let $F = \{(\mathbf{x}, \mathbf{y}) \in P : x_{11} + x_{1j} + x_{21} + x_{2j} - y_{11}^{2j} = 0\}$. Since $(\mathbf{u}_{1p} + \mathbf{u}_{2r}) \in P$ for some $2 \leq p < r \leq n, p \neq j \neq r$, but not in F, F is a proper face of P. Suppose there exists a valid inequality $\boldsymbol{\alpha}\mathbf{x} + \boldsymbol{\beta}\mathbf{y} \leq \gamma$ for P such that every $(\mathbf{x}, \mathbf{y}) \in F$ satisfies $\boldsymbol{\alpha}\mathbf{x} + \boldsymbol{\beta}\mathbf{y} = \gamma$.

Locally Ideal LP Formulations II 127

(i) Since $(\mathbf{u}_{1j} + \mathbf{u}_{2p}) \in F$ for $2 \leq p \neq j \leq n$, $\alpha_{2p} = \alpha_{2r}$ for $2 \leq p, r \leq n, p \neq j \neq r$.

(ii) Since $(\mathbf{u}_{1p} + \mathbf{u}_{2j}) \in F$ for $2 \leq p \neq j \leq n$, $\alpha_{1p} = \alpha_{1r}$ for $2 \leq p, r \leq n, p \neq j \neq r$.

(iii) Since $(\mathbf{u}_{1p} + \mathbf{u}_{21} + \mathbf{v}_{11}^{2p}) \in F$ for $2 \leq p \leq n$, $\alpha_{1p} - \alpha_{1r} = \beta_{11}^{2r} - \beta_{11}^{2p}$ for all $2 \leq p, r \leq n$. Moreover, by (ii), $\beta_{11}^{2p} = \beta_{11}^{2r}$ for all $2 \leq p, r \leq n, p \neq j \neq r$.

(iv) Since $(\mathbf{u}_{11} + \mathbf{u}_{2p} + \mathbf{v}_{11}^{2p}), (\mathbf{u}_{1p} + \mathbf{u}_{21} + \mathbf{v}_{11}^{2p}) \in F$ for $2 \leq p \leq n$, $\alpha_{11} - \alpha_{1p} = \alpha_{21} - \alpha_{2p}$ for all $2 \leq p \leq n$.

(v) Since $(\mathbf{u}_{1p} + \mathbf{u}_{2j}), (\mathbf{u}_{11} + \mathbf{u}_{2j} + \mathbf{v}_{11}^{2j}) \in F$ for $2 \leq p \leq n, p \neq j$, $\alpha_{11} - \alpha_{1p} = -\beta_{11}^{2j}$ for all $2 \leq p \leq n, p \neq j$.

Consequently, $\boldsymbol{\alpha}\mathbf{x} + \boldsymbol{\beta}\mathbf{y} = \gamma$ becomes $\sum_{i=1}^{2} \alpha_{ir} \sum_{p=1}^{n} x_{ip} + \beta_{11}^{2r}(-x_{11} - x_{21} + \sum_{p=2}^{n} y_{11}^{2p}) + (\alpha_{11} - \alpha_{1r} + \beta_{11}^{2r})(x_{11} + x_{1j} + x_{21} + x_{2j} - y_{11}^{2j}) = \alpha_{11} + \alpha_{2r} + \beta_{11}^{2r}$; equivalently, $(\alpha_{11} - \alpha_{1r} + \beta_{11}^{2r})(x_{11} + x_{1j} + x_{21} + x_{2j} - y_{11}^{2j}) = (\alpha_{11} - \alpha_{1r} + \beta_{11}^{2r})$ for all $(\mathbf{x}, \mathbf{y}) \in F$ where $2 \leq r \leq n$. $\square$

Proposition 5.28 *Inequality (5.38) defines a facet of P for all $1 \leq i \leq 2, \emptyset \neq S \subset N - \{1\}, |S| \leq n - 3$.*

Proof. By Remark (5.8), (5.38) is valid for P. WROG assume $i = 1$ and let $F = \{(\mathbf{x}, \mathbf{y}) \in P : -x_{11} - \sum_{j \in S}(x_{1j} - y_{11}^{2j}) = 0\}$. Since $(\mathbf{u}_{11} + \mathbf{u}_{2r} + \mathbf{v}_{11}^{2r}) \in P$ for some $r \notin S$, but not in F, F is a proper face of P. Suppose there exists a valid inequality $\boldsymbol{\alpha}\mathbf{x} + \boldsymbol{\beta}\mathbf{y} \leq \gamma$ for P such that every $(\mathbf{x}, \mathbf{y}) \in F$ satisfies $\boldsymbol{\alpha}\mathbf{x} + \boldsymbol{\beta}\mathbf{y} = \gamma$.

(i) Since $(\mathbf{u}_{1r} + \mathbf{u}_{2p}) \in F$ for $p \in S, r \notin S$, $\alpha_{1p} = \alpha_{1r}$ for $p, r \notin S$ and $\alpha_{2p} = \alpha_{2r}$ for all $p, r \in S$.

(ii) Since $(\mathbf{u}_{1r} + \mathbf{u}_{2p}), (\mathbf{u}_{11} + \mathbf{u}_{2p} + \mathbf{v}_{11}^{2p}) \in F$ for $p \in S, r \notin S$, $\alpha_{1r} = \alpha_{11} + \beta_{11}^{2p}$, i.e. $\alpha_{1r} - \alpha_{11} = \beta_{11}^{2p}$ for all $p \in S, r \notin S$. By (i), $\beta_{11}^{2p} = \beta_{11}^{2r}$ for all $p, r \in S$.

(iii) Since $(\mathbf{u}_{1p} + \mathbf{u}_{21} + \mathbf{v}_{11}^{2p}), (\mathbf{u}_{11} + \mathbf{u}_{2p} + \mathbf{v}_{11}^{2p}) \in F$ for $p \in S$, $\alpha_{1p} + \alpha_{21} = \alpha_{11} + \alpha_{2p}$. By (i), $\alpha_{1p} = \alpha_{1r}$ for all $p, r \in S$.

(iv) Since $(\mathbf{u}_{1r} + \mathbf{u}_{21} + \mathbf{v}_{11}^{2r}), (\mathbf{u}_{11} + \mathbf{u}_{2p} + \mathbf{v}_{11}^{2p}) \in F$ for $p \in S, r \notin S$, $\alpha_{1r} + \alpha_{21} + \beta_{11}^{2r} = \alpha_{11} + \alpha_{2p} + \beta_{11}^{2p}$ for all $p \in S, r \notin S$. By (iii), $\beta_{11}^{2r} = \alpha_{2p} - \alpha_{21}$ and $\beta_{11}^{2p} = \beta_{11}^{2r}$ for all $p, r \notin S$.

(v) Since $(\mathbf{u}_{1p} + \mathbf{u}_{2r}), (\mathbf{u}_{1p} + \mathbf{u}_{21} + \mathbf{v}_{11}^{2p}) \in F$ for $p, r \notin S$, $\alpha_{2r} = \alpha_{21} + \beta_{11}^{2p}$ for all $p, r \notin S$. By (iv), $\beta_{11}^{2r} = \alpha_{2r} - \alpha_{21}$ for all $r \notin S$. Moreover, by (i), $\alpha_{2p} = \alpha_{2r}$ for all $2 \leq p, r \leq n$.

(vi) Since $(\mathbf{u}_{11} + \mathbf{u}_{2p} + \mathbf{v}_{11}^{2p}), (\mathbf{u}_{1r} + \mathbf{u}_{21} + \mathbf{v}_{11}^{2r}) \in F$ for $p \in S, r \notin S$, $\alpha_{11} + \alpha_{2p} + \beta_{11}^{2p} = \alpha_{1r} + \alpha_{21} + \beta_{11}^{2r}$ for all $p \in S, r \notin S$. By (v), $\beta_{11}^{2p} = \alpha_{1r} - \alpha_{11}$ for all $p \in S, r \notin S$. Moreover, by (ii), $\alpha_{1p} - \alpha_{11} = \alpha_{2r} - \alpha_{21}$ for all $p \in S, 2 \leq r \leq n$.

Consequently, $\alpha \mathbf{x} + \beta \mathbf{y} = \gamma$ becomes $(\alpha_{11} - \alpha_{1p})x_{11} + (\alpha_{1p} - \alpha_{1r})\sum_{\ell \in S} x_{1\ell} + \alpha_{1r}\sum_{\ell \notin S} x_{1\ell} + (\alpha_{21} - \alpha_{2r})x_{21} + \alpha_{2r}\sum_{\ell=2}^{n} x_{2\ell} + (\alpha_{1r} - \alpha_{1p})\sum_{\ell \in S} y_{11}^{2\ell} + (\alpha_{1p} - \alpha_{11})\sum_{\ell=2}^{n} y_{11}^{2\ell} = \sum_{i=1}^{2} \alpha_{ir} + (\alpha_{1p} - \alpha_{11})(-x_{11} - x_{21} + \sum_{\ell=2}^{n} y_{11}^{2\ell}) + (\alpha_{1r} - \alpha_{1p})(-x_{11} - \sum_{\ell \in S}(x_{1\ell} - y_{11}^{2\ell})) = \sum_{i=1}^{2} \alpha_{ir}$; equivalently, $(\alpha_{1r} - \alpha_{1p})(-x_{11} - \sum_{\ell \in S}(x_{1\ell} - y_{11}^{2\ell})) = 0$ for all $(\mathbf{x}, \mathbf{y}) \in F$ where $p \in S, r \notin S$. □

We have the following conjecture for $P = conv(D)$ which is true for $3 \leq n \leq 9$.

Conjecture 5.1 *The system of equations and inequalities* (5.34), ..., (5.40) *is an ideal linear description of the local polytope P for $n \geq 5$.*

The linear system of the conjecture, though complete, is not minimal for $n = 3$ and $n = 4$. For $n = 4$, by dropping all of the three inequalities (5.36) for $2 \leq j \leq n$ and for $n = 3$, by dropping any one of the two inequalities (5.36), let us say, the inequality (5.36) for $j = 3$, we obtain an ideal description of P from the above system of equations and inequalities. Moreover, for $n = 3$, since the inequality (5.37) for $j = 3$ given by $x_{11} + x_{13} + x_{21} + x_{23} - y_{11}^{23} \leq 1$ is equivalent to $x_{11} + x_{12} + x_{21} + x_{22} - y_{11}^{22} \geq 1$, these inequalities (5.37) for $j = 2$ and 3 can be replaced by an equation $x_{11} + x_{12} + x_{21} + x_{22} - y_{11}^{22} = 1$.

Using the conjecture, we consider the following equations and inequalities to linearize $y_{rj}^{s\ell} = x_{rj}x_{s\ell} + x_{r\ell}x_{sj}$ for all $1 \leq j < \ell \leq n$ and a pair of indices r and s with $1 \leq r < s \leq n$:

$$\sum_{j=1}^{n} x_{ij} = 1 \quad \text{for } i \in \{r, s\} \quad (5.41)$$

$$-x_{rj} - x_{sj} + \sum_{\ell=1}^{j-1} y_{r\ell}^{sj} + \sum_{\ell=j+1}^{n} y_{rj}^{s\ell} = 0 \quad \text{for } 1 \leq j \leq n \quad (5.42)$$

$$-x_{rj} - x_{sj} + y_{rj}^{s\ell} \leq 0 \quad \text{for } 1 \leq j < \ell \leq n \quad (5.43)$$

$$x_{rj} + x_{r\ell} + x_{sj} + x_{s\ell} - y_{rj}^{s\ell} \leq 1 \quad \text{for } 1 \leq j < \ell \leq n \quad (5.44)$$

$$-x_{ij} - \sum_{\ell \in S} x_{i\ell} + \sum_{j > \ell \in S} y_{r\ell}^{sj} + \sum_{j < \ell \in S} y_{rj}^{s\ell} \leq 0 \quad \text{for } \begin{array}{l} \emptyset \neq S \subset N - \{j\}, j \in N, \\ |S| \leq n - 3, i \in \{r, s\} \end{array} \quad (5.45)$$

$$x_{ij} \geq 0 \quad \text{for } i \in \{r, s\}, 1 \leq j \leq n \quad (5.46)$$

$$y_{rj}^{s\ell} \geq 0 \quad \text{for } 1 \leq j < \ell \leq n. \quad (5.47)$$

Using symmetry and similar arguments as done previously, we consider the following system of equations and inequalities to linearize the variables y_{ir}^{ks} for all $1 \leq i < k \leq n$ and a pair of indices r and s with $1 \leq r < s \leq n$:

$$\sum_{i=1}^{n} x_{ij} = 1 \qquad \text{for } j \in \{r, s\} \tag{5.48}$$

$$-x_{ir} - x_{is} + \sum_{k=1}^{i-1} y_{kr}^{is} + \sum_{k=i+1}^{n} y_{ir}^{ks} = 0 \qquad \text{for } 1 \le i \le n \tag{5.49}$$

$$-x_{ir} - x_{is} + y_{ir}^{ks} \le 0 \qquad \text{for } 1 \le i < k \le n \tag{5.50}$$

$$x_{ir} + x_{kr} + x_{is} + x_{ks} - y_{ir}^{ks} \le 1 \qquad \text{for } 1 \le i < k \le n \tag{5.51}$$

$$-x_{ij} - \sum_{k \in S} x_{kj} + \sum_{i > k \in S} y_{kr}^{is} + \sum_{i < k \in S} y_{ir}^{ks} \le 0 \qquad \text{for } \begin{matrix} \emptyset \ne S \subset N - \{i\}, \\ |S| \le n-3, i \in N, j \in \{r,s\} \end{matrix} \tag{5.52}$$

$$x_{ij} \ge 0 \qquad \text{for } j \in \{r, s\}, 1 \le i \le n, \tag{5.53}$$

$$y_{ir}^{ks} \ge 0 \qquad \text{for } 1 \le i < k \le n. \tag{5.54}$$

Remark 5.9 *The inequalities (5.43), (5.44), (5.50) and (5.51) are a linear combination of (5.41), (5.42), (5.48) and (5.49) and a nonnegative linear combination of (5.46), (5.47), (5.53) and (5.54) and thus redundant.*

Proof. (i) For some $1 \le g \le n$, summing (5.47) for $1 \le j < \ell \le n$ and $j \ne g, \ell \ne h$ where $g < h \le n$, we obtain $-\sum_{j=1}^{g-1} y_{rj}^{sg} - \sum_{j=g+1}^{h-1} y_{rg}^{sj} - \sum_{j=h+1}^{n} y_{rg}^{sj} \le 0$. Adding this inequality to (5.42) for g, we obtain that $-x_{rg} - x_{sg} + y_{rg}^{sh} \le 0$. Hence, (5.43) for all $1 \le j < \ell \le n$ are redundant. By a similar argument, it follows that (5.50) for all $1 \le i < k \le n$ are redundant.

(ii) For some fixed $1 \le g < h \le n$, the linear combination of (5.41) and (5.42) for $1 \le j \le n$ given by $\sum_{j=1}^{n} x_{rj} + \sum_{j=1}^{n} x_{sj} - (-x_{rg} - x_{sg} + \sum_{\ell=1}^{g-1} y_{r\ell}^{sg} + \sum_{\ell=g+1}^{n} y_{rg}^{s\ell}) - (-x_{rh} - x_{sh} + \sum_{\ell=1}^{h-1} y_{r\ell}^{sh} + \sum_{\ell=h+1}^{n} y_{rh}^{s\ell}) + \sum_{\{g,h\} \ne j=1}^{n} (-x_{rj} - x_{sj} + \sum_{\ell=1}^{j-1} y_{r\ell}^{sj} + \sum_{\ell=j+1}^{n} y_{rj}^{s\ell}) = 2(x_{rg} + x_{rh} + x_{sg} + x_{sh} - y_{rg}^{sh} + \sum_{\{g,h\} \ne j=1}^{n} \sum_{\{g,h\} \ne \ell=j+1}^{n} y_{rj}^{s\ell}) = 2$. Dividing by two, we get $x_{rg} + x_{rh} + x_{sg} + x_{sh} - y_{rg}^{sh} + \sum_{\{g,h\} \ne j=1}^{n-1} \sum_{\{g,h\} \ne \ell=j+1}^{n} y_{rj}^{s\ell} = 1$. Adding an appropriate nonnegative linear combination of (5.47) as done in (i), we obtain $x_{rg} + x_{rh} + x_{sg} + x_{sh} - y_{rg}^{sh} \le 1$. Hence, (5.44) for all $1 \le j < \ell \le n$ are redundant. By a similar argument, it follows that (5.51) for all $1 \le i < k \le n$ are redundant. □

Considering all equations and inequalities resulting from our conjecture on the locally ideal linearization of the variables giving rise to quadratic terms in the objective function of the SQP, except the inequalities shown to be redundant in Remark (5.9) and inequalities (5.45) and (5.52), we formulate the SQP as the LP problem given by:

$$\min \left\{ \sum_{i,j \in N} c_{ij} x_{ij} + \sum_{i<k \in N} \sum_{j,\ell \in N} q_{ij}^{k\ell} y_{ij}^{k\ell} : (\mathbf{x},\mathbf{y}) \in SQP_n \right\}, \quad (\mathcal{O}SQP_n)$$

where SQP_n is the polytope defined by the convex hull of solutions $(\mathbf{x},\mathbf{y}) \in \mathbb{R}^{n^2+n^2(n-1)^2/4}$ to the following equations and inequalities in zero-one variables:

$$\sum_{j=1}^{n} x_{ij} = 1 \quad \text{for } i \in N \quad (5.55)$$

$$\sum_{i=1}^{n} x_{ij} = 1 \quad \text{for } j \in N \quad (5.56)$$

$$-x_{ij} - x_{i\ell} + \sum_{k=1}^{i-1} y_{kj}^{i\ell} + \sum_{k=i+1}^{n} y_{ij}^{k\ell} = 0 \quad \text{for } i \in N, j < \ell \in N \quad (5.57)$$

$$-x_{ij} - x_{kj} + \sum_{\ell=1}^{j-1} y_{i\ell}^{kj} + \sum_{\ell=j+1}^{n} y_{ij}^{k\ell} = 0 \quad \text{for } i < k \in N, j \in N \quad (5.58)$$

$$x_{ij} \geq 0 \quad \text{for } i,j \in N \quad (5.59)$$

$$y_{ij}^{k\ell} \geq 0 \quad \text{for } i < k \in N, j < \ell \in N \quad (5.60)$$

$$x_{ij} \in \{0,1\} \quad \text{for } i,j \in N. \quad (5.61)$$

We show now that a formulation of the SQP has been obtained. Inequalities (5.45) and (5.52) are not needed for a formulation. The study as to their possible facet-defining properties is left for future work.

Proposition 5.29 $\mathcal{O}SQP_n$ *is a formulation of the Symmetric Quadratic Assignment Problem with* $2n + n^2(n-1)$ *equations where* $n \geq 3$.

Proof. By a similar argument as in Remark (5.8), $(\mathbf{x},\mathbf{y}) \in DSQP_n$ satisfies (5.55), ..., (5.61); hence, $DSQP_n \subseteq SQP_n$. Let $(\mathbf{x},\mathbf{y}) \in SQP_n$. We show that $y_{ij}^{k\ell} = x_{ij} x_{k\ell} + x_{i\ell} x_{kj}$ for all $1 \leq i < k \leq n, 1 \leq j < \ell \leq n$. Suppose that there exist $1 \leq p < r \leq n$ and $1 \leq d < s \leq n$ such that $y_{pd}^{rs} \neq x_{pd} x_{rs} + x_{ps} x_{rd}$. If $x_{pd} = x_{ps} = 0$ then from (5.57) we conclude using (5.61) that $y_{pd}^{rs} = 0$; likewise, we conclude $y_{pd}^{rs} = 0$ when $x_{pd} = x_{rd} = 0$. Next, assume $x_{pd} = x_{rs} = 1$. Since, $x_{pd} = 1$ implies $x_{pg} = x_{hd} = 0$ for $1 \leq d \neq g \leq n, 1 \leq p \neq h \leq n$ and $x_{rs} = 1$ implies $x_{rg} = x_{hs} = 0$ for $1 \leq s \neq g \leq n, 1 \leq r \neq h \leq n$. But, then by a similar argument as above, we have $y_{pg}^{rd} = y_{pd}^{rh}$ for $1 \leq g < d < h \leq n, h \neq s$, $y_{pd}^{rs} = y_{ps}^{rh} = 0$ for

$1 \leq g < s < h \leq n, g \neq d$, $y_{gd}^{ps} = y_{pd}^{hs} = 0$ for $1 \leq g < p < h \leq n, h \neq r$ and $y_{gd}^{rs} = y_{rd}^{hs} = 0$ for $1 \leq g < p < h \leq s, g \neq p$. Hence, by (5.57) for $i = p, j = d$ and $\ell = s$ and by (5.58) $i = p, k = r$ and $j = d$, $y_{pd}^{rs} = 0$. So necessarily (x_{pd} and $x_{rs} = 0$) or (x_{ps} and $x_{rd} = 0$); WROG assume $x_{pd} = 1$ and $x_{rs} = 0$. But, then there exists $1 \leq g \neq s \leq n$ such that $x_{rd} = 1$, which implies, following a similar argument as above, that $y_{pd}^{rg} = 1$ if $d < g$ and $y_{pg}^{rd} = 1$ otherwise. Using (5.58), we obtain $y_{pd}^{rh} = 0$ for $1 \leq h \neq g \leq n$ and in particular, $y_{pd}^{rs} = 0$, a contradiction to the assumption that $y_{pd}^{rs} \neq x_{pd}x_{rs} + x_{ps}x_{rd}$. Thus $y_{ij}^{k\ell} = x_{ij}x_{k\ell} + x_{i\ell}x_{kj}$ and every zero-one point of SQP_n is in $DSQP_n$. The rest follows by counting. □

In Chapter 7.3 we address the issue of the *minimality* of our formulation.

6

QUADRATIC SCHEDULING PROBLEMS

As noted in Chapter 4.2, the operations scheduling problem (OSP) with machine independent quadratic interaction costs is identical with the graph partitioning problem (GPP). We compare in this chapter these alternative formulations for the OSP in this special case. By comparing the two formulations we do *not* mean an *empirical* comparison, but rather an *analytical* comparison such as the one carried out by Padberg and Sung [1991] for four different formulations of the traveling salesman problem. This guarantees that our results have validity for any numerical calculations based on the formulations that we propose in Chapters 4.2 and 4.3. In the second half of this chapter we derive some results on the facial structure of the OSP.

6.1 Alternative Formulations of the OSP

Though the OSP permits more general cost functions, in the special case where the quadratic interaction costs are machine independent, we have the option of working with either the OSP formulation or the GPP formulation. The OSP formulation is in a larger space of variables while the GPP formulation is in a smaller space of variables. We are interested in comparing the quality of the two linear programming relaxations analytically. Given two different formulations A and B of the same problem in the same space of variables and associated polyhedra $\mathcal{X}_A$ and $\mathcal{X}_B$ respectively, formulation A is superior to formulation B if $\mathcal{X}_A \subset \mathcal{X}_B$. However, since the alternative formulations of the OSP with machine independent interaction cost that we have presented are stated in terms of different sets of variables, we have to map the linear description of the polyhedron in the higher dimensional space of the OSP onto the lower dimensional space of the GPP in order to analytically compare the two formulations. Let A and C be alternative formulations of a problem where the formulation C models the problem in a higher dimensional space while the formulation A models the problem in a lower dimensional space. Likewise, let $\mathcal{X}_A$ and $\mathcal{Z}_C$ be the respective polyhedra associated with the formulations A and C. Let $\mathbf{T}$ be an affine transformation that maps the polyhedron $\mathcal{Z}_C$ onto the space of variables where the polyhedron $\mathcal{X}_A$ resides. If $\mathbf{T}(\mathcal{Z}_C) \supset \mathcal{X}_A$ then formulation A is evidently better than formulation C since no additional polyhedral information is provided for by the formulation C. On the other hand, formulation C is better than formulation A if $\mathbf{T}(\mathcal{Z}_C) \subset \mathcal{X}_A$.

It is well-known that every affine transformation from $\mathbb{R}^n$ to $\mathbb{R}^m$ with $m \leq n$ maps a polyhedron $\mathcal{Z} \subseteq \mathbb{R}^n$ onto another polyhedron $\mathcal{X} \subseteq \mathbb{R}^m$. Let $\mathbf{x} = \mathbf{f} + \mathbf{L}\mathbf{z}$ be an affine transformation of full rank from $\mathbb{R}^n$ into $\mathbb{R}^m$, i.e., $\mathbf{f} \in \mathbb{R}^m$ and $\mathbf{L}$ is an $m \times n$ matrix having full row rank. Since $rank(\mathbf{L}) = m$, we can partition the matrix $\mathbf{L}$ into two parts $\mathbf{L}_1$ and $\mathbf{L}_2$ such that $\mathbf{L}_1$ is an $m \times m$ nonsingular matrix corresponding to the first m columns of $\mathbf{L}$. Given a linear description of some polyhedron $\mathcal{Z} \subseteq \mathbb{R}^n$ we are interested in finding a linear description of its image under the affine transformation and so we next state a theorem from Padberg and Sung [1991], see also Chapter 7.3 of Padberg [1995], which lets us do that.

Theorem 6.1 *Let* $\mathcal{Z} = \{\mathbf{z} \in \mathbb{R}^n : \mathbf{A}\mathbf{z} = \mathbf{b}, \mathbf{D}\mathbf{z} \leq \mathbf{d}, \mathbf{z} \geq \mathbf{0}\}$, *where* $\mathbf{A}$ *is a* $p \times n$ *matrix and* $\mathbf{D}$ *is a* $q \times n$ *matrix. Set* $\mathcal{X} = \{\mathbf{x} \in \mathbb{R}^m : \exists\, \mathbf{z} \in \mathcal{Z}\ \text{such that}\ \mathbf{x} = \mathbf{f} + \mathbf{L}\mathbf{z}\}$ *and* $t = p + q + m$. *Then* $\mathcal{X} = \mathcal{X}_\mathcal{C}$, *where* $\mathcal{X}_\mathcal{C}$ *and* $\mathcal{C}$ *are given by*

$$\mathcal{X}_\mathcal{C} = \{\mathbf{x} \in \mathbb{R}^m :$$
$$(\boldsymbol{\alpha}\mathbf{A}_1 + \boldsymbol{\beta}\mathbf{D}_1 - \boldsymbol{\gamma})\mathbf{L}_1^{-1}(\mathbf{x} - \mathbf{f}) \leq \boldsymbol{\alpha}\mathbf{b} + \boldsymbol{\beta}\mathbf{d}\ \text{for all}\ (\boldsymbol{\alpha}, \boldsymbol{\beta}, \boldsymbol{\gamma}) \in \mathcal{C}\}, \quad (6.1)$$
$$\mathcal{C} = \{(\boldsymbol{\alpha}, \boldsymbol{\beta}, \boldsymbol{\gamma}) \in \mathbb{R}^t :$$
$$\boldsymbol{\alpha}(\mathbf{A}_2 - \mathbf{A}_1\mathbf{L}_1^{-1}\mathbf{L}_2) + \boldsymbol{\beta}(\mathbf{D}_2 - \mathbf{D}_1\mathbf{L}_1^{-1}\mathbf{L}_2) + \boldsymbol{\gamma}\mathbf{L}_1^{-1}\mathbf{L}_2 \geq \mathbf{0},\ \boldsymbol{\beta}, \boldsymbol{\gamma} \geq \mathbf{0}\}. (6.2)$$

The set $\mathcal{C}$ defined in (6.2) is a convex polyhedral cone. Since every $(\boldsymbol{\alpha}, \boldsymbol{\beta}, \boldsymbol{\gamma}) \in \mathcal{C}$ can be written as the sum of a linear combination of the elements of a basis of the lineality space $L_\mathcal{C}$ of the cone $\mathcal{C}$ and a non-negative combination of the conical generators of $\mathcal{C}$, we can replace the requirement "for all $(\boldsymbol{\alpha}, \boldsymbol{\beta}, \boldsymbol{\gamma}) \in \mathcal{C}$" in the linear description of the polyhedron $\mathcal{X}$ by the requirement "for all $(\boldsymbol{\alpha}, \boldsymbol{\beta}, \boldsymbol{\gamma})$ in a minimal generator system of $\mathcal{C}$". Polyhedral cones have finite generator systems. Thus we get a finite system of inequalities for $\mathcal{X}$. Furthermore, if the linear programs over $\mathcal{Z}$ and $\mathcal{X}$ are *comparable* in the sense that $\mathbf{c} = \mathbf{d}\mathbf{L}$, then $min\{\mathbf{c}\mathbf{z} : \mathbf{z} \in \mathcal{Z}\} = min\{\mathbf{d}\mathbf{x} : \mathbf{x} \in \mathcal{X}\} - \mathbf{d}\mathbf{f}$.

Quadratic Scheduling Problems

As noted in Chapter 4.3, the OSP with machine independent quadratic interaction cost can also be formulated as a GPP. For ease of reference we restate these alternative formulations of the OSP. Letting $N = \{1, \ldots, n\}$ we formulate the GPP in Chapter 4.2 as the linear program

$$\min \left\{ \sum_{i=1}^{m}\sum_{j=1}^{n} c_{ij}x_{ij} + \sum_{i=1}^{m-1}\sum_{k=i+1}^{m} q_{ik}z_{ik} : (\mathbf{x}, \mathbf{z}) \in GPP_n^m \right\}, \quad (\mathcal{O}GPP_n^m)$$

where GPP_n^m is the polytope defined by the convex hull of solutions $(\mathbf{x}, \mathbf{z}) \in \mathbb{R}^{mn+m(m-1)/2}$ to the following system of equations and inequalities in zero-one variables:

$$\sum_{j=1}^{n} x_{ij} = 1 \quad \text{for } 1 \leq i \leq m \quad (6.3)$$

$$x_{ij} + x_{kj} - z_{ik} \leq 1 \quad \text{for } 1 \leq i < k \leq m, 1 \leq j \leq n \quad (6.4)$$

$$\sum_{j \in S} x_{ij} - \sum_{j \in S} x_{kj} + z_{ik} \leq 1 \quad \text{for } 1 \leq i < k \leq m, \emptyset \neq S \subset N \quad (6.5)$$

$$x_{ij} \geq 0 \quad \text{for } 1 \leq i \leq m, 1 \leq j \leq n \quad (6.6)$$

$$z_{ik} \geq 0 \quad \text{for } 1 \leq i < k \leq m \quad (6.7)$$

$$x_{ij} \in \{0, 1\} \quad \text{for } 1 \leq i \leq m, 1 \leq j \leq n. \quad (6.8)$$

As shown in Chapter 4.2, the linear programming relaxation $(6.3), \ldots, (6.7)$ of GPP_n^m is solvable in polynomial time despite the exponentiality of its constraint set. In Chapter 4.3 we formulate the OSP as the linear program

$$\min \left\{ \sum_{i=1}^{m}\sum_{j=1}^{n} c_{ij}x_{ij} + \sum_{i=1}^{m-1}\sum_{k=i+1}^{m}\sum_{j=1}^{n} q_{ikj}y_{ikj} : (\mathbf{x}, \mathbf{y}) \in QSP_n^m \right\}, \quad (\mathcal{O}QSP_n^m)$$

where QSP_n^m is the polytope defined by the convex hull of solutions $(\mathbf{x}, \mathbf{y}) \in \mathbb{R}^{mn(m+1)/2}$ to the following system of equations and inequalities in zero-one variables:

$$\sum_{j=1}^{n} x_{ij} = 1 \quad \text{for } 1 \leq i \leq m \quad (6.9)$$

$$x_{ij} + x_{kj} - y_{ikj} + \sum_{j \neq \ell=1}^{n} y_{ik\ell} \leq 1 \quad \text{for } 1 \leq i < k \leq m,\ 1 \leq j \leq n \quad (6.10)$$

$$-x_{ij} + y_{ikj} \leq 0 \quad \text{for } 1 \leq i < k \leq m,\ 1 \leq j \leq n \quad (6.11)$$

$$-x_{kj} + y_{ikj} \leq 0 \quad \text{for } 1 \leq i < k \leq m,\ 1 \leq j \leq n \quad (6.12)$$

$$y_{ikj} \geq 0 \quad \text{for } 1 \leq i < k \leq m,\ 1 \leq j \leq n \quad (6.13)$$

$$x_{ij} \in \{0, 1\} \quad \text{for } 1 \leq i \leq m,\ 1 \leq j \leq n. \quad (6.14)$$

To carry out the comparison we define two polytopes P_S and P_T as follows:

$$P_S = \{(\mathbf{x}, \mathbf{y}) \in \mathbb{R}^{mn(m+1)/2} : (\mathbf{x}, \mathbf{y}) \text{ satisfies } (6.9), \ldots, (6.13)\},$$
$$P_T = \{(\mathbf{x}, \mathbf{z}) \in \mathbb{R}^{mn+m(m-1)/2} : \exists\, (\mathbf{x}, \mathbf{y}) \in P_S \text{ such that } (\mathbf{x}, \mathbf{z}) = \mathbf{L}(\mathbf{x}, \mathbf{y})\},$$

where the linear transformation matrix $\mathbf{L}$ is defined below, see (6.15). P_S is the linear relaxation of the polytope QSP_n^m obtained by dropping the integrality requirements (6.14) and P_T its linear transformation. Likewise we define the linear relaxation of the graph partitioning polytope, obtained by dropping the integrality requirements (6.8), as follows:

$$P_G = \{(\mathbf{x}, \mathbf{z}) \in \mathbb{R}^{mn+m(m-1)/2} : (\mathbf{x}, \mathbf{z}) \text{ satisfies } (6.3), \ldots, (6.7)\}.$$

To compare the GPP formulation with the standard OSP formulation we have to calculate the linear description of the polytope P_T. The linear transformation that maps P_S into P_T consists of the identity for the $\mathbf{x}$-variables, while the $\mathbf{z}$-variables are obtained from the $\mathbf{y}$-variables via the transformation

$$z_{ik} = \sum_{j \in N} y_{ikj} \quad \text{for all } 1 \leq i < k \leq m. \tag{6.15}$$

From (6.9),...(6.13) it follows that $0 \leq z_{ik} \leq 1$ for $1 \leq i < k \leq m$ and moreover, zero-one points are mapped into zero-one points under this transformation. Letting

$$\mathbf{x}^j = (x_{1j}, \ldots, x_{mj})^T, \quad \mathbf{y}^j = (y_{12j}, \ldots, y_{1mj}, y_{23j}, \ldots, y_{2mj}, \ldots, y_{m-1,mj})^T$$

for $1 \leq j \leq n$ and $\mathbf{z} = (z_{12}, \ldots, z_{1m}, z_{23}, \ldots, z_{2m}, \ldots, z_{m-1,m})^T$, the linear transformation is

$$\mathbf{x}^j = \mathbf{x}^j \text{ for } 1 \leq j \leq n, \quad \mathbf{z} = \sum_{j=1}^n \mathbf{y}^j.$$

To apply Theorem 6.1 we write the matrix $\mathbf{L}$ corresponding to this transformation in partitioned form as $(\mathbf{L}_1, \mathbf{L}_2)$ where

$$\mathbf{L}_1 = \begin{pmatrix} \mathbf{I}_m & \ldots & \mathbf{O} & \mathbf{O} \\ \vdots & \ddots & \vdots & \vdots \\ \mathbf{O} & \ldots & \mathbf{I}_m & \mathbf{O} \\ \mathbf{O} & \ldots & \mathbf{O} & \mathbf{I}_s \end{pmatrix}, \quad \mathbf{L}_2 = \begin{pmatrix} \mathbf{O} & \ldots & \mathbf{O} \\ \vdots & \ddots & \vdots \\ \mathbf{O} & \ldots & \mathbf{O} \\ \mathbf{I}_s & \ldots & \mathbf{I}_s \end{pmatrix},$$

$s = m(m-1)/2$ and $\mathbf{I}_k$ for any $k \geq 1$ is the $k \times k$ identity matrix. The matrix $\mathbf{L}_1$ is nonsingular and of the required size. Thus Theorem 6.1 applies. Denote

Quadratic Scheduling Problems

$$\mathbf{D}_1 = \begin{pmatrix} \mathbf{A}_G^T & \mathbf{O} & \cdots & \mathbf{O} & -\mathbf{I}_s \\ \mathbf{O} & \mathbf{A}_G^T & \cdots & \mathbf{O} & \mathbf{I}_s \\ \vdots & \vdots & \ddots & \vdots & \vdots \\ \mathbf{O} & \mathbf{O} & \cdots & \mathbf{A}_G^T & \mathbf{I}_s \\ -\mathbf{K} & \mathbf{O} & \cdots & \mathbf{O} & \mathbf{I}_s \\ \mathbf{O} & -\mathbf{K} & \cdots & \mathbf{O} & \mathbf{O} \\ \vdots & \vdots & \ddots & \vdots & \vdots \\ \mathbf{O} & \mathbf{O} & \cdots & -\mathbf{K} & \mathbf{O} \\ -\mathbf{H} & \mathbf{O} & \cdots & \mathbf{O} & \mathbf{I}_s \\ \mathbf{O} & -\mathbf{H} & \cdots & \mathbf{O} & \mathbf{O} \\ \vdots & \vdots & \ddots & \vdots & \vdots \\ \mathbf{O} & \mathbf{O} & \cdots & -\mathbf{H} & \mathbf{O} \end{pmatrix} \quad \mathbf{D}_2 = \begin{pmatrix} \mathbf{I}_s & \mathbf{I}_s & \cdots & \mathbf{I}_s \\ -\mathbf{I}_s & \mathbf{I}_s & \cdots & \mathbf{I}_s \\ \vdots & \vdots & \ddots & \vdots \\ \mathbf{I}_s & \mathbf{I}_s & \cdots & -\mathbf{I}_s \\ \mathbf{O} & \mathbf{O} & \cdots & \mathbf{O} \\ \mathbf{I}_s & \mathbf{O} & \cdots & \mathbf{O} \\ \vdots & \vdots & \ddots & \vdots \\ \mathbf{O} & \mathbf{O} & \cdots & \mathbf{I}_s \\ \mathbf{O} & \mathbf{O} & \cdots & \mathbf{O} \\ \mathbf{I}_s & \mathbf{O} & \cdots & \mathbf{O} \\ \vdots & \vdots & \ddots & \vdots \\ \mathbf{O} & \mathbf{O} & \cdots & \mathbf{I}_s \end{pmatrix}$$

Figure 6.1 The partitioning of the inequalities (6.10), ..., (6.12)

by $\mathbf{A}_G$ the node-edge incidence matrix of the complete undirected graph having m nodes and by $\mathbf{e}_k$ the **column** vector having k components equal to one. We let

$$\mathbf{F} = (\mathbf{I}_m \ \cdots \ \mathbf{I}_m), \quad \mathbf{K} = \begin{pmatrix} \mathbf{e}_{m-1} & \mathbf{0} & \cdots & \mathbf{0} & \mathbf{0} \\ \mathbf{0} & \mathbf{e}_{m-2} & \cdots & \mathbf{0} & \mathbf{0} \\ \vdots & \vdots & \ddots & \vdots & \vdots \\ \mathbf{0} & \mathbf{0} & \cdots & 1 & 0 \end{pmatrix}, \quad \mathbf{H} = \begin{pmatrix} \mathbf{H}_1 \\ \mathbf{H}_2 \\ \vdots \\ \mathbf{H}_{m-1} \end{pmatrix},$$

where $\mathbf{F}$ is of size $m \times mn$ and $\mathbf{H}_i = (\mathbf{0} \ldots \mathbf{0} \ \mathbf{I}_{m-i})$ is of size $(m-i) \times m$ for $1 \le i \le m-1$. Note that in this notation $\mathbf{A}_G^T = \mathbf{K} + \mathbf{H}$. Let $\mathbf{d}^j = (d_1^j, \ldots, d_n^j)$ denote the vectors with components $d_j^j = -1, d_\ell^j = 1$ for $1 \le \ell \ne j \le n$ where $1 \le j \le n$. We write the constraint set of OSP in matrix/vector form as follows, where the constraints (6.9), ..., (6.12) are listed in the order implied by the above and the indexing of the variables of the problem.

$$\begin{aligned} \mathbf{F}\mathbf{x} &= \mathbf{e}_m \\ \mathbf{A}_G^T \mathbf{x}^j + \sum_{\ell=1}^n d_\ell^j \mathbf{y}^\ell &\le 1 \quad \text{for } 1 \le j \le n \\ -\mathbf{K}\mathbf{x}^j + \mathbf{y}^j &\le 0 \quad \text{for } 1 \le j \le n \\ -\mathbf{H}\mathbf{x}^j + \mathbf{y}^j &\le 0 \quad \text{for } 1 \le j \le n \\ \mathbf{y}^j &\ge 0 \quad \text{for } 1 \le j \le n. \end{aligned}$$

To determine the linear description of the cone (6.2) we calculate in the notation of Theorem 6.1 that $\mathbf{A}_2 - \mathbf{A}_1 \mathbf{L}_1^{-1} \mathbf{L}_2 = \mathbf{O}$ and the corresponding calculation of

$\mathbf{D}_2 - \mathbf{D}_1\mathbf{L}_1^{-1}\mathbf{L}_2$ is carried out in the notation given above. In Figure 6.1 we display the matrices $\mathbf{D}_1$ and $\mathbf{D}_2$. It follows that in the case of mapping (6.15) the associated cone (6.2) is given by

$$C = \left\{ \begin{array}{l} (\alpha, \beta, \gamma, \delta, \omega) \in \mathbb{R}^\phi : \\ 2\beta^1 - 2\beta^j - \gamma^1 + \gamma^j - \delta^1 + \delta^j + \omega^{n+1} \geq 0 \text{ for } 2 \leq j \leq n, \\ \beta \geq 0, \gamma \geq 0, \delta \geq 0, \omega \geq 0 \end{array} \right\},$$

where $\phi = m+3ns+mn+s$ and $s = m(m-1)/2$. Moreover, $\alpha \in \mathbb{R}^m$, $\beta^j, \gamma^j, \delta^j \in \mathbb{R}^s$ for $1 \leq j \leq n$, $\omega^j \in \mathbb{R}^m$ for $1 \leq j \leq n$, $\omega^{n+1} \in \mathbb{R}^s$ and $\beta = (\beta^1, \ldots, \beta^n)$, $\gamma = (\gamma^1, \ldots, \gamma^n)$, $\delta = (\delta^1, \ldots, \delta^n)$, $\omega = (\omega^1, \ldots, \omega^n, \omega^{n+1})$. The lineality space of the cone C is generated by

(i) $\alpha = \pm \mathbf{u}^i$ for $1 \leq i \leq m, \beta = \gamma = \delta = 0, \omega = 0$,

where $\mathbf{u}^i \in \mathbb{R}^m$ is the i-th unit vector. Intersecting C with the orthogonal complement of its lineality space we obtain a pointed cone. Using the intersection property of cones; see e.g. Proposition 1 of Padberg and Sung [1991], we find that

(ii) $\alpha = 0, \beta = \gamma = \delta = 0, \omega^{n+1} = 0, \omega^j = \mathbf{u}^k$ for $1 \leq k \leq m$ and $1 \leq j \leq n$,

are extreme rays of the corresponding cone. Moreover, we can simplify the cone C of our linear transformation and using the substitution $\tilde{\beta}^j = 2\beta^j$ for $1 \leq j \leq n$ we are left with the task of finding the extreme rays of the pointed cone

$$C' = \left\{ \begin{array}{l} (\tilde{\beta}, \gamma, \delta, \omega^{n+1}) \in \mathbb{R}^\psi : \\ \tilde{\beta}^1 - \tilde{\beta}^j - \gamma^1 + \gamma^j - \delta^1 + \delta^j + \omega^{n+1} \geq 0 \text{ for } 2 \leq j \leq n, \\ \tilde{\beta} \geq 0, \gamma \geq 0, \delta \geq 0, \omega^{n+1} \geq 0 \end{array} \right\},$$

where $\psi = 3ns + s$ and $s = m(m-1)/2$. From the definition of an extreme ray of a pointed cone and the symmetry of the constraint set of C' it follows that $(\tilde{\beta}, \gamma, \delta, \omega^{n+1})$ is an extreme ray of C', if and only if $(\tilde{\beta}, \delta, \gamma, \omega^{n+1})$ is an extreme ray of C'. Moreover, for every extreme ray $(\tilde{\beta}, \gamma, \delta, \omega^{n+1})$ of C' we have $\gamma_u^j \delta_u^j = 0$ for all $1 \leq u \leq s$ and $2 \leq j \leq n$. (To see this, suppose $\gamma_u^j > 0$ and $\delta_u^j > 0$ for some u and j. Set e.g. $\tilde{\gamma}_u^j = \gamma_u^j + \delta_u^j$, $\delta_u^j = 0$ and leave all other components unchanged. Then the rank of the corresponding equation system is increased by 1, which contradicts the assumption that $(\tilde{\beta}, \gamma, \delta, \omega^{n+1})$ is an extreme ray of C'.) From the symmetry of the constraint set it follows that the corresponding statements are correct for the vectors $\tilde{\beta}^1$ and ω^{n+1} as well. Consequently, we can simplify the cone C' further and it suffices to determine the extreme rays of the pointed cone

$$C'' = \{(\tilde{\beta}, \gamma) \in \mathbb{R}^\rho : \tilde{\beta}^1 - \tilde{\beta}^j - \gamma^1 + \gamma^j \geq 0 \text{ for } 2 \leq j \leq n, \tilde{\beta} \geq 0, \gamma \geq 0\},$$

where $\rho = 2ns$ and $s = m(m-1)/2$.

Claim 6.1 *The extreme rays of C'' are given by*

(a1) $\tilde{\beta}^1 = \mathbf{v}^\ell$, $\tilde{\beta}^j = \mathbf{v}^\ell$ *for* $j \in T \subseteq \{2,\ldots,n\}$, $\tilde{\beta}^k = \mathbf{0}$ *for* $k \notin T, k \geq 2, \gamma = 0$,

(a2) $\tilde{\beta}^j = \gamma^j = \mathbf{v}^\ell$ *for some* $j \in \{1,\ldots,n\}$, $\tilde{\beta}^k = \gamma^k = \mathbf{0}$ *for* $1 \leq k \neq j \leq n$,

(a3) $\tilde{\beta} = \mathbf{0}$, $\gamma^j = \mathbf{v}^\ell$ *for* $1 \leq j \leq n$,

(a4) $\tilde{\beta} = \mathbf{0}$, $\gamma^j = \mathbf{v}^\ell$ *for some* $j \in \{2,\ldots,n\}$, $\gamma^k = \mathbf{0}$ *for* $1 \leq k \neq j \leq n$,

where $1 \leq \ell \leq s$, $\mathbf{v}^\ell \in \mathbb{R}^s$ *is the ℓ-th unit vector and* $s = m(m-1)/2$.

Proof. Every vector $(\tilde{\beta}, \gamma) \in \mathbb{R}^p$ defined by $(a1), \ldots, (a4)$ belongs to C'' and satisfies exactly $2ns - 1$ linearly independent rows of the constraint set of C'' at equality, i.e. it defines an extreme ray of C''. It remains to show that every $(\tilde{\beta}, \gamma) \in C''$ is a nonnegative combination of the extreme rays $(a1), \ldots, (a4)$ of C''. Listing the extreme rays in the order implied by $(a1), \ldots, (a4)$ this is equivalent to showing that for $(\tilde{\beta}, \gamma) \in C''$ the equation system

$$\sum_T \boldsymbol{\lambda}^T + \boldsymbol{\mu}^1 = \tilde{\beta}^1, \quad \boldsymbol{\mu}^1 + \boldsymbol{\mu}^{n+1} = \gamma^1, \quad \sum_{j \in T} \boldsymbol{\lambda}^T + \boldsymbol{\mu}^j = \tilde{\beta}^j, \quad \boldsymbol{\mu}^j + \boldsymbol{\mu}^{n+1} + \boldsymbol{\mu}^{n+j} = \gamma^j$$

for $2 \leq j \leq n$ has a nonnegative solution, where $\boldsymbol{\lambda}^T \in \mathbb{R}^s$ for $T \subseteq \{2, \ldots, n\}$ and $\boldsymbol{\mu}^j \in \mathbb{R}^s$ for $1 \leq j \leq 2n$. Eliminating the $\boldsymbol{\mu}$-variables from this system the assertion is equivalent to showing that the system of inequalities

$$\sum_T \boldsymbol{\lambda} \leq \tilde{\beta}^1, \quad \sum_{j \notin T} \boldsymbol{\lambda}^T \leq \tilde{\beta}^1 - \tilde{\beta}^j - \gamma^1 + \gamma^j \text{ for } 2 \leq j \leq n, \quad -\sum_T \boldsymbol{\lambda}^T \leq \gamma^1 - \tilde{\beta}^1,$$

has a nonnegative solution for $(\tilde{\beta}, \gamma) \in C''$. Suppose not. Then by Farkas' lemma

$$\sum_{k \notin T} \mathbf{u}^k - \mathbf{u}^{n+1} \geq \mathbf{0} \text{ for } T \subseteq \{2, \ldots, n\},$$

$$\mathbf{u}^1 \tilde{\beta}^1 + \sum_{j=2}^n \mathbf{u}^j (\tilde{\beta}^1 - \tilde{\beta}^j - \gamma^1 + \gamma^j) + \mathbf{u}^{n+1}(\gamma^1 - \tilde{\beta}^1) < 0$$

has a solution $\mathbf{u}^k \geq \mathbf{0}$ where $\mathbf{u}^k \in \mathbb{R}^s$ and $1 \leq k \leq n+1$. Note that the summations include $T = \emptyset$ and that in this case $k \notin T$ is to be read as $k = 1, \ldots, n$. Since $(\tilde{\beta}, \gamma) \in C''$, we have $\tilde{\beta}^1 \geq \mathbf{0}$, $\gamma^1 \geq \mathbf{0}$ and $\tilde{\beta}^1 - \tilde{\beta}^j - \gamma^1 + \gamma^j \geq \mathbf{0}$ for $2 \leq j \leq n$. For $T = \{2, \ldots, n\}$ we get $\mathbf{u}^1 - \mathbf{u}^{n+1} \geq \mathbf{0}$, thus $(\mathbf{u}^1 - \mathbf{u}^{n+1})\tilde{\beta}^1 + \sum_{j=2}^n \mathbf{u}^j (\tilde{\beta}^1 - \tilde{\beta}^j - \gamma^1 + \gamma^j) + \mathbf{u}^{n+1} \gamma^1 \geq 0$ for all $\mathbf{u}^1 \geq \mathbf{0}, \ldots, \mathbf{u}^{n+1} \geq \mathbf{0}$ which is a contradiction. $\square$

Now we are ready to derive the extreme rays of the cone C' and to complete the minimal generator system of the cone C of the linear transformation that we are analyzing. From the remarks preceding the claim we get precisely the following additional generators for the conical part of C. In this listing we assume that the vectors $\alpha, \beta, \gamma, \delta$ and ω that are *not* shown must all equal zero. Moreover, we state each class of generators pairwise as suggested by the symmetry of the constraints of C' and let $1 \leq \ell \leq s$ be arbitrary.

(iii) $\beta^1 = \mathbf{v}^\ell$, $\beta^j = \mathbf{v}^\ell$ for $j \in T \subseteq \{2, \ldots, n\}$, $\beta^k = \mathbf{0}$ otherwise and $\beta^j = \mathbf{v}^\ell$ for $j \in T \subseteq \{2, \ldots, n\}$, $\beta^k = \mathbf{0}$ otherwise, $\omega^j = \mathbf{0}$ for $1 \leq j \leq n$, $\omega^{n+1} = 2\mathbf{v}^\ell$,

(iv) $\beta^1 = \mathbf{v}^\ell$, $\gamma^1 = 2\mathbf{v}^\ell$, $\beta^k = \gamma^k = \mathbf{0}$ for $2 \leq k \leq n$ and $\gamma^1 = \mathbf{v}^\ell, \gamma^k = \mathbf{0}$ for $2 \leq k \leq n$, $\omega^j = \mathbf{0}$ for $1 \leq j \leq n$, $\omega^{n+1} = \mathbf{v}^\ell$,

(v) $\beta^1 = \mathbf{v}^\ell, \delta^1 = 2\mathbf{v}^\ell, \beta^k = \delta^k = \mathbf{0}$ for $2 \leq k \leq n$ and $\delta^1 = \mathbf{v}^\ell, \delta^k = \mathbf{0}$ for $2 \leq k \leq n$, $\omega^j = \mathbf{0}$ for $1 \leq j \leq n$, $\omega^{n+1} = \mathbf{v}^\ell$,

(vi) $\beta^j = \mathbf{v}^\ell,, \gamma^j = 2\mathbf{v}^\ell$ for some $j \in \{2, \ldots, n\}$, $\beta^k = \gamma^k = \mathbf{0}$ for $1 \leq k \neq j \leq n$ and $\beta^j = \mathbf{v}^\ell, \delta^j = 2\mathbf{v}^\ell$ for some $j \in \{2, \ldots, n\}, \beta^k = \delta^k = \mathbf{0}$ for $1 \leq k \neq \leq j \leq n$,

(vii) $\gamma^j = \mathbf{v}^\ell$ for $1 \leq j \leq n$ and $\delta^j = \mathbf{v}^\ell$ for $1 \leq j \leq n$,

(viii) $\gamma^j = \mathbf{v}^\ell$ for some $j \in \{2, \ldots, n\}, \gamma^k = \mathbf{0}$ for $1 \leq k \neq j \leq n$ and $\delta^j = \mathbf{v}^\ell$ for some $j \in \{2, \ldots, n\}, \delta^k = \mathbf{0}$ for $1 \leq k \neq j \leq n$.

We apply Theorem 6.1 again and calculate the linear description of the image P_T of the OSP polytope QSP_n^m under the transformation (6.15). In the calculation of (6.1) we use the fact that the index ℓ with $1 \leq \ell \leq s$ corresponds to some index pair i, k with $1 \leq i < k \leq m$.

The generators (i) give the equations (6.3) and the generators (ii) the inequalities (6.6).

For $T = \emptyset$ the generators (iii) give $s = m(m-1)/2$ inequalities (6.4) for $j = 1$ and the s inequalities (6.7). For $T = \{j\}$ we get $x_{i1} + x_{k1} + x_{ij} + x_{kj} \leq 2$ for some $j \geq 2$, which are redundant by (6.3), and the remaining $s(n-1)$ inequalities (6.4) for $2 \leq j \leq n$. For $2 \leq |T| \leq n-1$ we get the inequalities

$$x_{i1}+x_{k1}+\sum_{j \in T}(x_{ij}+x_{kj})+(|T|-1)z_{ik} \leq |T|+1, \quad \sum_{j \in T}(x_{ij}+x_{kj})+(|T|-2)z_{ik} \leq |T|.$$

The generators (iv) give s inequalities $-x_{i1} + x_{k1} + z_{ik} \leq 1$ and inequalities $-x_{i1} \leq 0$, which we have already. Using (6.3) the first inequalities are equivalent to (6.5) for $S = N - \{1\}$.

The generators (v) give all inequalities (6.5) for $S = \{1\}$ and $-x_{k1} \leq 0$. The generators (vi) give all remaining inequalities (6.5) for $S = N - \{j\}$ and $S = \{j\}$ where $2 \leq j \leq n$ and the generators (vii) and (viii) give redundant inequalities.

Quadratic Scheduling Problems 141

The inequalities (6.5) for $S = \{j\}$ and $S = N - \{j\}$ imply that $z_{ik} \leq 1$. Consequently, using (6.3) we find that the inequalities that were obtained from the generators (iii) for $2 \leq |T| \leq n-1$ are redundant.

Summarizing the preceding material we have proven the following proposition.

Proposition 6.1 *Let P_T be the image of the linear relaxation P_S of the polytope QSP_n^m under the linear transformation (6.15). Then*

$$P_T = \left\{ (\mathbf{x}, \mathbf{z}) \in \mathbb{R}^{mn+s} : \begin{array}{l} (\mathbf{x}, \mathbf{z}) \text{ satisfies } (6.3), (6.4), (6.6), (6.7) \text{ and } (6.5) \\ \text{for } S = \{j\} \text{ and } S = N - \{j\} \text{ where } j \in N \end{array} \right\}$$

and $P_T \supset P_G$, where P_G is the linear relaxation of the polytope GPP_n^m and $s = m(m-1)/2$.

Denote by $\mathbf{c} \in \mathbb{R}^{mn}$ the row vector of the c_{ij} and by $\mathbf{q}^* \in \mathbb{R}^{ns}$ the row vector of the q_{ikj} of the objective function of the OSP in the appropriate indexing. Machine independence of the quadratic interaction cost means that $q_{ikj} = q_{ik}$ for all $1 \leq i < k \leq m$ and $1 \leq j \leq n$. Let $\mathbf{q} \in \mathbb{R}^s$ be the vector of the q_{ik} in the usual indexing. The assumption of the machine independence then implies that

$$(\mathbf{c}, \mathbf{q}^*) = (\mathbf{c}, \mathbf{q})\mathbf{L},$$

where $\mathbf{L}$ is the matrix of the linear transformation (6.15). Thus the linear programs over P_S and P_T are comparable. Writing $\mathbf{x} = (\mathbf{x}^1, \ldots, \mathbf{x}^n)$ and $\mathbf{y} = (\mathbf{y}^1, \ldots, \mathbf{y}^n)$ it follows that

$$\begin{aligned} min\{\mathbf{cx} + \mathbf{q}^*\mathbf{y} : (\mathbf{x}, \mathbf{y}) \in P_S\} &= min\{\mathbf{cx} + \mathbf{qz} : (\mathbf{x}, \mathbf{z}) \in P_T\} \\ &\leq min\{\mathbf{cx} + \mathbf{qz} : (\mathbf{x}, \mathbf{z}) \in P_G\} \end{aligned}$$

since $P_G \subset P_T$. This is true no matter what (machine independent) objective function coefficients are used. It means that in the case of machine independent interaction cost the lower bound obtained from the linear relaxation (6.9), ..., (6.13) of the OSP is in all cases *worse* than the lower bound obtained from the LP relaxation (6.3), ..., (6.7) of the GPP.

On one hand this shows that additional information – such as the machine independence of the interaction cost – should be utilized at the modeling stage, especially in this case where many superfluous variables can be avoided. More precisely, the explicit consideration of the additional variables "hurts," rather than "helps" the linear programming relaxation of the problem. On the other hand, the preceding shows that the detailed analysis of the graph partitioning problem via the locally ideal linearization of Chapter 4.2 yields a better result than what can be obtained from the OSP formulation of Chapter 4.3 via the linear transformation (6.15). It is interesting to note that the weaker formulation of the OSP with machine independent interaction cost obtained *via* (6.15) agrees fully with the formulation of the graph partitioning problem due to Chopra and Rao [1989a, 1993]; see also Chapter 4.2 on this point.

While in the case of the OSP the outcome of the linear transformation technique – the mapping of a polyhedron from a higher-dimensional space into a lower dimensional space – is negative in the sense that a weaker formulation is obtained, this is, of course, not always the case. To give a concrete example consider the case of the general model of Chapter 4.6 which generalizes all preceding formulations of Chapter 4. In Remark (5.5) we show that by eliminating certain variables the general model is reduced to the seemingly less general VLSI circuit layout design problem (CLDP). More precisely, we show that by eliminating the variables y_{ij}^{kj} for $1 \leq i < k \leq m$ and $1 \leq j \leq n$ and appropriately modifying the objective function of the general model the formulation $(5.6), \ldots, (5.10)$ of the CLDP is obtained.

The same result can be obtained by *projecting out* the corresponding ns y-variables from the linear formulation $(5.15), \ldots, (5.18)$ of the general model where $s = m(m-1)/2$. Indexing the variables of the general model to be retained in the order of the variables of the CLDP, see Chapter 4.5, and the variables y_{ij}^{kj} to be projected out as the last variables, we thus have a linear transformation $(\mathbf{x}, \mathbf{z}) = \mathbf{L}(\mathbf{x}, \mathbf{y})$ where

$$\mathbf{L} = (\mathbf{I}_{mn+t} \ \mathbf{O}),$$

$t = n(n-1)s$ and the zero matrix is of size $(mn+t) \times ns$. Denote

$$P_{GM} = \{(\mathbf{x}, \mathbf{y}) \in \mathbb{R}^{mn+n^2s} : (\mathbf{x}, \mathbf{y}) \text{ satisfies } (5.15), \ldots, (5.18)\},$$
$$P_{IM} = \{(\mathbf{x}, \mathbf{z}) \in \mathbb{R}^{mn+t} : \exists (\mathbf{x}, \mathbf{y}) \in P_{GM} \text{ such that } (\mathbf{x}, \mathbf{z}) = \mathbf{L}(\mathbf{x}, \mathbf{y})\},$$
$$P_{CL} = \{(\mathbf{x}, \mathbf{z}) \in \mathbb{R}^{mn+t} : (\mathbf{x}, \mathbf{z}) \text{ satisfies } (5.6), \ldots, (5.9)\}.$$

P_{GM} is the linear relaxation of the polytope QGP_n^m of the general model, P_{IM} its image under the projection $\mathbf{L}$ and P_{CL} the linear relaxation of the polytope QDP_n^m of the circuit layout design problem CLDP. We apply Theorem 6.1 with $\mathbf{L}$ partitioned into $\mathbf{L}_1 = \mathbf{I}_{mn+t}$ and $\mathbf{L}_2 = \mathbf{O}$. Denote the system of equations (5.15), (5.16) and (5.17) by $\mathbf{A}(\mathbf{x}, \mathbf{y})^T = \mathbf{b}$, partition $\mathbf{A} = (\mathbf{A}_1, \mathbf{A}_2)$ according to $(\mathbf{L}_1, \mathbf{L}_2)$ and let $r = (n-1)s$. We calculate

$$\mathbf{A}_2 - \mathbf{A}_1 \mathbf{L}_1^{-1} \mathbf{L}_2 = \begin{pmatrix} \mathbf{O} & \mathbf{O} \\ \mathbf{I}_r & \mathbf{O} \\ \mathbf{O} & \mathbf{I}_s \\ \mathbf{I}_r & \mathbf{O} \end{pmatrix}.$$

Since the matrix **D** of Theorem 6.1 is void, we get the cone
$$C = \{(\alpha, \beta, \gamma, \delta, \omega) \in \mathbb{R}^\phi : \beta + \delta \geq 0, \gamma \geq 0, \omega \geq 0\},$$
where $\phi = m + 2r + s + mn + t$ with $t = n(n-1)s$ and $\alpha \in \mathbb{R}^m$, $\beta, \delta \in \mathbb{R}^r$, $\gamma \in \mathbb{R}^s$, $\omega \in \mathbb{R}^{mn+t}$. The lineality space of C is generated by
(b1) $\alpha = \pm \mathbf{u}^i$ for $1 \leq i \leq m$, $\beta = \delta = 0$, $\gamma = 0$, $\omega = 0$,
(b2) $\alpha = 0$, $\beta = \pm \mathbf{v}^i$, $\delta = \mp \mathbf{v}^i$ for $1 \leq i \leq r$, $\gamma = 0$, $\omega = 0$,
where $\mathbf{u}^i \in \mathbb{R}^m$, $\mathbf{v}^i \in \mathbb{R}^r$ are unit vectors and $r = (n-1)s$. From the intersection property of cones we find the following generators of C
(b3) $\alpha = 0, \beta = \delta = 0$, $\gamma = \mathbf{r}^i$ for $1 \leq i \leq s$, $\omega = 0$,
(b4) $\alpha = 0, \beta = \delta = 0$, $\gamma = 0$, $\omega = \mathbf{t}^i$ for $1 \leq i \leq mn + t$,
where $\mathbf{r}^i \in \mathbb{R}^s$, $\mathbf{t}^i \in \mathbb{R}^{mn+t}$ are unit vectors, $s = m(m-1)/2$ and $t = n(n-1)s$. Moreover, the cone C simplifies and after intersecting it with the orthogonal complement of the lineality space, we are left with determining the extreme rays of the pointed cone
$$C' = \{(\beta, \delta) \in \mathbb{R}^{2r} : \beta + \delta \geq 0, \beta - \delta = 0\},$$
which are easily determined. This gives the remaining generators of C
(b5) $\alpha = 0, \beta = \delta = \mathbf{v}^i$ for $1 \leq i \leq r$, $\gamma = 0$, $\omega = 0$.
From the derivation it follows that $(b1), \ldots, (b5)$ is a minimal generator system of the polyhedral cone C of the mapping from the space of variables of the general model to the one of CLDP. It remains to calculate the linear description of the image P_{IM} of P_{GM} by (6.1).

The generators (b1) of C give the equations (5.6) and the generators (b2) the equations (5.7) when we replace the $y_{i\ell}^{kj}$ by the $z_{i\ell}^{kj}$ of our linear transformation. The generators (b3) give the inequalities (5.8) for $j = n$. The generators (b4) give the inequalities (5.9) and the redundant inequalities $x_{ij} \geq 0$ for $1 \leq i \leq m, 1 \leq j \leq n$. The generators (b5) yield

$$-x_{ij} + \sum_{j \neq \ell = 1}^n y_{ij}^{k\ell} - x_{kj} + \sum_{j \neq \ell = 1}^n y_{i\ell}^{kj} \leq 0 \text{ for } 1 \leq i < k \leq m, 1 \leq j \leq n-1,$$

where we have simply written $y_{ij}^{k\ell}$ rather than $z_{ij}^{k\ell}$ as required by our transformation. Using (5.7) to eliminate the second half of this inequality we thus find all remaining inequalities (5.8) multiplied by a factor of two, which is immaterial because the right-hand equals zero.

It follows that the projection P_{IM} of the polytope P_{GM} obtained by the linear transformation technique is exactly the polytope P_{CL}. To get comparability of the linear programs over P_{GM} and P_{IM}, respectively, the objective function of the general model has to be changed so as to produce zero coefficients for the variables that are projected out. Thus – except for the slightly more general objective function of the general model – the CLDP and the general model are the same. Of course, *you* should have inferred this without the analysis that we just went through: the general model has equations only except for the nonnegativities (5.18) which we have preserved in the elimination process of Remark (5.5). Variable elimination corresponds in this case exactly to projection and so the result was predictable. The linear (or affine) transformation technique confirmed in this case the obvious. The technique is, however, much more widely applicable and as we have seen before, the results are not always predictable. Indeed, a much more frequent use of this technique is desirable to the end of *analytically comparing* formulations proposed by different authors for the same problem. Historically, such comparisons were carried out *empirically* by testing different formulations on numerical data. Besides wasting computer time and journal paper – not to speak of refereeing time – this approach can and should be replaced by the more profound analysis of the type done here; see also Padberg and Sung [1991].

6.2 Quadratic Scheduling Polytopes

From among the scheduling problems described in Chapter 1, we will study the facial strucure of the OSP only, because it permits the most general cost function. For special cases of the OSP a substantial body of literature already exists; see Grötschel and Wakabayashi [1989, 1990] for the *clique partitioning* problem and Chopra and Rao [1989a, 1993] for the graph partitioning problem. We denote the convex hull of solutions to the OSP by QSP_n^m as before and refer to it as the quadratic scheduling polytope. Let $\overline{\mathbf{u}}_{ij} \in \mathbb{R}^{mn}, \overline{\mathbf{v}}_{ikj} \in \mathbb{R}^{mn(m-1)/2}$, $\mathbf{u}_{ij}, \mathbf{v}_{ikj} \in \mathbb{R}^{mn+mn(m-1)/2}$ be as defined in Chapter 4.3 and define $\mathbf{z}_I(j) = (\sum_{i \in I} \mathbf{u}_{ij} + \sum_{i<k \in I} \mathbf{v}_{ikj})$ for $1 \leq j \leq n$ where $I \subseteq M = \{1, \ldots, m\}$. We set $N = \{1, \ldots, n\}$ and assume $m \geq n \geq 3$.

Proposition 6.2 *The dimension of QSP_n^m is $dim(QSP_n^m) = mn(m+1)/2 - m$.*

Proof. Since the m equations (6.9) are linearly independent, $dim(QSP_n^m) \leq mn(m+1)/2 - m$. We establish $dim(QSP_n^m) \geq mn(m+1)/2 - m$ by showing that every equation $\alpha \mathbf{x} + \beta \mathbf{y} = \gamma$ that is satisfied by all $(\mathbf{x}, \mathbf{y}) \in QSP_n^m$ is a linear combination of (6.9).

Quadratic Scheduling Problems

(i) Since $(\mathbf{z}_{M\setminus\{s\}}(r) + \mathbf{u}_{s\ell}) \in QSP_n^m$ for every $s \in M$ and $r \neq \ell \in N$, $\alpha_{sr} = \alpha_{s\ell} = \omega_s$ for all $r \neq \ell \in N$ where ω_s are constants for $s \in M$.

(ii) Since $(\mathbf{z}_{is}(\ell) + \mathbf{z}_{M\setminus\{i,s\}}(k))$, $(\mathbf{u}_{i\ell} + \mathbf{u}_{sr} + \mathbf{z}_{M\setminus\{i,s\}}(k)) \in QSP_n^m$ for $k \neq \ell \neq r$, $k, \ell, r \in N$, comparing these solutions with the ones used in (i), we get $\beta_{is\ell} = 0$ for $i < s$ and $\beta_{si\ell} = 0$ for $s < i$ and $\ell \in N$.

Hence $\boldsymbol{\alpha}\mathbf{x} + \boldsymbol{\beta}\mathbf{y} = \gamma$ becomes $\sum_{s \in M} \sum_{k \in N} \omega_s x_{sk} = \sum_{s \in M} \omega_s = \gamma$, which is a linear combination of the equations (6.9) and the proposition follows. $\square$

Proposition 6.3 *The inequalities $y_{pgr} \geq 0$ define facets of QSP_n^m for all $p, g \in M$ and $r \in N$.*

Proof. Let $F = \{(\mathbf{x}, \mathbf{y}) \in QSP_n^m : y_{pgr} = 0\}$. Since $(\mathbf{z}_{M\setminus\{p,g\}}(k) + \mathbf{z}_{pg}(r)) \in QSP_n^m$ for $k \in N \setminus \{r\}$ but not in F, F is a proper face of QSP_n^m. Suppose there exists a valid inequality $\boldsymbol{\alpha}\mathbf{x} + \boldsymbol{\beta}\mathbf{y} \leq \gamma$ for QSP_n^m such that every $(\mathbf{x}, \mathbf{y}) \in F$ satisfies $\boldsymbol{\alpha}\mathbf{x} + \boldsymbol{\beta}\mathbf{y} = \gamma$. To prove the proposition we need to show that $(\boldsymbol{\alpha}, \boldsymbol{\beta}, \gamma) = (\sum_s \omega_s \mathbf{e}_s, \pi \overline{\mathbf{v}}_{pgr}, \sum_s \omega_s)$ where $\mathbf{e}_s \in \mathbb{R}^{mn}$ is a vector with one in its (s, ℓ) components for all $\ell \in N$ and zero elsewhere, $\pi \in \mathbb{R}^1$ and $\omega_s \in \mathbb{R}^1$ are constants for all $s \in M$.

(i) Since $(\mathbf{z}_{M\setminus\{s\}}(k) + \mathbf{u}_{s\ell}) \in F$ for $s \in M$, $\ell \in N$ and every $k \in N \setminus \{\ell\}$ with $k \neq r$ or if $k = r$ then $s = p$ or g, $\alpha_{sk} = \alpha_{s\ell} = \omega_s$ for all $k, \ell \in N$.

(ii) Since $(\mathbf{z}_{ij}(\ell) + \mathbf{z}_{M\setminus\{i,j\}}(k))$, $(\mathbf{u}_{ir} + \mathbf{u}_{j\ell} + \mathbf{z}_{M\setminus\{i,j\}}(k)) \in F$ for $k \neq \ell \in N \setminus \{r\}$, comparing these solutions with the ones used in (i), we get $\beta_{ij\ell} = 0$ for all $i < j \in M$ and $r \neq \ell \in N$.

(iii) Since $(\mathbf{z}_{ij}(r) + \mathbf{z}_{M\setminus\{i,j\}}(\ell))$, $(\mathbf{u}_{ir} + \mathbf{u}_{jk} + \mathbf{z}_{M\setminus\{i,j\}}(\ell)) \in F$ given at least one of $i < j \in M \setminus \{p, g\}$, $k \neq \ell \in N \setminus \{r\}$, comparing these solutions with the ones used in (ii), we get $\beta_{ijr} = 0$ where at least one of $i < j \in M \setminus \{p, g\}$.

Hence $\boldsymbol{\alpha}\mathbf{x} + \boldsymbol{\beta}\mathbf{y} \leq \gamma$ becomes $\sum_{s \in M} \sum_{k \in N} \omega_s x_{sk} + \beta_{pgr} y_{pgr} \leq \gamma = \sum_s \omega_s$ or equivalently, $\beta_{pgr} y_{pgr} \leq 0$ for $p, g \in M$ and $r \in N$. Since F is a proper face of QSP_n^m and $y_{pgr} \geq 0$ valid for QSP_n^m, $\beta_{pgr} \leq 0$ and hence $\beta_{pgr} < 0$. Taking $\pi = \beta_{pgr}$ the proposition follows. $\square$

Proposition 6.4 *Inequalities $-x_{pr} + y_{pgr} \leq 0$ define facets of QSP_n^m for $p, g \in M$ and $r \in N$.*

Proof. Let $F = \{(\mathbf{x}, \mathbf{y}) \in QSP_n^m : -x_{pr} + y_{pgr} = 0\}$. Since $(\mathbf{z}_{M\setminus\{p\}}(r) + \mathbf{u}_{pk}) \in QSP_n^m$ for $k \in N \setminus \{r\}$ but not in F, F is a proper face of QSP_n^m. Suppose there exists a valid inequality $\boldsymbol{\alpha}\mathbf{x} + \boldsymbol{\beta}\mathbf{y} \leq \gamma$ for QSP_n^m such that every $(\mathbf{x}, \mathbf{y}) \in F$ satisfies $\boldsymbol{\alpha}\mathbf{x} + \boldsymbol{\beta}\mathbf{y} = \gamma$. To prove the proposition we need to show that $(\boldsymbol{\alpha}, \boldsymbol{\beta}, \gamma) = (\sum_s \omega_s \mathbf{e}_s + \pi \overline{\mathbf{u}}_{pr}, -\pi \overline{\mathbf{v}}_{pgr}, \sum_s \omega_s)$ where $\mathbf{e}_s \in \mathbb{R}^{mn}$ is a vector with one in its (s, ℓ) components for all $\ell \in N$ and zero elsewhere, $\pi \in \mathbb{R}^1$ and $\omega_s \in \mathbb{R}^1$ are constants for all $s \in M$.

(i) Since $(\mathbf{z}_{M\setminus\{p\}}(r)+\mathbf{u}_{p\ell}) \in F$ for all $\ell \in N\setminus\{r\}$, $\alpha_{pk} = \alpha_{p\ell}$ for $k, \ell \in N\setminus\{r\}$.

(ii) Since $(\mathbf{z}_{M\setminus\{p,g\}}(k) + \mathbf{z}_{pg}(r)), (\mathbf{z}_{M\setminus\{p,g\}}(k) + \mathbf{u}_{p\ell} + \mathbf{u}_{gr}) \in F$ for $k \neq \ell \in N \setminus \{r\}$, $\alpha_{pr} = \alpha_{p\ell} - \beta_{pgr}$ for $\ell \in N \setminus \{r\}$.

(iii) Since $(\mathbf{z}_{M\setminus\{i\}}(k) + \mathbf{u}_{ir}), (\mathbf{z}_{M\setminus\{i\}}(k) + \mathbf{u}_{i\ell}) \in F$ for $k \neq \ell \in N \setminus \{r\}$ and $i \in M \setminus \{p\}$, $\alpha_{ir} = \alpha_{i\ell}$ for $k \neq \ell \in N \setminus \{r\}$ and $i \in M \setminus \{p\}$.

(iv) $(\mathbf{z}_{M\setminus\{p,g\}}(k) + \mathbf{z}_{pg}(\ell))$, $(\mathbf{z}_{M\setminus\{p,g\}}(k) + \mathbf{u}_{p\ell} + \mathbf{u}_{gr})$, $(\mathbf{z}_{M\setminus\{g,i\}}(k) + \mathbf{u}_{gr} + \mathbf{u}_{i\ell})$, $(\mathbf{z}_{M\setminus\{p,g,r\}}(k) + \mathbf{z}_{pgi}(r))$, $(\mathbf{z}_{M\setminus\{p,g,i\}}(k) + \mathbf{z}_{pg}(r) + \mathbf{u}_{i\ell}) \in F$ for $k \neq \ell \in N \setminus \{r\}$. Thus except for β_{pgr} all other β's are equal to zero.

Hence $\boldsymbol{\alpha}\mathbf{x} + \boldsymbol{\beta}\mathbf{y} \leq \gamma$ becomes $\sum_{s \in M} \sum_{k \in N} \omega_s x_{sk} - \beta_{pgr}(x_{pr} - y_{pgr}) \leq \sum_{s \in M} \omega_s$ or equivalently, $-\beta_{pgr}(x_{pr} - y_{pgr}) \leq 0$. Since F is a proper face of QSP_n^m and $-x_{pr} + y_{pgr} \leq 0$ valid for QSP_n^m, $\beta_{pgr} > 0$. Taking $\pi = -\beta_{pgr}$, the proposition follows. $\square$

Proposition 6.5 *The inequalities* $x_{pr} + x_{gr} - y_{pgr} + \sum_{h \in N\setminus\{r\}} y_{pgh} \leq 1$ *define facets of QSP_n^m for all $p, g \in M$ and $r \in N$.*

Proof. Let $F = \{(\mathbf{x}, \mathbf{y}) \in QSP_n^m : x_{pr} + x_{gr} - y_{pgr} + \sum_{h \in N\setminus\{r\}} y_{pgh} = 1\}$. Since $(\mathbf{z}_{M\setminus\{p,g\}}(r) + \mathbf{u}_{pk} + \mathbf{u}_{g\ell}) \in QSP_n^m$ for $(i, r) \in S_r$ and $k \neq \ell \in N \setminus \{r\}$ but not in F, F is a proper face of QSP_n^m. Suppose there exists a valid inequality $\boldsymbol{\alpha}\mathbf{x} + \boldsymbol{\beta}\mathbf{y} \leq \gamma$ for QSP_n^m such that every $(\mathbf{x}, \mathbf{y}) \in F$ satisfies $\boldsymbol{\alpha}\mathbf{x} + \boldsymbol{\beta}\mathbf{y} = \gamma$. To prove the proposition we need to show that $(\boldsymbol{\alpha}, \boldsymbol{\beta}, \gamma) = (\sum_s \omega_s \mathbf{e}_s + \pi(\overline{\mathbf{u}}_{pr} + \overline{\mathbf{u}}_{gr}), -\pi(\overline{\mathbf{V}}_{pgr} - \sum_{h \in N\setminus\{r\}} \overline{\mathbf{V}}_{pgh}), \sum_s \omega_s + \pi)$ where $\mathbf{e}_s \in \mathbb{R}^{mn}$ is a vector with one in its (s, ℓ) components for all $\ell \in N$ and zero elsewhere, $\pi \in \mathbb{R}^1$ and $\omega_s \in \mathbb{R}^1$ are constants for $s \in M$.

(i) Since $(\mathbf{z}_{M\setminus\{p,g\}}(k) + \mathbf{z}_{pg}(r)), (\mathbf{z}_{M\setminus\{p,g\}}(k) + \mathbf{u}_{p\ell} + \mathbf{u}_{gr}) \in F$ for $k \neq \ell \in N \setminus \{r\}$, $\alpha_{pr} = \alpha_{p\ell} - \beta_{pgr}$ for $\ell \in N \setminus \{r\}$ and likewise, $\alpha_{gr} = \alpha_{g\ell} - \beta_{pgr}$ for $\ell \in N \setminus \{r\}$.

(ii) Since $(\mathbf{z}_{M\setminus\{i\}}(k) + \mathbf{u}_{ir}), (\mathbf{z}_{M\setminus\{i\}}(k) + \mathbf{u}_{i\ell}) \in F$ for $i \in M \setminus \{p, g\}$, $k \neq \ell \in N \setminus \{r\}$, $\alpha_{ir} = \alpha_{i\ell}$ for $i \in M \setminus \{p, g\}$, $\ell \in N \setminus \{r\}$.

(iii) Since $(\mathbf{z}_{M\setminus\{i,j\}}(r) + \mathbf{z}_{ij}(k)), (\mathbf{z}_{M\setminus\{i,j\}}(r) + \mathbf{u}_{i\ell} + \mathbf{u}_{ik}) \in F$ for $i < j \in M \setminus \{p, g\}$, $k \neq \ell \in N \setminus \{r\}$, $\beta_{ijk} = 0$ for $i < j \in M \setminus \{p, g\}$, $k \in N \setminus \{r\}$.

(iv) Since $(\mathbf{z}_{M\setminus\{pg\}}(k) + \mathbf{z}_{pg}(\ell)), (\mathbf{z}_{M\setminus\{pg\}}(k) + \mathbf{z}_{pg}(r)) \in F$ for $k \neq \ell \in N \setminus \{r\}$, $\beta_{pg\ell} = -\beta_{pgr}$ for $\ell \in N \setminus \{r\}$.

Hence $\boldsymbol{\alpha}\mathbf{x} + \boldsymbol{\beta}\mathbf{y} \leq \gamma$ becomes $\sum_{s \in M} \sum_{k \in N} \omega_s x_{sk} + \beta_{pgr}(x_{pr} + x_{gr} - y_{pgr} + \sum_{h \in N\setminus\{r\}} y_{pgh}) \leq \sum_{s \in M} \omega_s + \beta_{pgr}$ or equivalently, $\beta_{pgr}(x_{pr} + x_{gr} - y_{pgr} + \sum_{h \in N\setminus\{r\}} y_{pgh}) \leq \beta_{pgr}$. Since F is a proper face of QSP_n^m and $x_{pr} + x_{gr} - y_{pgr} + \sum_{h \in N\setminus\{r\}} y_{pgh} \leq 1$ valid for QSP_n^m, $\beta_{pgr} > 0$. Taking $\pi = \beta_{pgr}$, the proposition follows. $\square$

Quadratic Scheduling Problems

To analyze QSP_n^m, we associate to our problem an undirected graph $G = (V, E)$ with mn vertices and $mn(m-1)/2$ edges. Every vertex $(i, j) \in V$ corresponds to a variable x_{ij} and an edge between a pair of nodes (i, j) and (k, j) to a variable y_{ikj} for $1 \leq i < k \leq m$ and $1 \leq j \leq n$; i.e. there is an edge between nodes (i, j) and (k, ℓ) if and only if $i \neq k$ and $j = \ell$. For $r \in N$, let $V_r = \{(i, r) : i \in M\}$. Evidently, $\bigcup_{r \in N} V_r = V$. For any valid inequality $\alpha x + \beta y \leq \gamma$ of QSP_n^m we denote by $G(\alpha, \beta) = (V(\alpha, \beta), E(\alpha, \beta))$ its *minimal support graph* where $E(\alpha, \beta) = \{e \in E : \beta_e \neq 0\}$ and $V(\alpha, \beta)$ is the subset of vertices of G spanned by $E(\alpha, \beta)$. The following lemma states two elementary properties of the support graph of facet inducing inequalities of QSP_n^m.

Lemma 6.1 *If $\alpha x + \beta y \leq \gamma$ defines a facet of QSP_n^m, then*
(i) $\beta_e \neq 0$ for at least one $e \in E$.
(ii) $\alpha x + \beta y \leq \gamma$ is of the form (6.11), (6.12) or (6.13) if $|V(\alpha, \beta)| \leq 2$.

Proof. (i) Suppose not. Then $\alpha x + \beta y \leq \gamma$ becomes $\alpha x \leq \gamma$. Since $max\{\alpha x : (x, y) \in QSP_n^m\} = \sum_{i \in M} max\{\alpha_{ij} : j \in N\}$, it follows that $\gamma \geq max\{\alpha_{ij} : j \in N\}$. Hence $\alpha x + \beta y \leq \gamma$ is implied by a linear combination of the inequalities $x_{ij} \leq 1$ for $i \in M$, $j \in N$. These are implied by (6.9), ..., (6.13) and hence, so is $\alpha x + \beta y \leq \gamma$.
(ii) By (i) $|V(\alpha, \beta)| \neq 1$. Assume $|V(\alpha, \beta)| = 2$. By (i) $|E(\alpha, \beta)| \geq 1$ and $\alpha x + \beta y = \alpha_{ij} x_{ij} + \alpha_{kj} x_{kj} + \beta_{ikj} y_{ikj}$ with $\beta_{ikj} \neq 0$. Suppose the lemma is not true. Since $m \geq n \geq 3$ there exist $(x, y) \in QSP_n^m$ with $x_{ij} = x_{kj} = y_{ikj} = 0$ and thus $\gamma \geq 0$. By assumption $\alpha x + \beta y \leq \gamma$ is different from (6.13) and thus there exists $(x, y) \in QSP_n^m$ with $y_{ikj} = 1$ and $\alpha x + \beta y = \gamma$. By (6.11) and (6.12) $x_{ij} = x_{kj} = 1$ for such $(x, y) \in QSP_n^m$ and thus $\alpha_{ij} + \alpha_{kj} + \beta_{ikj} = \gamma$. Likewise, since $\alpha x + \beta y \leq \gamma$ is different from (6.11) and (6.12) we conclude that $\alpha_{ij} = \alpha_{ik} = \gamma$ and thus $\beta_{ikj} = -\gamma$ with $\gamma > 0$. Consequently, $\alpha x + \beta y \leq \gamma$ is a positive multiple of the inequality $x_{ij} + x_{kj} - y_{ikj} \leq 1$, which is dominated by (6.10) and thus not a facet of QSP_n^m. □

To show that the facet-defining clique and cut inequalities of the Boolean quadric polytope, see Padberg [1989], extend naturally to the quadratic scheduling polytope QSP_n^m we introduce some notation. For $S_r \subseteq V_r$ and $T_r \subseteq V - S_r$ we let
$$E(S_r) = \{((i, r), (j, r)) : (i, r) \in S_r, (j, r) \in S_r\},$$
$$(S_r : T_r) = \{((i, r), (j, r)) : (i, r) \in S_r, (j, r) \in T_r\},$$
$$x(S_r) = \sum_{(i,r) \in S_r} x_{ir}, \quad y(E(S_r)) = \sum_{e \in E(S_r)} y_e.$$

Lemma 6.2 *For $S_r \subseteq V_r$ and integer α the* clique inequality
$$\alpha x(S_r) - y(E(S_r)) \leq \alpha(\alpha + 1)/2 \tag{6.16}$$

is valid for QSP_n^m, where $r \in N$ is arbitrary.

Proof. For any zero-one point $(\mathbf{x},\mathbf{y}) \in QSP_n^m$ let $\mu = |S_r \cap \{(i,r) \in V_r : x_{ir} = 1\}|$. We calculate $\alpha\mathbf{x}(S_r) - \mathbf{y}(E(S_r)) - \alpha(\alpha+1)/2 = \alpha\mu - \mu(\mu-1)/2 - \alpha(\alpha+1)/2 = -(\alpha-\mu)(\alpha+1-\mu)/2 \leq 0$ for all integer α and μ. Since all extreme points of the polytope QSP_n^m are zero-one, it follows that (6.16) is valid for QSP_n^m, no matter what $r \in N$. □

For $|S_r| = 2$ and $\alpha = 1$ the clique inequality (6.16) is *dominated* by (6.10).

Proposition 6.6 *The clique inequality (6.16) with $\alpha = 1$ defines a facet of QSP_n^m for any $r \in N$ and $S_r \subseteq V_r$ with $|S_r| \geq 3$.*

Proof. Let $F = \{(\mathbf{x},\mathbf{y}) \in QSP_n^m : \mathbf{x}(S_r) - \mathbf{y}(E(S_r)) = 1\}$. Since $(\mathbf{z}_M(k)) \in QSP_n^m$ for $k \in N\setminus\{r\}$ but not in F, F is a proper face of QSP_n^m. Suppose there exists a valid inequality $\alpha\mathbf{x} + \beta\mathbf{y} \leq \gamma$ for QSP_n^m such that every $(\mathbf{x},\mathbf{y}) \in F$ satisfies $\alpha\mathbf{x} + \beta\mathbf{y} = \gamma$. To prove the theorem we need to show that $(\boldsymbol{\alpha}, \boldsymbol{\beta}, \gamma) = (\sum_s \omega_s \mathbf{e}_s + \pi \sum_{(p,r) \in S_r} \overline{\mathbf{u}}_{pr}, -\pi \sum_{(p,g,r) \in E(S_r)} \overline{\mathbf{v}}_{pgr}, \sum_s \omega_s + \pi)$ where $\mathbf{e}_s \in \mathbb{R}^{mn}$ a vector with one in its (s,r) components for $r \in N$ and zero elsewhere, $\pi \in \mathbb{R}^1$ and $\omega_s \in \mathbb{R}^1$ are constants for $s \in M$.

(i) Since $(\mathbf{z}_{pg}(r) + \mathbf{z}_{M\setminus\{p,g\}}(k))$, $(\mathbf{u}_{pr} + \mathbf{u}_{g\ell} + \mathbf{z}_{M\setminus\{p,g\}}(k)) \in F$ for $(p,r), (g,r) \in S_r$, $k \neq \ell \in N \setminus \{r\}$, $\alpha_{gr} + \beta_{pgr} = \alpha_{g\ell}$ for $(p,r), (g,r) \in S_r$, and $k \neq \ell \in N \setminus \{r\}$.

(ii) Since $(\mathbf{z}_{pi}(r) + \mathbf{z}_{M\setminus\{p,i\}}(k))$, $(\mathbf{u}_{pr} + \mathbf{u}_{i\ell} + \mathbf{z}_{M\setminus\{p,i\}}(k)) \in F$ for $(p,r) \in S_r$, $(i,r) \notin S_r$, $k \neq \ell \in M \setminus \{r\}$, $\alpha_{ir} + \beta_{pir} = \alpha_{i\ell}$ for $(p,r) \in S_r$, $(i,r) \notin S_r$, and $k \neq \ell \in N \setminus \{r\}$.

(iii) Since $(\mathbf{z}_{pij}(r) + \mathbf{z}_{M\setminus\{p,i,j\}}(k))$, $(\mathbf{z}_{pj}(r) + \mathbf{u}_{i\ell} + \mathbf{z}_{M\setminus\{p,i,j\}}(k)) \in F$ for $(p,r) \in S_r$, $(i,r), (j,r) \notin S_r$, $k \neq \ell \in M \setminus \{r\}$, $\alpha_{ir} + \beta_{pir} + \beta_{ijr} = \alpha_{i\ell}$ and hence $\beta_{ijr} = 0$ for $(p,r) \in S_r$, $(i,r), (j,r) \notin S_r$, and $k \neq \ell \in N \setminus \{r\}$.

(iv) Since $(\mathbf{z}_{pgi}(r) + \mathbf{z}_{M\setminus\{p,g,i\}}(k))$, $(\mathbf{z}_{pg}(r) + \mathbf{u}_{i\ell} + \mathbf{z}_{M\setminus\{p,g,i\}}(k)) \in F$ for $(p,r), (g,r) \in S_r$, $(i,r) \notin S_r$, $k \neq \ell \in M \setminus \{r\}$, $\alpha_{ir} + \beta_{pir} + \beta_{gir} = \alpha_{i\ell}$ and hence $\beta_{gir} = 0$ for $(p,r), (g,r) \in S_r$, $(i,r) \notin S_r$, and $k \neq \ell \in N \setminus \{r\}$.

(v) Since $(\mathbf{z}_{gi}(\ell) + \mathbf{u}_{pr} + \mathbf{z}_{M\setminus\{p,g,i\}}(k))$, $(\mathbf{z}_{pg}(r) + \mathbf{u}_{i\ell} + \mathbf{z}_{M\setminus\{p,g,i\}}(k))$ for $(p,r), (g,r) \in S_r$, $(i,r) \notin S_r$, $k \neq \ell \in M \setminus \{r\}$, $\alpha_{g\ell} + \beta_{gi\ell} = \alpha_{gr} + \beta_{pgr}$ and hence $\beta_{gi\ell} = 0$ for (p,r) $(g,r) \in S_r$, $(i,r) \notin S_r$, and $k \neq \ell \in N \setminus \{r\}$.

(vi) Since $(\mathbf{z}_{ij}(\ell) + \mathbf{u}_{gr} + \mathbf{z}_{M\setminus\{g,i,j\}}(k))$, $(\mathbf{z}_{gi}(r) + \mathbf{u}_{j\ell} + \mathbf{z}_{M\setminus\{g,i,j\}}(k)) \in F$ for $(g,r) \in S_r$, $(i,r), (j,r) \notin S_r$, $k \neq \ell \in M \setminus \{r\}$, $\alpha_{i\ell} + \beta_{ij\ell} = \alpha_{ir} + \beta_{gir}$ and hence $\beta_{ij\ell} = 0$ for $(g,r) \in S_r$, $(i,r), (j,r) \notin S_r$, and $k \neq \ell \in N \setminus \{r\}$.

Hence $\alpha\mathbf{x} + \beta\mathbf{y} \leq \gamma$ becomes $\sum_{s \in M} \sum_{k \in N} \omega_s x_{sk} + \beta_{pgr}(\mathbf{x}(S_r) - \mathbf{y}(E(S_r))) \leq \omega_s + \beta_{pgr}$ or equivalently, $\beta_{pgr}(\mathbf{x}(S_r) - \mathbf{y}(E(S_r))) \leq \beta_{pgr}$. Since F is a proper face of QSP_n^m and $\mathbf{x}(S_r) - \mathbf{y}(E(S_r)) \leq 1$ valid for QSP_n^m, $\beta_{pgr} > 0$. Taking $\pi = \beta_{pgr}$, the theorem follows. □

Lemma 6.3 For $S_r \subseteq V_r$ with $|S_r| \geq 1$ and $T_r \subseteq V_r - S_r$ with $T_r \geq 2$ the cut inequality
$$-\mathbf{x}(S_r) - \mathbf{y}(E(S_r)) + \mathbf{y}(S_r : T_r) - \mathbf{y}(E(T_r)) \leq 0 \tag{6.17}$$
is valid for QSP_n^m, where $r \in N$ is arbitrary.

Proof. For any zero-one point $(\mathbf{x}, \mathbf{y}) \in QSP_n^m$ let $\mu = |S_r \cap \{(i,r) \in V_r : x_{ir} = 1\}|$ and $\nu = |T_r \cap \{(i,r) \in V_r : x_{ir} = 1\}|$. We calculate $-\mathbf{x}(S_r) - \mathbf{y}(E(S_r)) + \mathbf{y}(S_r : T_r) - \mathbf{y}(E(T_r)) = -\mu - \mu(\mu-1)/2 + \mu\nu - \nu(\nu-1)/2 = -(\nu-\mu)(\nu-\mu-1)/2 \leq 0$ for all integer μ and ν. Validity of (6.17) for the polytope QSP_n^m follows like in the proof of Lemma 6.2. $\square$

Proposition 6.7 The cut inequality (6.17) defines a facet of QSP_n^m for any $r \in N$ and $S_r \subseteq V_r$, $T_r \subseteq V_r - S_r$ with $|S_r| \geq 1$ and $|T_r| \geq 2$.

Proof. Let $F = \{(\mathbf{x},\mathbf{y}) \in QSP_n^m : -\mathbf{x}(S_r) - \mathbf{y}(E(S_r)) + \mathbf{y}(S_r : T_r) - \mathbf{y}(E(T_r)) = 0\}$. Since $(\mathbf{z}_{M \setminus \{i\}}(k) + \mathbf{u}_{ir}) \in QSP_n^m$ for $k \in N \setminus \{r\}$ but not in F, F is a proper face of QSP_n^m. Suppose there exists a valid inequality $\boldsymbol{\alpha}\mathbf{x} + \boldsymbol{\beta}\mathbf{y} \leq \gamma$ for QSP_n^m satisfied at equality by all $(\mathbf{x},\mathbf{y}) \in F$. To prove the theorem we need to show that $(\boldsymbol{\alpha}, \boldsymbol{\beta}, \gamma) = (\sum_s \omega_s \mathbf{e}_s + \pi \sum_{(p,r) \in S_r} \overline{\mathbf{u}}_{pr}, -\pi(\sum_{(p,r),(j,r) \in S_r} \overline{\mathbf{v}}_{pjr} + \sum_{(p,r) \in S_r (g,r) \in T_r} \overline{\mathbf{v}}_{pgr} - \sum_{(g,r),(i,r) \in T_r} \overline{\mathbf{v}}_{gir}), \sum_s \omega_s)$ where $\mathbf{e}_s \in \mathbb{R}^{mn}$ a vector with one in its (s, ℓ) components for all $\ell \in N$ and zero elsewhere, $\pi \in \mathbb{R}^1$ and $\omega_s \in \mathbb{R}^1$ are constants for $s \in M$.

(i) Since $(\mathbf{z}_{pg}(r) + \mathbf{z}_{M \setminus \{p,g\}}(k))$, $(\mathbf{u}_{p\ell} + \mathbf{u}_{gr} + \mathbf{z}_{M \setminus \{p,g\}}(k)) \in F$ for $(p,r) \in S_r$, $(g,r) \in T_r$, $k \neq \ell \in N \setminus \{r\}$, $\alpha_{pr} + \beta_{pgr} = \alpha_{p\ell}$ for $(p,r) \in S_r$, $(g,r) \in T_r$, and $k \neq \ell \in N \setminus \{r\}$.

(ii) Since $(\mathbf{u}_{g\ell} + \mathbf{z}_{M \setminus \{g\}}(k))$, $(\mathbf{u}_{gr} + \mathbf{z}_{M \setminus \{g\}}(k)) \in F$ for $(g,r) \in T_r$, $k \neq \ell \in N \setminus \{r\}$, $\alpha_{gr} = \alpha_{g\ell}$ for $(g,r) \in T_r$, and $k \neq \ell, \in N \setminus \{r\}$.

(iii) Since $(\mathbf{z}_{pgi}(r) + \mathbf{z}_{M \setminus \{p,g,i\}}(k))$, $(\mathbf{z}_{pi}(r) + \mathbf{u}_{g\ell} + \mathbf{z}_{M \setminus \{p,g,i\}}(k)) \in F$ for $(p,r) \in S_r$, (g,r), $(i,r) \in T_r$, $k \neq \ell \in N \setminus \{r\}$, $\alpha_{gr} + \beta_{pgr} + \beta_{gir} = \alpha_{g\ell}$ and hence $\beta_{pgr} = -\beta_{gir}$ for $(p,r) \in S_r$, (g,r), $(i,r) \in T_r$, and $k \neq \ell \in N \setminus \{r\}$.

(iv) Since $(\mathbf{z}_{pgij}(r) + \mathbf{z}_{M \setminus \{p,g,i,j\}}(k))$, $(\mathbf{z}_{gij}(r) + \mathbf{u}_{p\ell} + \mathbf{z}_{M \setminus \{p,g,i,j\}}(k)) \in F$ for (p,r), $(j,r) \in S_r$, (g,r), $(i,r) \in T_r$ and $k \neq \ell \in N \setminus \{r\}$, $\alpha_{pr} + \beta_{pgr} + \beta_{pir} + \beta_{pjr} = \alpha_{p\ell}$ and hence $\beta_{pjr} = -\beta_{pir}$ for (p,r), $(j,r) \in S_r$, (g,r), $(i,r) \in T_r$, and $k \neq \ell \in N \setminus \{r\}$.

(v) Since $(\mathbf{z}_{pgh}(r) + \mathbf{z}_{M \setminus \{p,g,h\}}(k))$, $(\mathbf{z}_{gh}(r) + \mathbf{u}_{p\ell} + \mathbf{z}_{M \setminus \{p,g,h\}}(k)) \in F$ for $(p,r) \in S_r$, $(g,r) \in T_r$, $(h,r) \notin (S_r \cup T_r)$ and $k \neq \ell \in N \setminus \{r\}$, $\beta_{phr} = 0$ for $(p,r) \in S_r$, $(h,r) \notin (S_r \cup T_r)$ and $k \neq \ell \in N \setminus \{r\}$.

(vi) Since $(\mathbf{u}_{h\ell} + \mathbf{z}_{M \setminus \{h\}}(k))$, $(\mathbf{u}_{hr} + \mathbf{z}_{M \setminus \{h\}}(k)) \in F$ for $(h,r) \notin (S_r \cup T_r)$ and $k \neq \ell \in N \setminus \{r\}$, $\alpha_{hr} = \alpha_{h\ell}$ for $(h,r) \notin (S_r \cup T_r)$, and $k \neq \ell \in N \setminus \{r\}$.

(vii) Since $(\mathbf{z}_{dh}(r) + \mathbf{z}_{M\setminus\{d,h\}}(k))$, $(\mathbf{u}_{dr} + \mathbf{u}_{h\ell} + \mathbf{z}_{M\setminus\{d,h\}}(k)) \in F$ for (d,r), $(h,r) \notin (S_r \cup T_r)$ and $k \neq \ell \in N\setminus\{r\}$, $\beta_{dhr} = 0$ for (d,r), $(h,r) \notin (S_r \cup T_r)$, and $k \neq \ell \in N\setminus\{r\}$.

(viii) Since $(\mathbf{z}_{gh}(r) + \mathbf{z}_{M\setminus\{g,h\}}(k))$, $(\mathbf{u}_{gr} + \mathbf{u}_{h\ell} + \mathbf{z}_{M\setminus\{g,h\}}(k)) \in F$ for $(g,r) \in T_r$, $(h,r) \notin (S_r \cup T_r)$ and $k \neq \ell \in N\setminus\{r\}$, $\beta_{ghr} = 0$ for $(g,r) \in T_r$, $(h,r) \notin (S_r \cup T_r)$, and $k \neq \ell \in N\setminus\{r\}$.

(ix) Since $(\mathbf{z}_{df}(\ell) + \mathbf{z}_{M\setminus\{d,f\}}(k))$, $(\mathbf{u}_{dr} + \mathbf{u}_{f\ell} + \mathbf{z}_{M\setminus\{d,f\}}(k)) \in F$ for $(d,r) \notin S_r$, $(f,r) \in V_r$ and $k \neq \ell \in N\setminus\{r\}$, $\beta_{df\ell} = 0$ for $(d,r) \notin S_r$, $(f,r) \in V_r$, and $k \neq \ell \in N\setminus\{r\}$.

(x) Since $(\mathbf{z}_{pgij}(r) + \mathbf{z}_{M\setminus\{p,g,i,j\}}(k))$, $(\mathbf{z}_{pij}(\ell) + \mathbf{u}_{gr} + \mathbf{z}_{M\setminus\{p,g,i,j\}}(k)) \in F$ for (p,r), $(j,r) \in S_r$, (g,r), $(i,r) \in T_r$ and $k \neq \ell \in N\setminus\{r\}$, $\beta_{pj\ell} = 0$ for (p,r), $(j,r) \in S_r$ and $k \neq \ell \in N\setminus\{r\}$.

Hence $\boldsymbol{\alpha}\mathbf{x} + \boldsymbol{\beta}\mathbf{y} \leq \gamma$ becomes $\sum_{s\in M}\sum_{k\in N}\omega_s x_{sk} + \beta_{pgr}(-\mathbf{x}(S_r) - \mathbf{y}(E(S_r)) + \mathbf{y}(S_r : T_r) - \mathbf{y}(E(T_r))) \leq \omega_s + \beta_{pgr}$ or equivalently, $\beta_{pgr}(\mathbf{x}(S_r) - \mathbf{y}(E(S_r))) \leq \beta_{pgr}$. Since F is a proper face of QSP_n^m and $-\mathbf{x}(S_r) - \mathbf{y}(E(S_r)) + \mathbf{y}(S_r : T_r) - \mathbf{y}(E(T_r)) \leq 0$ valid for QSP_n^m, $\beta_{pgr} > 0$. Taking $\pi = \beta_{pgr}$, the theorem follows. □

The facets that we have described in this section are – with the exception of inequalities (6.10) – "local" facets of the polytope QSP_n^m, because they correspond to configurations in a single connected component of the graph G associated with the OSP. While their number is important, see Padberg [1989] for a count of the clique and cut inequalities of the Boolean quadric polytope, different types of facets that like (6.10) tie the n components of the graph G together exist and can be expected to play a substantial role in numerical computations for this class of scheduling problems.

7

QUADRATIC ASSIGNMENT POLYTOPES

In this chapter we present various results and partial results on the facial structure of the quadratic assignment polytope QAP_n and its symmetric relative, the polytope SQP_n. We address primarily the questions of finding the affine hull and the dimension of the respective polytopes, but give also some valid inequalities for QAP_n. Some of these problems are left open and suggested in the form of conjectures for future work on this difficult, but interesting class of combinatorial optimization problems.

7.1 The Affine Hull and Dimension of QAP_n

In Chapter 5.3 we have formulated the quadratic assignment problem with $2n+n(n-1)(2n-1)$ equations in $n^2+n^2(n-1)^2/2$ nonnegative variables of which n^2 are required to be zero or one, see (5.26), ..., (5.32) and Proposition 5.22. Our formulation is related to, but shorter than the formulation of the QAP studied recently by Resende et al. [1994] which has $2n + 2n^2(n-1)$ equations. Their formulation is obtained from (5.26), ..., (5.32) by replacing $1 \leq k < i \leq n-1$ in (5.29) by $1 \leq k < i \leq n$. As we shall see in this section, their system of equations is *highly* redundant and even our formulation can be shortened somewhat by studying the *rank* of the system of equations. More precisely, $3n(n-1) + 2$ equations of the formulation due to Resende et al. [1994] can be dropped this way. The resulting smaller system of equations is an ideal, i.e. minimal and complete, linear description of the *affine hull* of the quadratic assignment polytope QAP_n for all $n \geq 3$. The case $n = 2$ is trivial.

Whenever one deals with a huge system of equations and seeks to find a minimal, linearly independent subsystem of it, there are typically many choices to take. The *art of research* consists in this case of finding a suitable subsystem that is tractable. We propose the following subset of equations in nonnegative/zero-one variables which we shall show to do the job.

$$\sum_{j=1}^{n} x_{ij} = 1 \qquad \text{for } 1 \leq i \leq n \qquad (7.1)$$

$$\sum_{i=1}^{n} x_{ij} = 1 \qquad \text{for } 1 \leq j \leq n-1 \qquad (7.2)$$

$$-x_{k\ell} + \sum_{i=1}^{k-1} y_{ij}^{k\ell} + \sum_{i=k+1}^{n} y_{k\ell}^{ij} = 0 \qquad \begin{array}{l} \text{for } 1 \leq j \neq \ell \leq n, 1 \leq k \leq n-1 \\ \text{and } 1 \leq \ell < j \leq n, k = n \end{array} \qquad (7.3)$$

$$-x_{ij} + \sum_{\ell=1}^{j-1} y_{ij}^{k\ell} + \sum_{\ell=j+1}^{n} y_{ij}^{k\ell} = 0 \qquad \begin{array}{l} \text{for } 1 \leq j \leq n, 1 \leq i \leq n-3, \\ i < k \leq n-1 \\ \text{and } 1 \leq j \leq n-1, i = n-2, \\ k = n-1 \end{array} \qquad (7.4)$$

$$-x_{kj} + \sum_{\ell=1}^{j-1} y_{i\ell}^{kj} + \sum_{\ell=j+1}^{n} y_{i\ell}^{kj} = 0 \qquad \begin{array}{l} \text{for } 1 \leq j \leq n-1, 1 \leq i \leq n-3, \\ i < k \leq n-1 \end{array} \qquad (7.5)$$

$$y_{ij}^{k\ell} \geq 0 \qquad \text{for } 1 \leq i < k \leq n, 1 \leq j \neq \ell \leq n \qquad (7.6)$$

$$x_{ij} \in \{0, 1\} \qquad \text{for } 1 \leq i, j \leq n, \qquad (7.7)$$

Counting the equations, we get $2n-1$ from (7.1) and (7.2), $n(n-1)^2+n(n-1)/2$ from (7.3), $n(n-1)^2/2 - n(n-1)/2 - 1$ from (7.4) and $n(n-1)^2/2 - n(n-1)$ from (7.5). Thus the total number of equations equals $2n(n-1)^2-(n-1)(n-2)$ and the number of variables appearing in (7.1),...,(7.5) is $n^2 + n^2(n-1)^2/2$.

Proposition 7.1 *The rank of* (7.1),...,(7.5) *equals* $2n(n-1)^2-(n-1)(n-2)$ *for all* $n \geq 3$.

Proof. For $n = 3$ we compute the rank of (7.1),...,(7.5) to be 22, for $n = 4$ we compute the rank to be 66 and thus the proposition is correct for $3 \leq n \leq 4$. Assume that $n \geq 5$. We partition (7.1),...,(7.5) into ten blocks (B1), ...,

(B10) as follows.

(B1) $$\sum_{j=1}^{n} x_{nj} = 1$$

$$-x_{k\ell} + \sum_{i=1}^{k-1} y_{ij}^{k\ell} + \sum_{i=k+1}^{n} y_{k\ell}^{ij} = 0 \quad \text{for } 1 \leq \ell < j \leq n, \; 1 \leq k \leq n-1$$

(B2) $$-x_{n-2,n} + \sum_{i=1}^{n-3} y_{ij}^{n-2,n} + \sum_{i=n-1}^{n} y_{n-2,n}^{ij} = 0 \quad \text{for } 1 \leq j \leq n-1$$

(B3) $$\sum_{j=1}^{n} x_{n-2,j} = 1$$

$$-x_{n\ell} + \sum_{i=1}^{n-1} y_{in}^{n\ell} = 0 \quad \text{for } 1 \leq \ell \leq n-1$$

$$-x_{ij} + \sum_{\ell=1}^{j-1} y_{ij}^{n-2,\ell} + \sum_{\ell=j+1}^{n} y_{ij}^{n-2,\ell} = 0 \quad \text{for } 1 \leq j \leq n-1, \; 1 \leq i \leq n-3$$

(B4) $$\sum_{i=1}^{n} x_{i,n-1} = 1$$

$$-x_{kn} + \sum_{i=1}^{k-1} y_{ij}^{kn} + \sum_{i=k+1}^{n} y_{kn}^{ij} = 0 \quad \text{for } 1 \leq j \leq n-1, \; 1 \leq k \leq n-1, \; k \neq n-2$$

(B5) $$\sum_{j=1}^{n} x_{n-1,j} = 1$$

$$-x_{ij} + \sum_{\ell=1}^{j-1} y_{i\ell}^{kj} + \sum_{\ell=j+1}^{n} y_{ij}^{k\ell} = 0 \quad \begin{array}{l} \text{for } 1 \leq j \leq n-1, \\ 1 \leq i < k \leq n-3 \\ \text{and } 1 \leq j \leq n-1, \\ 1 \leq i \leq n-2, \\ k = n-1 \end{array}$$

(B6) $$-x_{k\ell} + \sum_{i=1}^{k-1} y_{ij}^{k\ell} + \sum_{i=k+1}^{n} y_{k\ell}^{ij} = 0 \quad \text{for } 1 \leq j < \ell \leq n-1, \; n-2 \leq k \leq n-1$$

(B7) $$-x_{n\ell} + \sum_{i=1}^{n-1} y_{ij}^{n\ell} = 0 \quad \text{for } 1 \leq \ell < j \leq n-1$$

$$-x_{kj} + \sum_{\ell=1}^{j-1} y_{i\ell}^{kj} + \sum_{\ell=j+1}^{n} y_{i\ell}^{kj} = 0 \quad \begin{array}{l} \text{for } 2 \leq j \leq n-1, \\ 1 \leq i \leq n-3, \\ n-2 \leq k \leq n-1 \end{array}$$

(B8) $$\sum_{i=1}^{n} x_{ij} = 1 \quad \text{for } 1 \leq j \leq n-2$$

$$-x_{k\ell} + \sum_{i=1}^{k-1} y_{ij}^{k\ell} + \sum_{i=k+1}^{n} y_{k\ell}^{ij} = 0 \quad \begin{array}{l} \text{for } 1 \leq j < \ell \leq n-1, \\ 1 \leq k \leq n-3 \end{array}$$

$$-x_{in} + \sum_{\ell=1}^{n-1} y_{in}^{k\ell} = 0 \quad \begin{array}{l} \text{for } 1 \leq i \leq n-3, \\ n-2 \leq k \leq n-1 \end{array}$$

(B9) $$\sum_{j=1}^{n} x_{ij} = 1 \quad \text{for } i=1 \text{ and } i = n-3$$

$$-x_{kj} + \sum_{\ell=1}^{j-1} y_{i\ell}^{kj} + \sum_{\ell=j+1}^{n} y_{i\ell}^{kj} = 0 \quad \begin{array}{l} \text{for } 1 \leq j \leq n-1, \\ 1 \leq i < k \leq n-3 \\ \text{and } j=1, 1 \leq i \leq n-3, \\ n-2 \leq k \leq n-1 \end{array}$$

(B10) $$\sum_{j=1}^{n} x_{ij} = 1 \quad \text{for } 2 \leq i \leq n-4$$

$$-x_{in} + \sum_{\ell=1}^{n-1} y_{in}^{k\ell} = 0 \quad \text{for } 1 \leq i < k \leq n-3.$$

Checking (7.1) and (7.2) we find that these equations are listed exactly once in (B1), ..., (B10). There are precisely $n(n-1)^2 + n(n-1)/2$ distinct equations (7.3), $n(n-1)^2/2 - n(n-1)/2 - 1$ distinct equations (7.4) and $n(n-1)^2/2 - n(n-1)$ distinct equations in (7.5) in (B1), ..., (B10). The total number of equations (B1), ..., (B10) equals $2n(n-1)^2 - (n-1)(n-2)$ and thus (B1), ..., (B10) is a partitioning of (7.1),...,(7.5) into ten disjoint blocks. Likewise, we partition the $n^2+n^2(n-1)^2/2$ variables of $(7.1),\ldots,(7.5)$ into eleven classes.

(C1) x_{nn}, $y_{k\ell}^{nj}$ for $1 \leq \ell < j \leq n, 1 \leq k \leq n-1$
(C2) $y_{n-2,n}^{n-1,j}$ for $1 \leq j \leq n-1$
(C3) $x_{n-2,n}$, $y_{n-2,n}^{n\ell}$ for $1 \leq \ell \leq n-1$,
     $y_{ij}^{n-2,n}$ for $1 \leq j \leq n-1, 1 \leq i \leq n-3$
(C4) $x_{n,n-1}$, y_{kn}^{nj} for $1 \leq j \leq n-1, 1 \leq k \leq n-1, k \neq n-2$
(C5) $x_{n-1,n}$, y_{ij}^{kn} for $1 \leq j \leq n-1, 1 \leq i < k \leq n-3$,
     $y_{ij}^{n-1,n}$ for $1 \leq j \leq n-1, 1 \leq i \leq n-2$
(C6) $y_{n-2,\ell}^{n-1,j}$ for $1 \leq j \neq \ell \leq n-1$
(C7) $y_{ij}^{n\ell}$ for $n-2 \leq i \leq n-1, 1 \leq \ell < j \leq n-1$,
     $y_{i\ell}^{kj}$ for $1 \leq i \leq n-3, n-2 \leq k \leq n-1, 1 \leq \ell < j \leq n-1$
(C8) x_{ij} for $n-2 \leq i \leq n-1, 2 \leq j \leq n-2$, x_{nj} for $1 \leq j \leq n-2$,
     $y_{k\ell}^{n1}$ for $2 \leq \ell \leq n-1, 1 \leq k \leq n-3$,
     $y_{k\ell}^{ij}$ for $2 \leq j \leq \ell \leq n-1, n-2 \leq i \leq n, 1 \leq k \leq n-3$,
     $y_{in}^{k\ell}$ for $2 \leq \ell \leq n-1, 1 \leq i \leq n-3, n-2 \leq k \leq n-1$
(C9) x_{1j} for $1 \leq j \leq n-1$, $x_{n-3,n}$,
     $y_{i\ell}^{kj}$ for $1 \leq j \neq \ell \leq n-1, 1 \leq i < k \leq n-3$,
     $y_{i\ell}^{k1}$ for $2 \leq \ell \leq n, 1 \leq i \leq n-3, n-2 \leq k \leq n-1$
(C10) x_{ij} for $2 \leq i \leq n-4, 1 \leq j \leq n-1$,
     $y_{in}^{k\ell}$ for $1 \leq \ell \leq n-1, 1 \leq i < k \leq n-3$
(C11) x_{in} for $1 \leq i \leq n-4$, $x_{n-3,j}$ for $1 \leq j \leq n-1$,
     x_{i1}, $x_{i,n-1}$ for $n-2 \leq i \leq n-1$.

There are precisely n^2 variables x_{ij} in (C1), ..., (C11) and none is repeated. There are precisely $n^2(n-1)^2/2$ variables $y_{ij}^{k\ell}$ in (C1), ..., (C10) and none is repeated. Consequently, we have a partitioning of all variables occurring in $(7.1),\ldots,(7.5)$ into eleven disjoint classes. From a case-by-case analysis it follows that the variables in class (Ci) occur in block (Bi), but not in the blocks (Bk) for $k > i$, where $1 \leq i \leq 10$. Starting with (C1) and repeating with (C2), etc. we can thus eliminate all variables in (Ci) for $1 \leq i \leq 10$ and reduce the system (B1), ..., (B10) to zero rows. Hence the equations $(7.1),\ldots,(7.5)$ contain – modulo row and column permutations – an upper triangular matrix of size $(2n(n-1)^2 - (n-1)(n-2))^2$ having all entries equal to one on the main diagonal.   □

To give an outline of a proof that $(7.1),\ldots,(7.5)$ is an ideal description of the affine hull of QAP_n for $n \geq 3$, we introduce some notation. Let

$$\mathbf{y}^\ell = (y_{11}^{n+1,\ell}, \ldots, y_{n1}^{n+1,\ell}, \ldots, y_{1n}^{n+1,\ell}, \ldots, y_{nn}^{n+1,\ell}) \in \mathbb{R}^{n(n-1)},$$

Quadratic Assignment Polytopes

where $1 \leq \ell \leq n$. It is understood that the components $y_{1\ell}^{n+1,\ell}, \ldots, y_{n\ell}^{n+1,\ell}$ for $1 \leq \ell \leq n$ are missing from $\mathbf{y}^\ell$ because the corresponding variables do not exist in (7.1), ..., (7.5). For $1 \leq j \leq n$ we form the following vectors

$$\mathbf{z}_j = (y_{1j}^{2,n+1}, \ldots, y_{1j}^{n,n+1}, y_{2j}^{3,n+1}, \ldots, y_{2j}^{n,n+1}, \ldots, y_{n-1,j}^{n,n+1}) \in \mathbb{R}^{n(n-1)/2},$$
$$\mathbf{z}^j = (y_{1,n+1}^{2j}, \ldots, y_{1,n+1}^{nj}, y_{2,n+1}^{3j}, \ldots, y_{2,n+1}^{nj}, \ldots, y_{n-1,n+1}^{nj}) \in \mathbb{R}^{n(n-1)/2},$$
$$\mathbf{x}^{n+1} = (x_{1,n+1}, \ldots, x_{n,n+1}, x_{n+1,1}, \ldots, x_{n+1,n}) \in \mathbb{R}^{2n},$$

all of which, including $\mathbf{y}^\ell$ for $1 \leq \ell \leq n$, are subvectors of $(\mathbf{x}, \mathbf{y}) \in QAP_{n+1}$.

Proposition 7.2 *(i) The dimension of QAP_n equals $1+(n-1)^2+n(n-1)(n-2)(n-3)/2$ for all $n \geq 3$.*
(ii) The inequalities (7.6) define distinct facets of QAP_n for all $n \geq 4$.

Sketch of proof. (i) By Proposition (7.1) we have that $dim QAP_n \leq n^2 + n^2(n-1)^2/2 - 2n(n-1)^2 + (n-1)(n-2) = 1+(n-1)^2+n(n-1)(n-2)(n-3)/2$ for all $n \geq 3$. To prove that $dim QAP_n \geq 1+(n-1)^2+n(n-1)(n-2)(n-3)/2$ we use induction on $n \geq 3$. For $n = 3$ the 6×6 matrix

$$\begin{pmatrix} 1 & 0 & 0 & 0 & 0 & 0 \\ 1 & 0 & 0 & 1 & 0 & 0 \\ 0 & 1 & 0 & 0 & 0 & 0 \\ 0 & 1 & 0 & 1 & 1 & 0 \\ 0 & 0 & 1 & 0 & 0 & 0 \\ 0 & 0 & 1 & 0 & 1 & 1 \end{pmatrix}$$

is a submatrix of the list of the $n! = 6$ zero-one points in QAP_3 corresponding to the variables $x_{11}, x_{12}, x_{13}, x_{23}, x_{31}, y_{13}^{22}$. This matrix is nonsingular, thus $dim QAP_3 = 5$ and hence part (i) follows for $n = 3$. Suppose (i) is true for some $n \geq 3$. For $n+1$ we partition the list of all $(n+1)!$ zero-one points in QAP_{n+1} into two classes according to $x_{n+1,n+1} = 1$ and $x_{n+1,n+1} = 0$, respectively. Since every $(\tilde{\mathbf{x}}, \tilde{\mathbf{y}}) \in QAP_n$, say, can be completed to $(\mathbf{x}, \mathbf{y}) \in QAP_{n+1}$ by setting $x_{n+1,n+1} = 1$, the n^2 variables $y_{ij}^{n+1,n+1}$ with $1 \leq i, j \leq n$ according to $\tilde{\mathbf{x}}$ and the remaining variables equal to zero, it follows from the inductive hypothesis that the rank of the list of zero-one points in QAP_{n+1} with $x_{n+1,n+1} = 1$ is at least $1 + (n-1)^2 + n(n-1)(n-2)(n-3)/2$. Moreover, in the above notation $\mathbf{y}^\ell = 0$, $\mathbf{z}^\ell = \mathbf{z}_\ell = 0$ for $1 \leq \ell \leq n$ and $\mathbf{x}^{n+1} = 0$ for all $(\mathbf{x}, \mathbf{y}) \in QAP_{n+1}$ with $x_{n+1,n+1} = 1$. To prove the assertion it thus suffices to show that the rank of the submatrix of the list of all zero-one points in QAP_{n+1} with $x_{n+1,n+1} = 0$ corresponding to the variables $\mathbf{y}^\ell$, $\mathbf{z}_\ell$ and $\mathbf{z}^\ell$ for $1 \leq \ell \leq n$ is at least $2n - 1 + 2n(n-1)(n-2)$. This follows because the two variable sets are disjoint, thus the ranks are additive and we get $1 + (n-1)^2 + n(n-1)(n-2)(n-3)/2 + 2n - 1 + 2n(n-1)(n-2) = 1 + n^2 + (n+1)n(n-1)(n-2)/2$

as required by the induction. The proof then constructively provides a list of $2n - 1 + 2n(n - 1)(n - 2)$ points $(\mathbf{x},\mathbf{y}) \in QAP_{n+1}$ with $x_{n+1,n+1} = 1$ satisfying $y_{11}^{n2} = 0$ except for one point on the list such that the resulting $(2n - 1 + 2n(n - 1)(n - 2)) \times (2n + 2n^2(n - 1))$ matrix is of full rank. The details of the proof are too lengthy to be reproduced here; see Rijal [1995].

(ii) By the construction of part (i) the $(n!) \times (n^2 + n^2(n - 1)^2/2)$ list of all $n!$ points $(\mathbf{x},\mathbf{y}) \in QAP_n$ for all $n \geq 4$ contains a nonsingular submatrix of size $((n - 1)^2 + n(n - 1)(n - 2)(n - 3)/2)^2$ such that e.g. $y_{11}^{n2} = 0$. Thus $y_{11}^{n2} \geq 0$ defines a facet of QAP_n for all $n \geq 4$. Consequently, by permuting all indices $1 \leq i \leq n$ and $1 \leq j \leq n$ as required, the assertion follows for all $n \geq 4$. □

Remark 7.1 *For $n = 3$, the system of equations (7.1),...,(7.5) and inequalities (7.6) is a complete description of QAP_3; i.e., the integrality requirement (7.7) can be dropped from QAP_3. However, this system of equations and inequalities is not minimal because the system of equations (7.1),...,(7.5) implies that $y_{1j}^{2\ell} = y_{1j}^{3r} = y_{2\ell}^{3r}$ for $1 \leq j, \ell, r \leq 3$ and $j \neq \ell \neq r$ and $j \neq r$. Using this relationship, it follows that an ideal linear description of QAP_3 is given by $QAP_3 = \{(\mathbf{x},\mathbf{y}) \in \mathbb{R}^{27} : (\mathbf{x},\mathbf{y}) \text{ satisfies } (7.1),...,(7.5) \text{ and } y_{1j}^{2\ell} \geq 0 \text{ for } 1 \leq j \neq \ell \leq 3\}$. There are 22 equations (7.1),...,(7.5) and 6 inequalities (7.6) in an ideal description of QAP_3. For $n \geq 4$ many more inequalities are needed to describe the polytope QAP_n completely.*

It follows from Proposition 7.2 that the $3n(n-1)+2$ additional equations used e.g. by Resende et al. [1994] are linear combinations of the equations (7.1), ..., (7.5) and thus redundant for the linear program that they wish to solve. For $n = 30$ this means that 2,612 equations of their formulation can be dropped without affecting the outcome, which is a substantial saving given the number of 49,648 equations (7.1),...,(7.5) in this case.

The *assignment polytope* AP_n of the linear assignment problem, see Chapter 2.3, is the set of nonnegative solutions to (7.1) and (7.2). Its dimension equals $(n-1)^2$ for all $n \geq 3$ and we have n^2 variables. Thus from Proposition 7.2(i) we see that the $n^2(n-1)^2/2$ $\mathbf{y}$-variables of the QAP result in a "dimensional gain" of only $1+n(n-1)(n-2)(n-3)/2$. Interpreting this observation geometrically for large n this means that the polytope QAP_n becomes "flatter and flatter" relative to the space of variables in which it is embedded. This fact may explain asymptotic results on the QAP, such as those reported in Burkard [1990], where it is shown that the relative difference between a worst and an optimal solution to QAPs becomes arbitrarily small with a probability tending rapidly to 1 as the problem size tends to infinity.

7.2 Some Valid Inequalities for QAP_n

Like we did in Chapter 5.2 we can adapt the clique and the cut inequalities of the Boolean quadric polytope, see Padberg [1989], to the quadratic assignment polytope QAP_n. To do so we associate to our problem an undirected graph $G = (V, E)$ with n^2 vertices and $n^2(n-1)^2/2$ edges. Every vertex $(i,j) \in V$ corresponds to a variable x_{ij} and *vice versa*, an edge $((i,j),(k,\ell)) \in E$ between a pair of nodes $(i,j) \in V$ and $(k,\ell) \in V$ to a variable $y_{ij}^{k\ell}$ and *vice versa*, where $1 \leq i < k \leq n$ and $1 \leq j \neq \ell \leq n$. By construction an edge between nodes (i,j) and (k,ℓ) of G exists if and only if $i \neq k$ and $j \neq \ell$. A *clique* in a graph is any maximal subset of nodes of the graph such that every pair of nodes in the subset is connected by an edge of the graph. *Maximality* means that no node outside of the clique is connected to all nodes in the clique by the edges of the graph. For $S \subseteq V$ let

$$E(S) = \{((i,j),(k,\ell)) \in E : (i,j) \in S, (k,\ell) \in S\}.$$

If $(S, E(S))$ is a clique in G, then it follows from the construction of G that $\mathbf{x} \in \mathbb{R}^{n^2}$ defined by $x_{ij} = 1$ for all $(i,j) \in S$, $x_{ij} = 0$ otherwise is an *assignment*, i.e. $\mathbf{x}$ satisfies (7.1), (7.2) and (7.3). On the other hand, every assignment $\mathbf{x} \in \mathbb{R}^{n^2}$ gives rise to a clique in G and thus G has precisely $n!$ cliques all of which have exactly n nodes and $n(n-1)/2$ edges. For $S \subseteq V$ and $T \subseteq V - S$ we denote

$$(S:T) = \{((i,j),(k,\ell)) \in E : (i,j) \in S, (k,\ell) \in T\}, \quad \mathbf{x}(S) = \sum_{(i,j) \in S} x_{ij},$$

$$\mathbf{y}(E(S)) = \sum_{((i,j),(k,\ell)) \in E(S)} y_{ij}^{k\ell}, \quad \mathbf{y}(S:T) = \sum_{(i,j) \in S} \sum_{(k,\ell) \in T} y_{ij}^{k\ell}.$$

Lemma 7.1 *(i) For any $S \subseteq V$ and integer α the clique inequality*

$$\alpha \mathbf{x}(S) - \mathbf{y}(E(S)) \leq \alpha(\alpha+1)/2 \tag{7.8}$$

is satisfied by all $(\mathbf{x}, \mathbf{y}) \in QAP_n$. (ii) For any $S \subseteq V$ with $|S| \geq 1$ and $T \subseteq V - S$ with $|T| \geq 2$ the cut inequality

$$-\mathbf{x}(S) - \mathbf{y}(E(S)) + \mathbf{y}(S:T) - \mathbf{y}(E(T)) \leq 0 \tag{7.9}$$

is satisfied by all $(\mathbf{x}, \mathbf{y}) \in QAP_n$.

Proof. (i) For any zero-one point $(\mathbf{x}, \mathbf{y}) \in QAP_n$ let $\mu = |S \cap \{(i,j) \in V : x_{ij} = 1\}|$. Since $\mathbf{x}$ satisfies (7.1), (7.2) and (7.7) and $y_{ij}^{k\ell} = x_{ij} x_{k\ell}$ we calculate

$\alpha\mathbf{x}(S) - \mathbf{y}(E(S)) - \alpha(\alpha+1)/2 = \alpha\mu - \mu(\mu-1)/2 - \alpha(\alpha+1)/2 = -(\alpha-\mu)(\alpha+1-\mu)/2 \leq 0$ for all integer μ and integer α. Consequently, all extreme points of QAP_n satisfy (7.8) and thus (7.8) is valid for QAP_n.

(ii) For any zero-one point $(\mathbf{x}, \mathbf{y}) \in QAP_n$ we set $\mu = |S \cap \{(i,j) \in V : x_{ij} = 1\}|$ and $\nu = |T \cap \{(i,j) \in V : x_{ij} = 1\}|$. Since $\mathbf{x}$ satisfies (7.1), (7.2) and (7.7) we calculate as before $-\mathbf{x}(S) - \mathbf{y}(E(S)) + \mathbf{y}(S:T) - \mathbf{y}(E(T)) = -\mu - \mu(\mu-1)/2 + \mu\nu - \nu(\nu-1)/2 = -(\nu-\mu)(\nu-\mu-1)/2 \leq 0$ for all integer μ and ν. Validity of (7.9) for QAP_n follows like in the first part. $\square$

It is clear that not all clique and cut inequalities define facets of QAP_n. A complete study of when these inequalities define facets of the polytope is left for future work. For the cut inequalities we have derived conditions under which (7.9) does not define a facet of QAP_n. Let $N = \{1, \ldots, n\}$. For $T \subseteq V$ and $1 \leq i \leq n$ we define

$$T_i = \{j \in N : (i,j) \in T\}, \quad T^i = \{j \in N : (j,i) \in T\}.$$

Proposition 7.3 *The cut inequality (7.9) does not define a facet of QAP_n if any of the following conditions holds:*

(i) $S = \{(i,j)\}$ and $T \subseteq \{(k,\ell) : ((i,j),(k,\ell)) \in E, 1 \leq \ell \leq n\}$ for some $1 \leq k \leq n$ or $T \subseteq \{(k,\ell) : ((i,j),(k,\ell)) \in E, 1 \leq k \leq n\}$ for some $1 \leq \ell \leq n$ where $1 \leq i, j \leq n$.

(ii) $|T| = 2$.

(iii) $S = \{(i,j)\}$ and there exists $T' \subseteq T$ such that $T' = \{(k,\ell) : ((i,j),(k,\ell)) \in E, 1 \leq \ell \leq n\}$ for some $1 \leq i \neq k \leq n$ or $T' = \{(k,\ell) : ((i,j),(k,\ell)) \in E, 1 \leq k \leq n\}$ for some $1 \leq j \neq \ell \leq n$ where $1 \leq i, j \leq n$.

(iv) $|S| = 1$ and $T_i = T_k$ for all $1 \leq i \neq k \leq n$ and $T^j = T^\ell$ for all $1 \leq j \neq \ell \leq n$ such that $T_i \neq \emptyset \neq T_k$ and $T^j \neq \emptyset \neq T^\ell$ and $|T_i \cup T^j| \geq n$.

(v) There exist $S' \subseteq S$, $T' \subseteq T$ and $S' \cup T' \subset S \cup T$ such that $E(T') \cup (S' : T - T') \cup (T - T' : T') = \emptyset$ or $E(S') \cup (S - S' : S) \cup (S - S' : T') = \emptyset$ or $(S' : T - T') \cup (S - S' : T) = \emptyset$.

Proof. (i) If $i = k$ or $j = \ell$, then the cut inequality is of one of the forms

$$-\sum_{i \in N'} x_{ij} \leq 0, \quad -\sum_{j \in N'} x_{ij} \leq 0,$$

where $N' \subseteq N$. These inequalities can be obtained as a non-negative linear combination of $-x_{ig} \leq 0$ and $-x_{pj} \leq 0$ for $1 \leq p, g \leq n$ which are implied by (7.1), ..., (7.6). Hence, the cut inequalities satisfying the stated are not

Quadratic Assignment Polytopes   159

facet defining for QAP_n. Now assume $i \neq k$ and $j \neq \ell$. Then the cut inequality is of one of the following three forms

$$-x_{ij} + \sum_{k=1}^{i-1} y_{k\ell}^{ij} + \sum_{k=i+1}^{n} y_{ij}^{k\ell} \leq 0, \qquad -x_{ij} + \sum_{\ell=1}^{j-1} y_{k\ell}^{ij} + \sum_{\ell=j+1}^{n} y_{k\ell}^{ij} \leq 0,$$
$$-x_{ij} + \sum_{\ell=1}^{j-1} y_{ij}^{k\ell} + \sum_{\ell=j+1}^{n} y_{ij}^{k\ell} \leq 0,$$

or can be obtained as a non-negative linear combination of one of these inequalities with one or more of $-y_{ij}^{pg} \leq 0$ for $1 \leq i < p \leq n$ and $1 \leq j \neq g \leq n$. Since the inequalities given above are implied by (7.1),...,(7.6), it follows that cut inequalities satisfying the stated conditions do not define facets of QAP_n.
(ii) If T satisfies conditions (i), then there is nothing to be proved. So assume that T does not satisfy conditions (i) and WROG assume $T = \{(p,g),(r,s)\}$ and $1 \leq i \leq p < r \leq n$. Let $1 \leq j \neq g \leq n, 1 \leq j \neq s \leq n$. Then the cut inequality is of one of the following forms

$$-x_{ij} + y_{ij}^{rs} - y_{pg}^{rs} \leq 0, \qquad -x_{ij} + y_{ij}^{pg} + y_{ij}^{rs} - y_{pg}^{rs} \leq 0.$$

These inequalities are dominated by the cut inequality $-x_{ij} + y_{ij}^{pg} + y_{ij}^{rg} + y_{ij}^{rs} - y_{pg}^{rs} \leq 0$. Hence they do not define facets of QAP_n. A similar argument shows that if $1 \leq j = g \leq n$ or $1 \leq j = s \leq n$, then the cut inequality does not define a facet.
(iii) WROG assume $i = j = 1, T' = \{(2,2),(2,3),\ldots(2,n)\}$, and denote $R' \subseteq \{3,4,\ldots,n\}$ and $S_i = \{j : (i,j) \in T\}$. Then the cut inequality is given by

$$-x_{11} + \sum_{j=2}^{n} y_{11}^{2j} + \sum_{i \in R'} \sum_{1 \neq j \in S_i} y_{11}^{ij} - \sum_{j=2}^{n} \sum_{k \in R'} \sum_{j \neq \ell \in S_k} y_{2j}^{k\ell} - \sum_{i < k \in R'} \sum_{j \in S_i, j \neq \ell \in S_k} y_{ij}^{k\ell}$$

$$= \sum_{i \in R'} \sum_{1 \neq j \in S_i} y_{11}^{ij} - \sum_{j=2}^{n} \sum_{k \in R'} \sum_{j \neq \ell \in S_k} y_{2j}^{k\ell} - \sum_{i < k \in R'} \sum_{j \in S_i, j \neq \ell \in S_k} y_{ij}^{k\ell}$$

$$\leq \sum_{i \in R'} \sum_{1 \neq j \in S_i} y_{11}^{ij} - \sum_{j=2}^{n} \sum_{k \in R'} \sum_{j \neq \ell \in S_k} y_{2j}^{k\ell}$$

$$= \sum_{i \in R'} \sum_{j \in S_i} y_{11}^{ij} - \sum_{k \in R'} \sum_{j \neq \ell \in S_k} \left(x_{k\ell} - y_{21}^{k\ell} \right)$$

$$\leq - \sum_{k \in R'} \sum_{1 \neq j \in S_k} y_{11}^{k\ell}.$$

That is, the cut inequality satisfying conditions (iii) is dominated by a non-negative linear combination of a subset of $-y_{pg}^{rs} \leq 0$ for $1 \leq p < r$ and $1 \leq g \neq s \leq n$. Hence, it does not define facet of QAP_n. By a similar argument, if $i = j = 1, T' = \{(2,2),(3,2),\ldots(n,2)\}$, then the cut inequality does not define a facet of QAP_n.

(iv) WROG we assume $i = j = 1, T = \{(i,j) \in V : 2 \leq i \leq r, 2 \leq j \leq s\}$ and $r + s \geq n - 2$ to sketch the outline of the proof; see Rijal [1995] for detail. The cut inequality satisfies

$$-x_{11} + \sum_{k=2}^{r}\sum_{\ell=2}^{s} y_{11}^{k\ell} - \sum_{i=2}^{r-1}\sum_{k=i+1}^{r}\sum_{j=2}^{s}\sum_{j\neq\ell=2}^{s} y_{ij}^{k\ell}$$

$$= ((n-s)(n-s-1) - (r-2)(r-3))/2 + (r+s-n-2)(\sum_{i=2}^{r}\sum_{j=s+1}^{n} x_{ij} - \sum_{i=r+1}^{n} x_{i1})$$

$$- \sum_{k=r+1}^{n}\sum_{\ell=s+1}^{n}\sum_{\ell\neq j=s+1}^{n} y_{1\ell}^{kj} - \sum_{i=r+1}^{n-1}\sum_{k=i+1}^{n}\sum_{\ell=s+1}^{n}\sum_{\ell\neq j=s+1}^{n} y_{ij}^{k\ell}$$

$$\leq ((n-s)(n-s-1) - (r-2)(r-3))/2 + (r+s-n-2)\sum_{i=2}^{r}\sum_{j=s+1}^{n} x_{ij}$$

$$- \sum_{k=r+1}^{n}\sum_{\ell=s+1}^{n}\sum_{\ell\neq j=s+1}^{n} y_{1\ell}^{kj} - \sum_{i=r+1}^{n-1}\sum_{k=i+1}^{n}\sum_{\ell=s+1}^{n}\sum_{\ell\neq j=s+1}^{n} y_{ij}^{k\ell}$$

$$\leq ((n-s)(n-s-1) - (r-2)(r-3))/2 + (r+s-n-2)(n-s)$$

$$- \sum_{k=r+1}^{n}\sum_{\ell=s+1}^{n}\sum_{\ell\neq j=s+1}^{n} y_{1\ell}^{kj} - \sum_{i=r+1}^{n-1}\sum_{k=i+1}^{n}\sum_{\ell=s+1}^{n}\sum_{\ell\neq j=s+1}^{n} y_{ij}^{k\ell}$$

$$= -(n-s-r+2)(n-s-r+3)/2$$

$$- \sum_{k=r+1}^{n}\sum_{\ell=s+1}^{n}\sum_{\ell\neq j=s+1}^{n} y_{1\ell}^{kj} - \sum_{i=r+1}^{n-1}\sum_{k=i+1}^{n}\sum_{\ell=s+1}^{n}\sum_{\ell\neq j=s+1}^{n} y_{ij}^{k\ell}$$

$$\leq - \sum_{k=r+1}^{n}\sum_{\ell=s+1}^{n}\sum_{\ell\neq j=s+1}^{n} y_{1\ell}^{kj} - \sum_{i=r+1}^{n-1}\sum_{k=i+1}^{n}\sum_{\ell=s+1}^{n}\sum_{\ell\neq j=s+1}^{n} y_{ij}^{k\ell}.$$

That is, the cut inequality satisfying conditions (iv) is dominated by a nonnegative linear combination of a subset of $-y_{pg}^{rs} \leq 0$ for $1 \leq p < r$ and $1 \leq g \neq s \leq n$. Hence, it does not define a facet of QAP_n.

(v) Let $S_1 \subseteq S$, $T_1 \subseteq T$ and $S_1 \cup T_1 \subset S \cup T$; then the cut inequality can be written as:

$$\mathbf{x}(S) + \mathbf{y}(S:T) - \mathbf{y}(E(S)) - \mathbf{y}(E(T))$$
$$= \mathbf{x}(S_1) - \mathbf{x}(S - S_1) + \mathbf{y}(S_1:T_1) + \mathbf{y}(S - S_1:T_1)$$
$$+ \mathbf{y}(S:T - T_1) + \mathbf{y}(S - S_1:T - T_1) - \mathbf{y}(E(S_1)) - \mathbf{y}(E(S - S_1))$$
$$- \mathbf{y}(S_1:S - S_1) - \mathbf{y}(E(T_1)) - \mathbf{y}(E(T - T_1)) - \mathbf{y}(T_1:T - T_1)$$
$$\leq 0.$$

It follows that the inequality can either be obtained as or is dominated by a nonnegative linear combination of two cut inequalities defined on (i) S_1, T_1 and

Quadratic Assignment Polytopes

$S - S_1, T$ if $E(T_1) \cup (S_1 : T - T_1) \cup (T_1 : T - T_1) = \emptyset$; (ii) S_1, T_1 and $S : T - T_1$ if $E(S_1) \cup (S - S_1 : S_1) \cup (S - S_1 : T_1) = \emptyset$; and (iii) S_1, T_1 and $S - S_1, T - T_1$ if $(S_1 : T - T_1) \cup (S - S_1 : T_1) = \emptyset$. Hence it does not define a facet of QAP_n. □

We *conjecture* that all cut inequalities (7.9) except those shown not to be facet defining in Proposition 7.3 do indeed define facets of QAP_n.

Remark 7.2 *Dropping the integrality requirement from (7.1),...,(7.7) for $n = 4$ results in a polytope which has 148 fractional vertices in addition to the 24 integer vertices of QAP_4. For example, the non-zero components of a fractional vertex $(\mathbf{x}, \mathbf{y})$ to this system is given by $x_{ii} = .4$ for $1 \leq i \leq 4$, $x_{ij} = .2$ for $1 \leq i \neq j \leq 4$ and $y_{ik}^{ki} = y_{ii}^{kj} = y_{ij}^{kk} = .2$ for $1 \leq i < k \leq 4, 1 \leq j, \ell \leq n$ and $i \neq j \neq k$. The cut inequality $-x_{11} + y_{11}^{23} + y_{11}^{24} + y_{11}^{43} - y_{24}^{43} \leq 0$ cuts off this fractional vertex. Not only are the facet defining cut inequalities (7.9) sufficient to cut off all these 148 fractional vertices, but all of these cut inequalities together with (7.1),...,(7.6) also are a complete description of QAP_4. However, this system of equations and inequalities is not minimal because more than one cut inequality correspond to a facet of QAP_4. Let $T' = \{(i,j), (k,\ell), (p,r)\}$ for $2 \leq i < k \leq 4$ and $r = j$ if $p = k$ or $r = \ell$ if $p = i$ and $S' = \{(1,s)\}$ for $1 \leq s \leq 3, j \neq s \neq \ell$ and $s = 1$ if $j = 4$ or $\ell = 4$. Then the corresponding cut inequalities $-\mathbf{x}(S') + \mathbf{y}(S' : T') - \mathbf{y}(E(S')) - \mathbf{y}(E(T')) \leq 0$ suffice and together with (7.1),...,(7.6) an ideal description of QAP_4 is obtained. There are 66 equations (7.1),...,(7.5), 72 inequalities (7.6) and 72 such cut inequalities in an ideal description of QAP_4. An explicit listing of these cut inequalities is given in Table 7.1. For $n \geq 5$ many more inequalities (7.9) and many inequalities different from (7.9) are needed to describe QAP_n completely.*

7.3 The Affine Hull and Dimension of SQP_n

In Chapter 5.4 we have formulated the symmetric quadratic assignment problem as a mixed integer programming problem with $2n + n^2(n-1)$ equations in $n^2 + n^2(n-1)^2/4$ nonnegative variables of which n^2 must be zero-one valued, see Proposition 5.29. Now we address the issue of the minimality of the linear description of the affine hull of the associated polytope SQP_n. It appears that $n^2 + 1$ equations can be dropped from the formulation, which is considerable even for moderate values of n. Let $N = \{1, \ldots, n\}$. To support this statement

$$-x_{11} + y_{11}^{22} + y_{11}^{23} + y_{11}^{33} - y_{22}^{33} \le 0$$
$$-x_{11} + y_{11}^{22} + y_{11}^{23} + y_{11}^{43} - y_{22}^{43} \le 0$$
$$-x_{11} + y_{11}^{22} + y_{11}^{23} + y_{11}^{32} - y_{23}^{32} \le 0$$
$$-x_{11} + y_{11}^{22} + y_{11}^{23} + y_{11}^{42} - y_{23}^{42} \le 0$$
$$-x_{11} + y_{11}^{22} + y_{11}^{24} + y_{11}^{32} - y_{24}^{32} \le 0$$
$$-x_{11} + y_{11}^{22} + y_{11}^{24} + y_{11}^{42} - y_{24}^{42} \le 0$$
$$-x_{11} + y_{11}^{23} + y_{11}^{32} + y_{11}^{33} - y_{23}^{33} \le 0$$
$$-x_{11} + y_{11}^{32} + y_{11}^{33} + y_{11}^{43} - y_{32}^{43} \le 0$$
$$-x_{11} + y_{11}^{22} + y_{11}^{33} + y_{11}^{33} - y_{22}^{33} \le 0$$
$$-x_{11} + y_{11}^{32} + y_{11}^{33} + y_{11}^{42} - y_{33}^{42} \le 0$$
$$-x_{11} + y_{11}^{22} + y_{11}^{32} + y_{11}^{34} - y_{22}^{34} \le 0$$
$$-x_{11} + y_{11}^{32} + y_{11}^{34} + y_{11}^{42} - y_{34}^{42} \le 0$$
$$-x_{11} + y_{11}^{23} + y_{11}^{42} + y_{11}^{42} - y_{23}^{42} \le 0$$
$$-x_{11} + y_{11}^{33} + y_{11}^{42} + y_{11}^{43} - y_{33}^{42} \le 0$$
$$-x_{11} + y_{11}^{22} + y_{11}^{42} + y_{11}^{43} - y_{22}^{43} \le 0$$
$$-x_{11} + y_{11}^{32} + y_{11}^{42} + y_{11}^{43} - y_{32}^{43} \le 0$$
$$-x_{11} + y_{11}^{22} + y_{11}^{42} + y_{11}^{44} - y_{22}^{44} \le 0$$
$$-x_{11} + y_{11}^{32} + y_{11}^{42} + y_{11}^{44} - y_{32}^{44} \le 0$$
$$-x_{12} + y_{12}^{21} + y_{12}^{23} + y_{12}^{33} - y_{21}^{33} \le 0$$
$$-x_{12} + y_{12}^{21} + y_{12}^{23} + y_{12}^{43} - y_{21}^{43} \le 0$$
$$-x_{12} + y_{12}^{21} + y_{12}^{23} + y_{12}^{31} - y_{23}^{31} \le 0$$
$$-x_{12} + y_{12}^{21} + y_{12}^{24} + y_{12}^{31} - y_{24}^{31} \le 0$$
$$-x_{12} + y_{12}^{23} + y_{12}^{31} + y_{12}^{33} - y_{23}^{31} \le 0$$
$$-x_{12} + y_{12}^{31} + y_{12}^{33} + y_{12}^{43} - y_{31}^{43} \le 0$$
$$-x_{12} + y_{12}^{21} + y_{12}^{31} + y_{12}^{33} - y_{21}^{33} \le 0$$
$$-x_{12} + y_{12}^{21} + y_{12}^{31} + y_{12}^{34} - y_{21}^{34} \le 0$$
$$-x_{12} + y_{12}^{23} + y_{12}^{41} + y_{12}^{43} - y_{23}^{41} \le 0$$
$$-x_{12} + y_{12}^{33} + y_{12}^{41} + y_{12}^{43} - y_{33}^{41} \le 0$$
$$-x_{12} + y_{12}^{21} + y_{12}^{41} + y_{12}^{43} - y_{21}^{43} \le 0$$
$$-x_{12} + y_{12}^{21} + y_{12}^{41} + y_{12}^{44} - y_{21}^{44} \le 0$$
$$-x_{13} + y_{13}^{21} + y_{13}^{22} + y_{13}^{32} - y_{21}^{32} \le 0$$
$$-x_{13} + y_{13}^{21} + y_{13}^{22} + y_{13}^{31} - y_{22}^{31} \le 0$$
$$-x_{13} + y_{13}^{22} + y_{13}^{31} + y_{13}^{32} - y_{22}^{31} \le 0$$
$$-x_{13} + y_{13}^{21} + y_{13}^{31} + y_{13}^{32} - y_{21}^{32} \le 0$$
$$-x_{13} + y_{13}^{32} + y_{13}^{41} + y_{13}^{42} - y_{22}^{41} \le 0$$
$$-x_{13} + y_{13}^{21} + y_{13}^{41} + y_{13}^{42} - y_{21}^{42} \le 0$$

$$-x_{11} + y_{11}^{22} + y_{11}^{24} + y_{11}^{34} - y_{22}^{34} \le 0$$
$$-x_{11} + y_{11}^{22} + y_{11}^{24} + y_{11}^{44} - y_{22}^{44} \le 0$$
$$-x_{11} + y_{11}^{23} + y_{11}^{24} + y_{11}^{34} - y_{23}^{34} \le 0$$
$$-x_{11} + y_{11}^{23} + y_{11}^{24} + y_{11}^{44} - y_{23}^{44} \le 0$$
$$-x_{11} + y_{11}^{23} + y_{11}^{24} + y_{11}^{33} - y_{24}^{33} \le 0$$
$$-x_{11} + y_{11}^{23} + y_{11}^{24} + y_{11}^{43} - y_{24}^{43} \le 0$$
$$-x_{11} + y_{11}^{24} + y_{11}^{32} + y_{11}^{34} - y_{24}^{32} \le 0$$
$$-x_{11} + y_{11}^{32} + y_{11}^{34} + y_{11}^{44} - y_{32}^{44} \le 0$$
$$-x_{11} + y_{11}^{24} + y_{11}^{33} + y_{11}^{34} - y_{24}^{33} \le 0$$
$$-x_{11} + y_{11}^{33} + y_{11}^{34} + y_{11}^{44} - y_{33}^{44} \le 0$$
$$-x_{11} + y_{11}^{23} + y_{11}^{33} + y_{11}^{34} - y_{23}^{34} \le 0$$
$$-x_{11} + y_{11}^{33} + y_{11}^{34} + y_{11}^{43} - y_{34}^{43} \le 0$$
$$-x_{11} + y_{11}^{24} + y_{11}^{42} + y_{11}^{44} - y_{24}^{42} \le 0$$
$$-x_{11} + y_{11}^{34} + y_{11}^{42} + y_{11}^{44} - y_{34}^{42} \le 0$$
$$-x_{11} + y_{11}^{24} + y_{11}^{43} + y_{11}^{44} - y_{24}^{43} \le 0$$
$$-x_{11} + y_{11}^{34} + y_{11}^{43} + y_{11}^{44} - y_{34}^{43} \le 0$$
$$-x_{11} + y_{11}^{23} + y_{11}^{43} + y_{11}^{44} - y_{23}^{44} \le 0$$
$$-x_{11} + y_{11}^{33} + y_{11}^{43} + y_{11}^{44} - y_{33}^{44} \le 0$$
$$-x_{12} + y_{12}^{21} + y_{12}^{24} + y_{12}^{34} - y_{21}^{34} \le 0$$
$$-x_{12} + y_{12}^{21} + y_{12}^{24} + y_{12}^{44} - y_{21}^{44} \le 0$$
$$-x_{12} + y_{12}^{21} + y_{12}^{23} + y_{12}^{41} - y_{23}^{41} \le 0$$
$$-x_{12} + y_{12}^{21} + y_{12}^{24} + y_{12}^{41} - y_{24}^{41} \le 0$$
$$-x_{12} + y_{12}^{24} + y_{12}^{31} + y_{12}^{34} - y_{24}^{31} \le 0$$
$$-x_{12} + y_{12}^{31} + y_{12}^{34} + y_{12}^{44} - y_{31}^{44} \le 0$$
$$-x_{12} + y_{12}^{31} + y_{12}^{33} + y_{12}^{41} - y_{33}^{41} \le 0$$
$$-x_{12} + y_{12}^{31} + y_{12}^{34} + y_{12}^{41} - y_{34}^{41} \le 0$$
$$-x_{12} + y_{12}^{24} + y_{12}^{41} + y_{12}^{44} - y_{24}^{41} \le 0$$
$$-x_{12} + y_{12}^{34} + y_{12}^{41} + y_{12}^{44} - y_{34}^{41} \le 0$$
$$-x_{12} + y_{12}^{31} + y_{12}^{41} + y_{12}^{43} - y_{31}^{43} \le 0$$
$$-x_{12} + y_{12}^{31} + y_{12}^{41} + y_{12}^{44} - y_{31}^{44} \le 0$$
$$-x_{13} + y_{13}^{21} + y_{13}^{22} + y_{13}^{42} - y_{21}^{42} \le 0$$
$$-x_{13} + y_{13}^{21} + y_{13}^{22} + y_{13}^{41} - y_{22}^{41} \le 0$$
$$-x_{13} + y_{13}^{31} + y_{13}^{32} + y_{13}^{42} - y_{31}^{42} \le 0$$
$$-x_{13} + y_{13}^{31} + y_{13}^{32} + y_{13}^{41} - y_{32}^{41} \le 0$$
$$-x_{13} + y_{13}^{32} + y_{13}^{41} + y_{13}^{42} - y_{32}^{41} \le 0$$
$$-x_{13} + y_{13}^{31} + y_{13}^{41} + y_{13}^{42} - y_{31}^{42} \le 0$$

Table 7.1   All cut inequalities needed for a complete description of QAP_4

Quadratic Assignment Polytopes

$$\mathbf{F} = \begin{pmatrix} \mathbf{I}_{n-3} & 0 & 0 & 0 \\ 0 & 1 & 0 & 1 \\ 0 & 0 & 1 & 1 \\ \mathbf{e}_{n-3}^T & 1 & 1 & 0 \end{pmatrix}$$

Figure 7.1 The matrix $\mathbf{F}$ used in the proof of Proposition 7.3

we study the following subsystem of $(5.55), \ldots, (5.61)$ for $n \geq 3$.

$$\sum_{j=1}^{n} x_{ij} = 1 \quad \text{for } i \in N \tag{7.10}$$

$$\sum_{i=1}^{n} x_{ij} = 1 \quad \text{for } 1 \leq j \leq n-1 \tag{7.11}$$

$$-x_{ij} - x_{kj} + \sum_{\ell=1}^{j-1} y_{i\ell}^{kj} + \sum_{\ell=j+1}^{n} y_{ij}^{k\ell} = 0 \quad \text{for } j \in N, i < k \in N \tag{7.12}$$

$$-x_{kj} - x_{k\ell} + \sum_{i=1}^{k-1} y_{ij}^{k\ell} + \sum_{i=k+1}^{n} y_{kj}^{i\ell} = 0 \quad \begin{array}{l} \text{for } 1 \leq j < \ell \leq n-1, \\ 1 \leq j \leq n-3, k \in N \end{array} \tag{7.13}$$

$$y_{ij}^{k\ell} \geq 0 \quad \text{for } i < k \in N, j < \ell \in N \tag{7.14}$$

$$x_{ij} \geq 0 \quad \text{for } i,j \in N \tag{7.15}$$

$$x_{ij} \in \{0,1\} \quad \text{for } i,j \in N. \tag{7.16}$$

Counting the equations we find $2n - 1$ from (7.10) and (7.11), $n^2(n-1)/2$ from (7.12) and $n^2(n-3)/2$ from (7.13). The total number of equations equals $n^2(n-2) + 2n - 1$ and the number of variables appearing in $(7.10), \ldots, (7.13)$ is $n^2 + n^2(n-1)^2/4$. Thus by comparison to $(5.55), \ldots, (5.61)$ we have $n^2 + 1$ fewer equations.

Proposition 7.4 *The rank of* $(7.10), \ldots, (7.13)$ *is* $n^2(n-2) + 2n - 1$.

Proof. We start by partitioning $(7.10), \ldots, (7.13)$ into four disjoint classes.

(B1) $\quad -x_{ij} - x_{kj} + \sum_{\ell=1}^{j-1} y_{i\ell}^{kj} + \sum_{\ell=j+1}^{n} y_{ij}^{k\ell} = 0 \quad$ for $j \in N, i < k \in N$
(B2) $\quad \sum_{j=1}^{n} x_{ij} = 1 \quad$ for $i \in N$
(B3) $\quad -x_{kj} - x_{k\ell} + \sum_{i=1}^{k-1} y_{ij}^{k\ell} + \sum_{i=k+1}^{n} y_{kj}^{i\ell} = 0 \quad$ for $1 \leq j < \ell \leq n-1,$
$\qquad\qquad\qquad\qquad\qquad\qquad\qquad\qquad\qquad\qquad 1 \leq j \leq n-3, k \in N$
(B4) $\quad \sum_{i=1}^{n} x_{ij} = 1 \quad$ for $1 \leq i \leq n-1$.

(B1),...,(B4) is a reordering of (7.10),...,(7.13) and thus all equations are listed. We partition the variables as follows into four classes.

(C1) $y_{i1}^{kn},\ldots,y_{i,n-1}^{kn}, y_{i,n-2}^{k,n-1}$ for $i < k \in N$
(C2) x_{in} for $i \in N$
(C3) $y_{ij}^{k\ell}$ for $i < k \in N, 1 \le j < \ell \le n-2$,
$y_{ij}^{k,n-1}$ for $i < k \in N, 1 \le j \le n-3$
(C4) x_{ij} for $i \in N, 1 \le j \le n-1$.

All $n^2 + n^2(n-1)^2/4$ variables of (7.10),...,(7.13) are in (C1),...,(C4) and none is repeated. It follows that the variables in class (C1) occur only in (B1), but not in (B2), (B3) and (B4). Likewise, the variables in (C2) occur in (B2), but not in (B3) and (B4). Finally, the variables (C3) are all in (B3) but not in (B4). The variables (C1) are present in exactly two rows of (B1) and for every pair i, k with $1 \le i < k \le n$ the corresponding rows can be arranged so that the $n \times n$ matrix $\mathbf{F}$ shown in Figure 7.1 occurs in the columns corresponding to (C1). $\mathbf{F}$ is nonsingular, it is repeated $n(n-1)/2$ times and thus the rank of (B1) is exactly $n^2(n-1)/2$. Since the variables (C1) do not occur in (B2), (B3) and (B4) we can drop all rows in (B1) from further consideration. The variables (C2) form an $n \times n$ identity matrix in the rows (B2) which thus has a rank of n and we can drop (B2). For every pair j, ℓ with $1 \le j < \ell \le n-2$ the variables $y_{ij}^{k\ell}$ for $1 \le i < k \le n$ form the incidence matrix $\mathbf{K}_n$, say, of a complete graph on n nodes in the rows (B3) and so do the variables $y_{ij}^{k,n-1}$ for $1 \le i < k \le n$ and every j with $1 \le j \le n-3$. $\mathbf{K}_n$ has a rank n, it occurs exactly $n(n-3)/2$ times and thus (B3) has a rank of $n^2(n-3)/2$. Like before, we can drop all of (B3) from consideration. The remaining rows (B4) have rank $n-1$. By construction, we can add the ranks of (B1),..., (B4) and thus (7.10),...,(7.13) has full row rank. In Figure 7.2 we give an illustration of this proof where the asterix $*$ denotes a matrix of 0 or ± 1 as required by (B1),...,(B4) and the variables are ordered as suggested by (C1),...,(C4). □

It is not overly difficult to show that all equations of the formulation (5.55),..., (5.58) of the symmetric quadratic assignment problem are either members of the equation system (7.10),...,(7.13) or obtainable as linear combinations of (7.10),..., (7.13). Consequently, (7.10),...,(7.16) *formulates* the SKP correctly. To prove that (7.10),...,(7.13) defines the affine hull of SQP_n for all $n \ge 3$ there are several methods of achieving this result. We can provide a list of linearly independent zero-one points in SQP_n of size $n^2 + n^2(n-1)^2/4$ – either directly or inductively as done in the outline of the proof of Proposition 7.2. Alternatively, we can show that every equation that is satisfied by all zero-one points in SQP_n is a linear combination of the equations (7.10),...,(7.13). We

$$
\begin{array}{ccccccccccccc}
F & O & \ldots & O & * & * & * & \ldots & * & * & * & \ldots & * \\
O & F & \ldots & O & * & * & * & \ldots & * & * & * & \ldots & * \\
\vdots & \vdots & \ddots & \vdots & \vdots & \vdots & \vdots & \ddots & \vdots & \vdots & \vdots & \ddots & \vdots \\
O & O & \ldots & F & * & * & * & \ldots & * & * & * & \ldots & * \\
O & O & \ldots & O & I_n & * & * & \ldots & * & * & * & \ldots & * \\
O & O & \ldots & O & O & K_n & O & \ldots & O & * & * & \ldots & * \\
O & O & \ldots & O & O & O & K_n & \ldots & O & * & * & \ldots & * \\
\vdots & \vdots & \ddots & \vdots & \vdots & \vdots & \vdots & \ddots & \vdots & \vdots & \vdots & \ddots & \vdots \\
O & O & \ldots & O & O & O & O & \ldots & K_n & * & * & \ldots & * \\
O & O & \ldots & O & O & O & O & \ldots & O & I_{n-1} & * & \ldots & *
\end{array}
$$

Figure 7.2 Summary of the construction of the proof of Proposition 7.3

have encountered both proof techniques numerous times in this monograph. There are also other methods for proving the dimensionality result that are available in the literature. We leave this task for future work and formulate the following conjecture instead.

Conjecture 7.1 *The dimension of $SQPP_n$ equals $(n-1)^2 + n^2(n-3)^2/4$ for all $n \geq 4$.*

We have, of course, checked the conjecture by way of a computer and found it to be correct for $3 \leq n \leq 11$. So unless something unexpected happens in dimensions corresponding to $n = 12$ or higher, the conjecture will turn out to be correct. It is also very likely that the inequalities (7.14) and (7.15) define "trivial" facets of SQP_n for all $n \geq 3$. With this ground work completed, one can then look for more complicated facets of the polytope SQP_n which are surely going to be needed to solve larger-scale symmetric quadratic assignment problems successfully.

Another approach to SQP_n consists of exploiting the transformation (5.33), i.e.

$$y_{ij}^{k\ell} = x_{ij}x_{k\ell} + x_{i\ell}x_{kj} \quad \text{for } 1 \leq i < k \leq n, 1 \leq j < \ell \leq n.$$

To do so, you have to calculate the formulation of the symmetric quadratic assignment problem that results from the one of Chapter 7.1 for the quadratic assignment problem by way of the linear transformation technique. We have used this approach in Chapter 6.1 to compare the formulation of the operations scheduling problem with machine independent quadratic interaction costs with the one that results form the graph partitioning problem in this case. Information about the quadratic assignment problem can thus be "translated" into information about the symmetric quadratic assignment problem and there are many other meaningful ways to accumulate polyhedral knowledge about either problem. Such knowledge – without any doubt to the writers' mind – is necessary if you want to try the *exact* solution of these problems for any reasonable size.

8

SOLVING SMALL QAPs

Psychologically it is, of course, disadvantageous to start the last chapter with a disclaimer, but this is exactly what we are going to do. The software system that we are going to describe here is of a preliminary nature and our computational results should by no means be interpreted as limiting the potential of *branch-and-cut* algorithms for the solution of quadratic assignment and related quadratic zero-one optimization problems. The software system came about from our desire to write an interesting introductory chapter for this monograph dealing with the fascinating world of location, scheduling and design problems – see Chapters 1.3, 1.4 and 1.5. As an afterthought came then the idea to test the software system on a somewhat larger sample of the problems from QAPLIB. In spite of our reservations we have included this material in the book because it seems to fill a gap in the literature on how to solve quadratic assignment problems. Our efforts in locating suitable references notwithstanding and despite the fact that every author that we have read on quadratic assignment calls the problem a (mixed) zero-one programming problem, nobody seems to have taken the pain to solve QAPs via a standard mixed integer programming code using ordinary branch-and-bound. The development effort that is necessary to actually write such a software system for QAP is minimal – it took one of the authors about seven days of intense programming work to "string it all together and get the job done."

In terms of computation for the traveling salesman problem – which has known an explosive growth in the problem size now considered to be amenable to *exact* optimization – the software system that we have written does not even put us into the vicinity of Crowder and Padberg's 1980 article, where they reported the optimization of a 318-city traveling salesman problem. Here we consider the bare minimum ingredients for our solution approach: the formulation (1.8),...,(1.12) plus the inequalities (1.11a) and (1.11b). This permits us to invoke any branch-and-bound solver. Crowder and Padberg [1980], by contrast, utilized considerably more knowledge about the traveling salesman polytope in their work. The parallel between Crowder and Padberg [1980] and the work done here is given by the fact that in both cases a standard branch-and-bound code is utilized in the final optimization phase. Unfortunately, commercially available branch-and-bound codes – like in 1980 – are still much too inflexible to permit a sophisticated user to implement a branch-and-cut scheme easily. Moreover, we just do not have yet enough operational knowledge about the facial structure of QAPs, let alone suitable algorithms for separation and/or constraint identification. By consequence, we limited ourself to a very coarse implementation of cutting plane ideas and left lots of interesting work to be done for future efforts in this direction.

The software system has essentially four components. The top part – called QAPMIP – reads in the data and sets up the equations (1.8) and (1.9). The data input consists of the value n, the cost matrix of the c_{ij}'s, the flow matrix of the t_{ik}'s and the distance matrix d_{ij}. Several flags are read from a file called QAPSIZ. The flag SOLECH governs the output from the intermediate linear programs, BOUND is the upper bound of $+\infty$. If the program does not find the file VAR.in of variables indices for a starting solution it defaults to calling the second component of the solver, namely some heuristic algorithm to find a "reasonable" upper bound and an initial variable set to initialize the calculations. The heuristic is essentially inspired from Elshafei's [1977] combination of greedy ideas plus two-exchange and took a couple of hours to write.

The system then calls a routine QAPLOW to calculate lower bounds – including the Gilmore-Lawler bound – by solving $n^2 + 2$ linear programs. The best lower bound is used subsequently to govern row generation versus column generation.

The subroutine STRTEQ constructs a more complete initial variable set and an (infeasible) starting basis for the linear programming calculations. Included into the initial variable set are, in particular, all x_{ij} variables and all minimum-cost $y_{ij}^{k\ell}$ variables of the problem.

The subroutine LPSOLV is the interface of our FORTRAN routine with the CPLEX callable optimization routines of CPLEX, Inc., which is written in the language C. It goes without saying that any comparable LP solver can be used in lieu of the CPLEX routines. The initial linear program is solved. In the next step variables and/or constraints are added and/or dropped from the problem. This is done in the subroutines DRPVAR, ADDVAR, DRPROW and

Solving Small QAPs

	n	nz	nv	meq	mxv	mxr	z_{LP}	$mip1$	$mipv$	$mipr$	no	z_{IP}
chr 12a	12	11	870	155	397	219	9,552.0				0	9552
chr 12b	12	11	870	155	393	224	9,742.0				0	9,742
chr 12c	12	11	870	155	381	231	10,895.2	137	323	958	2	11,156
chr 15a	15	14	1,695	239	577	339	9,329.5	223	722	1,830	16	9,896
chr 15b	15	14	1,695	239	592	353	7,751.2	218	1,040	1,818	2	7,990
chr 15c	15	14	1,695	239	608	325	9,504.0				0	9,504
chr 18a	18	17	2,925	341	773	470	10,699.3	324	1,907	3,114	4	11,098
chr 18b	18	17	2,925	341	800	742	1,534.0				0	1,534
chr 20a	20	19	4,010	419	1,048	678	2,170.1	390	1,541	4,248	4	2,192
chr 20b	20	19	4,010	419	1,048	689	2,287.0	399	1,659	4,243	2	2,298
chr 20c	20	19	4,010	419	998	620	13,972.6	392	2,353	4,224	2	14,142
chr 22a	22	21	5,335	505	1,365	747	6,122.1	484	3,802	5,614	10	6,156
chr 22b	22	21	5,335	505	1,296	716	6,171.9	484	3,475	5,608	18	6,194
chr 25a	25	24	7,825	649	1,568	1,008	3,736.9	624	6,420	8,178	10	3,796

Table 8.1 Computational results for super sparse QAPLIB problems

ADDROW. They are invoked whenever necessary and e.g. variable/constraint dropping is performed to ensure convergence of the overall computation scheme. The overall set of variables/constraints is thus partitioned into an active set of variables/ constraints and an inactive one. The size of the linear program sent to the CPLEX routines changes from iteration to iteration. Whenever mathematically correct, the subroutine FIXRCO is invoked which – in the inner loop – fixes inactive variables to zero based on the linear programming reduced cost and the upper and lower bounds on the optimal solution value. This whole procedure is iterated until the linear programming relaxation of (1.8),...,(1.12) including all inequalities (1.11), (1.11a) and (1.11b) is optimized. A more complete version of the program should permit to add/drop *equations* of the formulation as well, but currently we add/drop only inequalities.

Having optimized the linear program the routine FIXRCO is called again to fix more variables both of the x_{ij} and the $y_{ij}^{k\ell}$ type. The subroutine SETMIP then sets up the mixed zero-one program to be sent to the branch-and-bound routine *mipoptimize* of the CPLEX routines. In this first implementation we generate all variables that have not yet been fixed plus all inequalities (1.11) that are missing, because they are required for the formulation of the problem. The result is a fairly large mixed zero-one programming problem that is subsequently subjected to branch-and-bound. We note that the routine *mipoptimize* of CPLEX, Inc., has incorporated many aspects of *branch-and-cut*. These features are, however, not used in the solution process because of the particular nature of our constraint sets. Evidently, from a problem solving point of view the generation of the entire problem as a mixed zero-one problem is wasteful

	n	nz	nv	meq	mxv	mxr	z_{LP}	mip1	mipv	mipr	no	z_{IP}
scr 10	10	22	1,090	239	660	406	26,384.5	100	852	694	6	26,992
scr 12	12	28	1,992	359	972	608	29,457.6	144	1,789	1,154	28	31,410
scr 15	15	42	4,635	659	2,437	1,158	48,714.9	225	4,464	2,238	18	51,140
els 19	19	56	9,937	1,101	5,064	1,909	16,276,915.9	360	9,086	4,373	14	17,212,548
scr 20	20	62	12,180	1,279	5,369	2,274	94,534.4	400	12,024	5,087	> 1,500	110,030

Table 8.2   Computational results for some selected QAPLIB problems

and simply not done in a proper *branch-and-cut* framework. We have permitted us to do so nevertheless as we were interested in getting some first results using the mixed zero-one formulation of QAPs quickly.

In Table 8.1 and 8.2 we summarize our findings on a selected group of test problems from the test problem file QAPLIB. As most of the problems in the file are randomly generated we have discarded most of them from consideration since we do not like *Monte Carlo* data sets. Whatever their origin, Table 8.1 reports on the super sparse problems from Christofides and Benavent [1989] which are solvable in polynomial time. They were solved without any problems by QAPMIP with solution times ranging from about one minute to 16 minutes of elapsed CPUtime on our computer; see Chapter 1.4. Given their polynomial time solvability a properly implemented branch-and-cut solver, using additional facet-defining inequalities that we do not have yet, should solve such problems without any branching at all. In the tables we use the following notation.

n = number of plants,
nz = number of nonzero t_{ik} with $1 \leq i < k \leq n$,
nv = number of variables of the overall problem,
meq = number of equations (1.8) and (1.9),
mxv = maximum number of variables of the linear program sent to LPSOLV,

mxr = maximum number of constraints sent to LPSOLV,
z_{LP} = the linear programming bound produced by QAPMIP,
$mip1$ = number of unfixed zero-one variables sent to MIPSOL,
$mipv$ = number of variables sent to MIPSOL,
$mipr$ = number of constraints sent to MIPSOL,
no = total number of nodes on the search tree produced by MIPSOL,
z_{IP} = optimal objective function of the mixed zero-one problem.

Table 8.2 shows similar results for five of the test problems from QAPLIB [1991].

Despite the preliminary nature of our numerical investigations – one might say justly that we used lots of *"hee-haw* and *chutzpah"* in even trying it this way

– we conclude that a direct attack on quadratic assignment problems is possible using the mixed zero-one programming formulation (1.8), ..., (1.12) which exploits the sparsity of real data sets. The size of the linear programs that a suitably developed *branch-and-cut* solver needs to solve appear to be reasonably small when compared to the overall number of variables and constraints of the problem. This is a first indication of the numerical success to be had by a more in-depth development effort using *branch-and-cut* for the solution of the kind of problems discussed in this monograph. The beauty of *branch-and-cut* lies in the fact that a common approach to all sorts of different combinatorial optimization problem is utilized, the differences in the problems necessitating in-depth mathematical studies of the different polytopes that are, of course, problem specific; see Chapter 10 of Padberg [1995] for an overview and further references on *branch-and-cut*.

A
FORTRAN PROGRAMS FOR SMALL SQPs

```
      PROGRAM QAPMIP
C Solver for (small) symmetric quadratic assignment problems.
C Written by Manfred Padberg, Oct. 1995.
      INTEGER MXM,MXM,MXNZA,MXNQ
      PARAMETER ( MXNQ  =          40)
      PARAMETER ( MXM   =      150000)
      PARAMETER ( MXM   =       75000)
      PARAMETER ( MXNZA=      2000000)
      INTEGER NVAR,NZA,I,K,MEQ,PRIMAL,TERMIN,OLDBA,MACT,FREQ,ECHO
      INTEGER SOLECH,FIRST,ROUND,NIO,NI,NACT,NZAC,NBLO,NQ,NDFLT
      INTEGER CUTOFF,BADD,NZF,MXADD,NWNAC,NWMAC,BSIZ,MXITH,MXVA,MXRO
      LOGICAL ADVA,ZEROE,FIXO1
      REAL*8  BOUND,Z,ZL,XCTOL,LOWBD,UPBND,BESTV,ZDIFF,ZOLD
      REAL    CPUTIM,TO
      INTEGER ARINX(MXNZA),ACINX(MXNZA),ROWPT(MXM+1),COLPT(MXM+1)
      INTEGER ROSTA(MXM),COSTA(MXM),VIND(MXM),PROFIT(MXM),FIXV(MXM)
      INTEGER NAMES(MXM),ACTPT(MXM+1),ACTRIX(MXNZA),BLPT(MXNQ+1)
      INTEGER BLKNO(MXNQ*(MXNQ-1)),PLANO(MXNQ*(MXNQ-1)),ACHAM(MXM)
      INTEGER APROFT(MXM),ACTCOF(MXNZA),B(MXM)
      LOGICAL INFLAG(MXM)
      REAL*8  UZERO(MXM),REDCO(MXM),XSOL(MXM),WORK(MXM)
      INTEGER   M01,M02,M03,M04,M05,M06,M07,M08,M09,M10
      COMMON/PT/ M01,M02,M03,M04,M05,M06,M07,M08,M09,M10
C
      WRITE(6,*)' Executing program qapmip_CPLX......'
      T0=CPUTIM()
      ZL=1.0D-4
      XCTOL=1.0D-3
      OPEN(15,FILE='QAPSIZ',STATUS='OLD',ERR=7000)
      READ(15,*) NQ,SOLECH,BOUND,BESTV
      CLOSE(15)
C Read data, set up equns in ARINX and PROFIT for all vars.
      CALL ZREADA( NQ, MXNZA )
      CALL READAT( NQ, MEQ, NVAR, NZA, NBLO, MXNZA,     MXM,    MXM,
     *             ACINX(M01),ACINX(M02),ACINX(M03),    ACINX(M04),
     *             ACINX(M05),ACINX(M06),ACINX(M07),    ACINX(M08),
     *             BLKNO,   PLANO,   BLPT,   ARINX,ROWPT,PROFIT)
      NZF=BLPT(NQ+1)
      BSIZ=NQ*(NQ-1)/2
      FREQ=2
C Max no of constraints generated on a most violated criterion.
      MXADD= 1 + MEQ/2
C Max no worst vars not priced out per block of vars.
      BADD=NQ
      IF (NBLO.GE.75) THEN
        MXADD=1 + MEQ/5
        BADD= 1 + NQ/2
      ENDIF
      OPEN(17,FILE='VAR.in',STATUS='OLD',ERR=1)
        READ(17,*) K,NIO,NI,I
        READ(17,*) (VIND(K),K=1,NI)
      CLOSE(17)
      FIRST=0
      DO K=1,NIO
        FIRST=FIRST+PROFIT(VIND(K))
      ENDDO
      WRITE(6,*)' File VAR.in read. No. vars=',NI,' Value=',FIRST
      GO TO 2
C Find heuristic solutions.
 1    WRITE(6,*)' File VAR.in does not exist. We are using QAPHEU.'
      MXITH=40
      CALL ZQAPHE( NQ, MXNZA )
      CALL QAPHEU(NVAR,   NQ,    NBLO,   NIO,NI,    NZF,MXITH,FIRST,
     *            VIND, PLANO,   BLKNO,    BLPT,          COLPT,COSTA,
     *            ACINX(M01),ACINX(M02),ACINX(M03),ACINX(M04),
     *            ACINX(M05),ACINX(M06),ACINX(M07),ACINX(M08),
     *            ACINX(M09),             REDCO,      UZERO,     XSOL)
C Set Cutoff, Best Value, the LP UPBND and compute LOWBD.
 2    IF (FIRST.LT.BESTV+ZL) BESTV=FIRST
      CUTOFF=1 + 0.01*BESTV
      UPBND=FIRST
      CALL QAPLOW( NQ, FIRST,   ZL,  BOUND,ACINX(M01),ACINX(M02),
     *             ACINX(M03),ACINX(M04),ACINX(M05), COLPT, ACTRIX,
     *                    ACTCOF, ACTPT,  B, COSTA,  REDCO,UZERO,XSOL)
      LOWBD=FIRST
      I=BESTV+ZL
      WRITE(6,*)' Initial Upbnd:',I,' Lowbd=',FIRST,' Cutoff=',CUTOFF,
     *                       ' CPUtim:',CPUTIM()-TO,' s.'
      CALL AIXCHG( ROWPT, COLPT, ARINX, ACINX, NZA, MEQ, NVAR )
      CALL ZSTRTE( NQ, MXM )
      CALL STRTEQ( NQ,MEQ, NVAR, NZA,  NIO, NI, NACT, NZAC,  MXM,
     *             NBLO, VIND, NAMES, ACTPT,ACTRIX,ACTCOF,ROWPT,
     *             ARINX, COLPT, ACINX, PROFIT,APROFT,INFLAG,  B,
     *             COSTA,ROSTA, ACHAM,FIXV,REDCO(M01),REDCO(M02),
     *             REDCO(M03), REDCO(M04),REDCO(M05),REDCO(M06),UZERO)

                                                    174
```

```
      ECHO=0
      OLDBA=1
      PRIMAL=1
      MACT=MEQ
      MXVA=NACT
      MXRO=NACT
      FIXO1=.FALSE.
      ZERONE=.FALSE.
      ZOLD=LOWBD
      ROUND=0
  10  ROUND=ROUND+1
      WRITE(6,*)' '
      WRITE(6,*) ' Round no. =',ROUND
      WRITE(6,*)' Act prob: Na=',NACT,' vrs, Ma=',MACT,' rws, Nzs=',
     *                ACTPT(NACT+1),',CPUtim:',CPUTIM()-TO,'s.'
      IF (NACT.GT.MXVA) MXVA=NACT
      IF (NACT.GT.MXRO) MXRO=NACT
C Solve the linear program.
      NDFLT=0          ! Set NoDeFauLT=1 for nondefault CPLEX settings.
      IF (ZERONE) NDFLT=1
      CALL LPSOLV( NACT, MEQ, MACT, BOUND,PRIMAL,TERMIN,    Z,    OLDBA,
     *             ECHO,NDFLT,APROFT,   B, ACTPT,ACTRIX, ACTCOF,
     *                              XSOL, UZERO, REDCO, ROSTA,   COSTA)
      CALL LPWRIT(NACT, MACT, Z, ZL, BESTV,SOLECH,TERMIN,TO,MEQ,NVAR,
     *    ZERONE,XSOL,UZERO, REDCO, VIND,  FIXV, NAMES, COSTA, ROSTA)
C Drop&add variables.
      ADVA=.FALSE.
      NWMAC=NACT
      IF (ZERONE) GO TO 25
      IF (Z.LT.LOWBD+ZL) GO TO 30
      IF (FREQ*(ROUND/FREQ).NE.ROUND) GO TO 30
      IF (DABS(ZOLD-Z).GT.ZL) THEN
         CALL DRPVAR(NACT, NZA, NQ, NVAR, ZL,CUTOFF, NWMAC,
     *      ACTRIX,ACTCOF, ACTPT, COSTA, NAMES,INFLAG,
     *              ACNAM, FIXV, REDCO,APROFT, XSOL)
         ZOLD=Z
      ENDIF
      NACT=NWMAC
      NZAC=ACTPT(NACT+1)
  25  CALL ADDVAR( NVAR,  NZA,    MACT,  NACT,   NZAC,   NQ,   BADD,
     *                     TO,   ZL, PROFIT, ACINX, COLPT,    NAMES,
     *           NBLO,      ZL, PROFIT, ACINX, ACTRIX,ACTCOF,  ACTPT,
     *           INFLAG,APROFT, ACNAM,  FIXV,ACTRIX,ACTCOF,  ACTPT,
     *                              VIND, COSTA, UZERO,    REDCO)
      ADVA=.TRUE.
      PRIMAL=1
      IF (NACT.GT.NWMAC) GO TO 10
      LOWBD=Z
      I=Z+1.0-ZL
      K=BESTV+ZL
      WRITE(6,*)'QAP: Lowbd=',I,' Bestv=',K,' Upbd LP=',UPBND
      ZDIFF=K - I
      CALL FIXRCO(NACT,  MACT,   NZAC,  NVAR,      NQ,   NBLO,    NZF,
     *             NZA,    ZL, ZDIFF, FIXO1,
     *            BLKNO,  PLANO,   BLPT,INFLAG,   FIXV,      B,
     *            ARINX,  ROWPT,  ACINX, COLPT, ACTRIX,ACTCOF,  ACTPT,
     *            NAMES,  ACNAM, COSTA, XSOL, UZERO,  REDCO, APROFT,
     *                           PROFIT, WORK(1),WORK(MXM/2+1))
      NZAC=ACTPT(NACT+1)
      NWMAC=MACT
  30  IF (DABS(Z-ZOLD).GT.ZL) THEN
         CALL DRPROW( MACT, NACT, NZAC,MEQ, NZA,NVAR, ZL,  NQ, NWMAC,
     *           ARINX, ROWPT, UZERO, ROSTA,  NAMES,   PROFIT,
     *           ACINX, COLPT,ACTRIX,ACTCOF,  ACTPT,   APROFT)
      ENDIF
      MACT=NWMAC
      NZAC=ACTPT(NACT+1)
      NZA=ROWPT(MACT+1)
      CALL ADDROW( MACT,     NZAC,   NVAR, MXNZA,MXM,NQ,
     *               NZA,   NBLO,    NZF, MXADD,TO, ZL, XCTOL,
     *           ARINX, ROWPT, ACINX, COLPT,ACTRIX,ACTCOF,
     *           ACTPT, NAMES, ROSTA, BLPT, PLANO, BLKNO,
     *                               VIND,  XSOL,  WORK)
      PRIMAL=0
      IF (MACT.GT.NWMAC) GO TO 10
      UPBND=Z
      IF (.NOT.ADVA) GO TO 20
      WRITE(6,*)' '
      I=LOWBD+1-ZL
      K=BESTV+ZL
      WRITE(6,*)'QAP: Lowbd=',I,' Bestv=',K,' Upbd LP=',UPBND
      WRITE(6,*)'Max no. vars=',MXVA,' max rows=',MXRO
      IF (I.EQ.K) THEN
         CALL MIPWRI( NQ, NACT, Z, ZL, TERMIN,     TO,     0,
     *                        SOLECH,  XSOL,  VIND, NAMES)
```

```fortran
              WRITE(6,*)' Done. CPUtim:',CPUTIM()-TO,' s.'
              STOP
            ENDIF
            FIXO1=.TRUE.
            ZDIFF= K - I
            CALL CALRCO(NQ,NACT,MACT,NZAC,APROFT,UZERO,REDCO,ACTRIX,ACTPT)
            CALL FIXRCO( NACT, MACT, NZAC, NVAR,    NQ,  NBLO,    NZF,
     *                    NZA,   ZL, ZDIFF, FIXO1,
     *                   BLKNO, PLANO, BLPT, INFLAG,  FIXV,    B,
     *                   ARINX, ROWPT, ACINX, COLPT, ACTRIX, ACTCOF, ACTPT,
     *                   NAMES, ACNAM, COSTA,  XSOL, UZERO,  REDCO, APROFT,
     *                   PROFIT, WORK(1), WORK(MXM/2+1))

            NZAC=ACTPT(MACT+1)
            CALL SETMIP(MACT,    MEQ,   MACT,  NZAC,   NVAR,     NQ,  NBLO,
     *                   NZF,    NZA,   MXM,  MXNZA,    ZL,      NI,
     *                  BLKNO, PLANO,  BLPT, INFLAG,  FIXV,      B,
     *                  ARINX, ROWPT, ACINX, COLPT, ACTRIX, ACTCOF, ACTPT,
     *                  NAMES, ACNAM, COSTA, ROSTA, APROFT, PROFIT, VIND)
C Solve the mixed zero-one linear program.
            CALL MIPSOL( NI,  NACT,    MEQ,  MACT, FIRST, TERMIN, Z, OLDBA, BESTV,
     *                  BOUND, APROFT, B,    ACTPT, ACTRIX, ACTCOF,  XSOL, ROSTA, COSTA,
            CALL MIPWRI(NQ,NACT,Z,ZL,TERMIN,TO,FIRST,SOLECH,XSOL,VIND, NAMES)
            WRITE(6,*)'Done. CPUtim:',CPUTIM()-TO,' s.'
            STOP
 7000     WRITE(6,*)' Input file QAPSIZ does not exist!'
            STOP
            END
C-----------------------------------------------------------------
            SUBROUTINE LPWRIT(NACT,    MACT,  Z,  ZL,  BESTV, SOLECH, TERMIN, TO, MEQ,
     *NVAR, ZEROONE, XSOL, UZERO, REDCO,  VIND,  FIXV,  NAMES, COSTA, ROSTA)
            INTEGER NVAR,NACT,MACT,SOLECH,TERMIN,K,COUNT,MEQ
            LOGICAL ZEROONE
            INTEGER VIND(NACT),NAMES(NACT),COSTA(NACT),ROSTA(MACT),FIXV(NVAR)
            REAL*8  Z,ZL,VARVAL,BESTV,XSOL(NACT),UZERO(MACT),REDCO(NACT)
            REAL    TO,CPUTIM
            IF (TERMIN.NE.1) THEN
              WRITE(6,1400) TERMIN,CPUTIM()-TO
 1400         FORMAT(' No feasible LP solution exists: Termin=',I10,
     *               '     CPUtime:',F12.0,'secs.')
              STOP
            ENDIF
            WRITE(6,1500) Z,CPUTIM()-TO
 1500       FORMAT(' Optimum LP solution: ObjValue=',F20.6,
     *             '     CPUtime:',F12.0,'secs.')
            ZEROONE=.TRUE.
            IF (SOLECH.GT.O) WRITE(6,1600)
 1600       FORMAT('   VarNo.       PValue              RedCo       RowNo.
     *=========================================================')
            *DValue')
            WRITE(6,*)'================================================'
     *       =========='
            COUNT=0
            DO K=1,MACT
              IF (COSTA(K).NE.0) THEN
                VARVAL=XSOL(K)
                IF (DABS(VARVAL).GT.ZL) THEN
                  IF (DABS(VARVAL - 1.0).GT. ZL) ZEROONE=.FALSE.
                  COUNT=COUNT+1
                  IF (COUNT.LE.MACT) THEN
                    IF (SOLECH.GT.0) WRITE(6,1700)
     *                 NAMES(K),VARVAL,REDCO(K),COUNT,UZERO(COUNT)
 1700               FORMAT(I8,1X,F12.6,1X,F12.4,1X,I8,F18.4)
                  ELSE
                    IF (SOLECH.GT.0)WRITE(6,1800) NAMES(K),VARVAL,REDCO(K)
 1800               FORMAT(I8,1X,F12.6,1X,F12.4)
                  ENDIF
                ENDIF
              ENDIF
            ENDDO
            DO K=COUNT+1,MACT
              IF (SOLECH.GT.1 .AND. DABS(UZERO(K)).GT.ZL)
     *           WRITE(6,1900) K,UZERO(K)
 1900       FORMAT(33X,I10,F18.4)
            ENDDO
            IF (ZEROONE) THEN
              WRITE(6,*)' Zero-one sol: Z=',Z
              IF (Z.LT.BESTV+ZL) THEN
                BESTV=Z
                COUNT=0
                DO K=1,NACT
                  IF (XSOL(K).GT.ZL) THEN
                    COUNT=COUNT+1
                    VIND(COUNT)=NAMES(K)
                    IF (FIXV(NAMES(K)).EQ.0) FIXV(NAMES(K))=1
                  ENDIF
                ENDDO
                OPEN(17,FILE='VAR.best',STATUS='UNKNOWN')
```

```fortran
            WRITE(17,*) COUNT,BESTV
            WRITE(17,*) (VIND(K),K=1,COUNT)
            CLOSE(17)
            IF (SOLECH.GT.O) WRITE(6,*)
     *      ' File VAR.best with best integer sol written.'
          ENDIF
        ENDIF
        IF (SOLECH.GT.O) WRITE(6,1500) Z,CPUTIM()-TO
        RETURN
        END
C------------------------------------------------------------------------
        SUBROUTINE MIPWRI( NQ, NACT, TERMIN, K, Z, ZL,TERMIN,    TO, FIRST,SOLECH,
     *                                               XSOL, VIND, NAMES)
        INTEGER NQ,NACT,TERMIN,K,COUNT,FIRST,J,NQSQ,SOLECH
        INTEGER VIND(NACT),NAMES(NACT)
        REAL*8  Z,ZL,XSOL(NACT)
        REAL    TO,CPUTIM
        IF (TERMIN.EQ.O) THEN
          WRITE(6,1400) TERMIN,CPUTIM()-TO
1400      FORMAT(' No feasible 0-1 solution exists: Termin=',I10,
     *                            ', CPUtime:',F12.0,'secs.')
          STOP
        ENDIF
        IF (SOLECH.GT.O)WRITE(6,1500) Z,FIRST,CPUTIM()-TO
1500    FORMAT(' Optimum 0-1 sol: ObjValue=',F10.1,
     *      ' , No nodes=',I6,' CPUtime:',F10.0,'secs.')
        IF (SOLECH.GT.O) WRITE(6,1600)
1600    FORMAT('  VarNo.       PValue         RedCo         RowNo.         DVa
     *lue')
        WRITE(6,*)'========================================================
     *======='
        COUNT=0
        DO K=1,NACT
          IF (DABS(XSOL(K)).GT.ZL) THEN
            COUNT=COUNT+1
            VIND(COUNT)=NAMES(K)
            IF (SOLECH.GT.O) WRITE(6,1700) NAMES(K),XSOL(K),O,COUNT,O
1700        FORMAT(I8,1X,F12.6,1X,F12.4,1X,I8,F18.4)
          ENDIF
        ENDDO
        CALL SRTUPI( COUNT, VIND, NAMES)
        NQSQ=NQ*NQ
        DO K=1,COUNT
          IF (VIND(K).GT.NQSQ) GO TO 100
          J=VIND(K)-(K-1)*NQ
          WRITE(6,*)' Plant i=',K,' is assigned to location j=',J
        ENDDO
100     WRITE(6,1500) Z,FIRST,CPUTIM()-TO
        RETURN
        END
C------------------------------------------------------------------------
        SUBROUTINE ZREADA( NQ, MXNZA )
        INTEGER NQ,MXNZA
        INTEGER    MO1,MO2,MO3,MO4,MO5,MO6,MO7,MO8,MO9,M10
        COMMON/PT/ MO1,MO2,MO3,MO4,MO5,MO6,MO7,MO8,MO9,M10
        MO1=1
        MO2=MO1+NQ*NQ
        MO3=MO2+NQ*NQ
        MO4=MO3+NQ*NQ
        MO5=MO4+NQ*NQ
        MO6=MO5+NQ
        MO7=MO6+NQ
        MO8=MO7+NQ
        MO9=MO8+NQ*(NQ-1)
        IF (MO9.GT.MXNZA) THEN
          WRITE(6,*)' Not enough workspace. Increase Mxnza to', MO9
          STOP
        ENDIF
        RETURN
        END
C------------------------------------------------------------------------
        SUBROUTINE READAT( NQ, MEQ, NVAR, NZA, NBLO, MXNZA,    MXM,    MXN,
     *            COST,   FLOW,   DIST,    AUX,ROWMIN,  DRSUM,  DCSUM,
     *            BLCK, BLKNO, PLANO,   BLPT,  ARINX,  ROWPT,   PROF)
        INTEGER NQ,NBLO,BSIZ,NVAR,MEQ,MXNZA,MXM,MXN
        INTEGER COST(NQ,NQ),FLOW(NQ,NQ),DIST(NQ,NQ),AUX(NQ,NQ)
        INTEGER ROWMIN(NQ),DRSUM(NQ),DCSUM(NQ),BLCK(NQ*(NQ-1)/2,2)
        INTEGER BLKNO(NQ*(NQ-1)),PLANO(NQ*(NQ-1)),BLPT(NQ+1)
        INTEGER ARINX(MXNZA),ROWPT(MXM+1),PROF(MXM)
        INTEGER I,J,K,L,CNT,BIG,P,Q,NQSQ,NZA,NQ1,MAX,MIN
        BIG=2*30
        OPEN(17,FILE='DATA.x', STATUS='OLD', ERR=9000)
        READ(17,*) NQ1
```

```
      IF (NQ1.NE.NQ) THEN
        WRITE(6,*)' Nq and Nq1 disagree. Wrong Data.'
        STOP
      ENDIF
      DO I=1,NQ
        READ(17,*) (COST(I,K),K=1,NQ)
      ENDDO
      DO I=1,NQ
        READ(17,*) (FLOW(I,K),K=1,NQ)
      ENDDO
      DO I=1,NQ
        READ(17,*) (DIST(I,K),K=1,NQ)
      ENDDO
      CLOSE(17)
      NQSQ=NQ*NQ
C Compute rowsums and colsums of DIST.
      DO K=1,NQ
        DRSUM(K)=0
        DCSUM(K)=0
      ENDDO
      DO K=1,NQ
        DO I=1,NQ
          DRSUM(K)=DRSUM(K)+DIST(K,I)
          DCSUM(K)=DCSUM(K)+DIST(I,K)
        ENDDO
      ENDDO
C Check the symmetry of FLOW.
 10   MAX=-BIG
      CNT=0
      DO K=1,NQ
        DO I=K+1,NQ
          IF (FLOW(I,K).NE.FLOW(K,I)) GO TO 9100
          IF (FLOW(I,K).GT.MAX) MAX=FLOW(I,K)
          CNT=CNT+FLOW(I,K)
        ENDDO
      ENDDO
C Get rowminima, reduce FLOW and change COST.
      DO K=1,NQ
        ROWMIN(K)=MAX+1
      ENDDO
      DO K=1,NQ
        DO I=1,NQ
          IF (K.NE.I) THEN
            IF (FLOW(K,I).LT.ROWMIN(K)) ROWMIN(K)=FLOW(K,I)
          ENDIF
        ENDDO
      ENDDO
      DO K=1,NQ
        IF (ROWMIN(K).GT.0) GO TO 15
      ENDDO
      WRITE(6,*)' Flow matrix is reduced. Total flow=',2*CNT
      GO TO 20
 15   DO K=1,NQ
        IF (ROWMIN(K).GT.0) THEN
          DO I=1,NQ
            IF (I.NE.K) THEN
              FLOW(K,I)=FLOW(K,I)-ROWMIN(K)
              FLOW(I,K)=FLOW(I,K)-ROWMIN(K)
            ENDIF
          ENDDO
          DO I=1,NQ
            COST(K,I)=COST(K,I)+ROWMIN(K)*(DRSUM(I)+DCSUM(I))
          ENDDO
        ENDIF
      ENDDO
      GO TO 10
C Construct list of indics I<K with Flow(i,k) > 0 in BLCK(*,1 or 2).
 20   NBLO=0
      DO K=1,NQ
        DO I=1,NQ
          AUX(K,I)=0
        ENDDO
      ENDDO
      DO I=1,NQ-1
        DO K=I+1,NQ
          IF (FLOW(I,K).GT.0) THEN
            NBLO=NBLO+1
            BLCK(NBLO,1)=I
            BLCK(NBLO,2)=K
            AUX(I,K)=NBLO
            AUX(K,I)=NBLO
          ENDIF
        ENDDO
      ENDDO
C Construct the list of blocknos and plantnos for each plant K.
```

```
          CNT=0
          BLPT(1)=0
          DO K=1,NQ
            DO I=1,NQ
              IF (AUX (K,I).GT.0) THEN
                CNT=CNT+1
                BLKNO(CNT)=AUX(K,I)
                PLANO(CNT)=I
              ENDIF
            ENDDO
            BLPT(K+1)=CNT
          ENDDO
          BSIZ=NQ*(NQ-1)/2
          NVAR=NQ*NQ+NBLO*BSIZ
          IF (NVAR.GT.MXN) THEN
            WRITE(6,*)' Parameter MXN too small. Need MXN=',NVAR
            STOP
          ENDIF
C Set the 2*NQ-1 assignment constraints.
          MEQ=0
          NZA=0
          ROWPT(1)=0
          DO K=1,NQ-1
            DO I=1,NQ
              NZA=NZA+1
              ARINX(NZA)=NZA
            ENDDO
            NZA=NZA+1
            ARINX(NZA)=(I-1)*NQ+K
            MEQ=MEQ+1
            ROWPT(MEQ+1)=NZA
          ENDDO
          MEQ=MEQ+1
          ROWPT(MEQ+1)=NZA
        ENDDO
C Set the equations for each pair with pos flow.
        DO L=1,NBLO
          I=BLCK(L,1)
          K=BLCK(L,2)
          DO P=1,NQ
            NZA=NZA+1
            ARINX(NZA)=(I-1)*NQ+P
            ARINX(NZA)=(K-1)*NQ+P
            DO Q=1,P-1
              NZA=NZA+1
              ARINX(NZA)=NQSQ+(L-1)*BSIZ+(Q-1)*NQ-Q*(Q-1)/2 +P - Q
            ENDDO
            DO Q=P+1,NQ
              NZA=NZA+1
              ARINX(NZA)=NQSQ+(L-1)*BSIZ+(P-1)*NQ-P*(P-1)/2 +Q - P
            ENDDO
            MEQ=MEQ+1
            ROWPT(MEQ+1)=NZA
          ENDDO
        ENDDO
C Set the profit array.
        CNT=0
        DO I=1,NQ
          DO K=1,NQ
            CNT=CNT+1
            PROF(CNT)=COST(I,K)
          ENDDO
        ENDDO
        MAX=-BIG
        MIN=BIG
        DO P=1,NBLO
          I=BLCK(P,1)
          K=BLCK(P,2)
          DO J=1,NQ-1
            DO L=J+1,NQ
              CNT=CNT+1
              PROF(CNT)=FLOW(I,K)*(DIST(J,L)+DIST(L,J))
              IF (PROF(CNT).GT.MAX) MAX=PROF(CNT)
              IF (PROF(CNT).LT.MIN) MIN=PROF(CNT)
            ENDDO
          ENDDO
        ENDDO
        WRITE(6,*)' Problem with Nq=',NQ,' plants and locations:'
        WRITE(6,*)' ',' No. vars=',NVAR,' no. equns=',MEQ,
     *              ' no. nonzs=',NZA,' no. blocks=',NBLO
        RETURN
 9000   WRITE(6,*)' File DATA.x does not exist.'
```

```
            STOP
 9100   WRITE(6,*)' FLOW matrix is not symmetric.'
        STOP
        END
C-----------------------------------------------------------------
C AIXCHG: flips col-major indcs to row-major and vice versa. From 1 to 2.
        SUBROUTINE AIXCHG(HEAD1,HEAD2,INDEX1,INDEX2,NZR,M1,M2)
        INTEGER M1,M2,NZR,I,K,J,L
        INTEGER INDEX1(NZR+1),INDEX2(NZR+1),HEAD1(M1+1),HEAD2(M2+1)
        DO I=1,M2
           HEAD2(I)=0
        ENDDO
C Count the number of nonzeroes in rows/cols.
        DO I=1,NZR
           K=INDEX1(I)
           HEAD2(K)=HEAD2(K) + 1
        ENDDO
C Reset the HEAD pointers.
        J=1
        K=HEAD2(1)
        HEAD2(1)=J
        DO I=2,M2
           J=J + K
           K=HEAD2(I)
           HEAD2(I)=J
        ENDDO
C Move the structure.
        L=0
        DO I=1,NZR
 10        IF (HEAD1(L+1).LT.I) THEN
              L=L+1
              GO TO 10
           ENDIF
           J=INDEX1(I)
           K=HEAD2(J)
           INDEX2(K)=L
           HEAD2(J)=HEAD2(J) + 1
        ENDDO
C Reset the HEAD pointers.
        HEAD2(M2+1)=HEAD1(M1+1)
        DO I=M2,2,-1
           HEAD2(I)=HEAD2(I-1) - 1
        ENDDO
        HEAD2(1)=0
        RETURN
        END
C-----------------------------------------------------------------
        SUBROUTINE ZSTRTE( NQ, MXN )
        INTEGER NQ,MXN
        INTEGER   M01,M02,M03,M04,M05,M06,M07,M08,M09,M10
        COMMON/PT/ M01,M02,M03,M04,M05,M06,M07,M08,M09,M10
        M01=1
        M02=M01+NQ*(NQ-1)/2
        M03=M02+NQ*(NQ-1)/2
        M04=M03+NQ
        M05=M04+NQ*NQ
        M06=M05+NQ
        M07=M06+NQ
        IF (M07.GT.MXN) THEN
           WRITE(6,*)' MXN is too small. Need MXN=',M07
           STOP
        ENDIF
        RETURN
        END
C-----------------------------------------------------------------
        SUBROUTINE STRTEQ(   NQ,MEQ, N,      NZA,NIO,NI,    NACT,    NZAC,    MXM,
     *                       NBLO,  VARIN,   NAMES, ACTPT,  ACTRIX,  ACTCOF,
     *                       ROWPT, ARINX,   COLPT, ACINX,  PROFIT,  APROFT,
     *                       INFLAG,         B,     COSTA,  ROSTA,   ACNAM,   FIXV,
     *                       AUX1,  AUX2,    COL,   ADJ,    STAR,    PRED,   SCRCH   )
        INTEGER   N,NZA,NIO,NI,NACT,NZAC,MEQ,NQ,CNT,R,MXM,NBLO
        INTEGER   NAMES(N),ACTPT(N+1),ACTRIX(NZA),ROWPT(MEQ+1),COLPT(N+1)
        INTEGER   ARINX(NZA),ACINX(NZA),VARIN(NI),COSTA(N),PROFIT(N)
        INTEGER   ROSTA(MEQ),AUX1(NQ*(NQ-1)/2),AUX2(NQ*(NQ-1)/2),ACNAM(N)
        INTEGER   COL(NQ),ADJ(NQ*NQ),STAR(NQ),PRED(NQ),SCRCH(MEQ)
        INTEGER   ACTCOF(NZA),APROFT(N),B(MXM),FIXV(N)
        LOGICAL   INFLAG(N)
        INTEGER   NQSQ,K,L,VAR,KV,IV,R1,R2,BSIZ,RNO,CNO,COUNT,BL
        NQSQ=NQ*NQ
        BSIZ=NQ*(NQ-1)/2
        DO K=1,N
           FIXV(K)=0
           COSTA(K)=0
           ACNAM(K)=0
           INFLAG(K)=.FALSE.
        ENDDO
```

```
      DO K=1,NIO
        FIXV(VARIN(K))=1
      ENDDO
      DO K=1,MEQ
        ROSTA(K)=0
      ENDDO
      DO K=1,2*NQ-1
        B(K)=1
      ENDDO
      DO K=2*NQ,MXM
        B(K)=0
      ENDDO
C Put the NQ*NQ first cols into active set.
      NACT=NQSQ
      NZAC=COLPT(NQSQ+1)
      DO VAR=1,NQSQ
        INFLAG(VAR)=.TRUE.
        NAMES(VAR)=VAR
        ACNAM(VAR)=VAR
        APROFT(VAR)=PROFIT(VAR)
      ENDDO
      DO K=1,COLPT(NQSQ+1)
        ACTRIX(K)=ACINX(K)
        IF (ACINX(K).LT.2*NQ) THEN
          ACTCOF(K)=1
        ELSE
          ACTCOF(K)=-1
        ENDIF
      ENDDO
      DO K=1,NQSQ+1
        ACTPT(K)=COLPT(K)
      ENDDO
      DO K=1,NIO
        IF (VARIN(K).LE.NQSQ) THEN
          COSTA(VARIN(K))=1
        ENDIF
      ENDDO
C Find a cheap tree (basis) in top rows.
      DO K=1,NQSQ
        ADJ(K)=PROFIT(K)
        SCRCH(K)=K
      ENDDO
      CALL SRTUPI( NQSQ, ADJ, SCRCH)
      DO K=1,NQSQ
        ADJ(K)=0
      ENDDO
      COUNT=0
      DO K=1,NQSQ
        IF (COSTA(K).EQ.1) THEN
          COUNT=COUNT+1
          IF (K.LE.NQ*(NQ-1)) THEN
            R1=ACINX(COLPT(K)+1)
            R2=ACINX(COLPT(K)+2)-NQ+1
          ELSE
            R1=NQ
            R2=ACINX(COLPT(K)+1)-NQ+1
          ENDIF
          ADJ(R1)=COUNT
          ADJ(NQ+R2)=COUNT
        ENDIF
      ENDDO
      DO K=1,NQSQ
        IV=SCRCH(K)
        IF (COSTA(IV).GT.0) GO TO 10
        IF (IV.LE.NQ*(NQ-1)) THEN
          R1=ACINX(COLPT(IV)+1)
          R2=ACINX(COLPT(IV)+2)-NQ+1
        ELSE
          R1=NQ
          R2=ACINX(COLPT(IV)+1)-NQ+1
        ENDIF
        IF (ADJ(R1) .EQ. ADJ(NQ+R2)) GO TO 10
        COSTA(IV)=1
        CNT=ADJ(R1)
        COUNT=ADJ(NQ+R2)
        IF (COUNT.LT.CNT) THEN
          CNT=ADJ(NQ+R2)
          COUNT=ADJ(R1)
        ENDIF
        DO L=1,2*NQ
          IF (ADJ(L) .EQ. COUNT) ADJ(L)=CNT
        ENDDO
 10   CONTINUE
C For each block put all shortest edges into the active set.
      DO BL=1,NBLO
        CNT=NQSQ+(BL-1)*BSIZ
```

```
      DO R1=1,NQ-1
        COUNT=CNT+(R1-1)*NQ - R1*(R1-1)/2 - R1
        VAR=2**30
        DO R2=R1+1,NQ
          IF (PROFIT(COUNT+R2).LT.VAR) VAR=PROFIT(COUNT+R2)
        ENDDO
        DO 50 R2=R1+1,NQ
          K=COUNT+R2
          IF (INFLAG(K) .OR. PROFIT(K).GT.VAR) GO TO 50
          NACT=NACT+1
          NAMES(NACT)=K
          ACNAM(K)=NACT
          INFLAG(K)=.TRUE.
          APROF(NACT)=PROFIT(K)
          DO L=COLPT(K)+1,COLPT(K+1)
            NZAC=NZAC+1
            ACTRIX(NZAC)=ACINX(L)
            ACTCOF(NZAC)=1
          ENDDO
          ACTPT(NACT+1)=NZAC
 50     CONTINUE
      ENDDO
C Now build the basic active set for flow vars.
      DO 200 BL=1,NBLO
        CNO=NQSQ+(BL-1)*BSIZ
        DO K=1,BSIZ
          AUX1(K)=PROFIT(CNO+K)
          AUX2(K)=CNO+K
        ENDDO
        CALL SRTUPI( BSIZ, AUX1, AUX2)
        RNO=2*NQ-1+(BL-1)*NQ
C Set a min spanning 1-tree in block BL.
        DO K=1,NQ
          COL(K)=K
        ENDDO
        DO K=1,NQSQ
          ADJ(K)=0
        ENDDO
        COUNT=0
        DO 70 K=1,BSIZ
          IV=AUX2(K)
          R1=ACINX(COLPT(IV)+1)-RNO
          R2=ACINX(COLPT(IV)+2)-RNO
          IF (R1.GT.R2) THEN
            L=R1
            R1=R2
            R2=L
          ENDIF
          IF (COL(R1) .EQ. COL(R2)) GO TO 70
          CNT=COL(R2)
          DO L=1,NQ
            IF (COL(L) .EQ. CNT) COL(L)=COL(R1)
          ENDDO
          AUX2(K)=-IV
          COUNT=COUNT+1
          ADJ((R1-1)*NQ+R2)=R2
          ADJ((R2-1)*NQ+R1)=R1
          IF (COUNT.GE.NQ-1) GO TO 75
 70     CONTINUE
C Now find a cheapest odd cycle.
 75     DO K=1,NQ
          COL(K)=0
          PRED(K)=0
        ENDDO
        CNT=0
C Build a layered tree with root at node 1.
        COL(1)=-1
        DO K=1,NQ
          IF (ADJ(K).GT.0) THEN
            CNT=CNT+1
            STAR(CNT)=ADJ(K)
            COL(ADJ(K))=1
            PRED(ADJ(K))=1
          ENDIF
        ENDDO
        COUNT=0
 80     COUNT=COUNT+1
        IF (COUNT.GT.CNT) GO TO 90
        R1=STAR(COUNT)
        DO 85 K=(R1-1)*NQ+1,R1*NQ
          IF (ADJ(K).EQ.0) GO TO 85
          IF (COL(ADJ(K)) .EQ.0) THEN
            CNT=CNT+1
            STAR(CNT)=ADJ(K)
            COL(ADJ(K))=COL(R1)+1
```

```
              PRED(ADJ(K))=R1
           ENDIF
 85     CONTINUE
        GO TO 80
 90     COL(1)=0
        DO 120 K=1,BSIZ
           IV=AUX2(K)
           IF (IV.LT.0) GO TO 120
           R1=ACINX(COLPT(IV)+1)-RNO
           R2=ACINX(COLPT(IV)+2)-RNO
           IF (COL(R2).GT.COL(R1)) THEN
              R1=ACINX(COLPT(IV)+1)-RNO
              R1=ACINX(COLPT(IV)+2)-RNO
           ENDIF
           R=R1
           COUNT=0
           STAR(COUNT)=PRED(R)
           R=PRED(R)
 95        IF (R.NE.1) GO TO 95
C Check: R2 on the path from R1 to node 1.
           DO L=1,COUNT
              IF (R2.EQ.STAR(L)) THEN
                 IF (2*((COL(R1)-COL(R2))/2) .NE. COL(R1)-COL(R2))
     *                                                 GO TO 120
                 AUX2(K)=-IV
                 GO TO 130
              ENDIF
           ENDDO
           R=PRED(R2)
 100       DO L=1,COUNT
              IF (R.EQ.STAR(L)) GO TO 110
           ENDDO
           R=PRED(R)
           GO TO 100
 110       L=COL(R1)-COL(R)+COL(R2)-COL(R)
           IF (2*(L/2).NE.L) GO TO 120
           AUX2(K)=-IV
           GO TO 130
 120    CONTINUE
C Put vars into the col-structure.
 130    DO 150 K=1,BSIZ
           IV=AUX2(K)
           IF (IV.GT.0) GO TO 150
           IV=-IV
           IF (INFLAG(IV)) THEN
              COSTA(ACNAM(IV))=1
              GO TO 150
           ENDIF
           NACT=NACT+1
           NAMES(NACT)=IV
           ACNAM(IV)=NACT
           COSTA(NACT)=1
           INFLAG(IV)=.TRUE.
           APROFT(NACT)=PROFIT(IV)
           DO L=COLPT(IV)+1,COLPT(IV+1)
              NZAC=NZAC+1
              ACTRIX(NZAC)=ACINX(L)
              ACTCOF(NZAC)=1
           ENDDO
           ACTPT(NACT+1)=NZAC
 150    CONTINUE
 200    CONTINUE
C Now put in the heuristic vars.
        DO 300 K=1,NIO
           KV=VARIN(K)
           IF (INFLAG(KV)) GO TO 300
           NACT=NACT+1
           NAMES(NACT)=KV
           ACNAM(KV)=NACT
           INFLAG(KV)=.TRUE.
           APROFT(NACT)=PROFIT(KV)
           DO L=COLPT(KV)+1,COLPT(KV+1)
              NZAC=NZAC+1
              ACTRIX(NZAC)=ACINX(L)
              ACTCOF(NZAC)=1
           ENDDO
           ACTPT(NACT+1)=NZAC
 300    CONTINUE
        RETURN
        END
C------------------------------------------------------------------
        SUBROUTINE ADDVAR( N, NZA, MACT, NACT, NZAC,   NQ, BADD,
     *                               NBLO,     ZL,PROFIT, ACINX, COLPT,
     *                        NAMES,INFLAG,APROFT, ACNAM,  FIXV,
```

```
*                 ACTRIX,ACTCOF,ACTPT, VIND, COSTA, UZERO, REDCO)
      INTEGER N,NZA,MACT,NACT,NZAC,NQ,BADD,NBLO
      REAL*8  ZL,REDC
      REAL    CPUTIM,TO
      INTEGER ACINX(NZA),COLPT(N+1),COSTA(N),VIND(N),ACNAM(N)
      INTEGER PROFIT(N),NAMES(N),ACTPT(N+1),ACTRIX(NZA),FIXV(N)
      INTEGER APROFT(N),ACTCOF(NZA)
      LOGICAL INFLAG(N)
      REAL*8  UZERO(MACT),REDCO(N)
      INTEGER I,K,COUNT,LL,BLO,NQSQ,BSIZ,KEEP,TCNT
C Check inactive cols that do not price out.
      KEEP=0
      TCNT=0
      NQSQ=NQ*NQ
      BSIZ=NQ*(NQ-1)/2
      DO BLO=1,NBLO
        COUNT=0
        DO 200 K=NQSQ+(BLO-1)*BSIZ+1, NQSQ+BLO*BSIZ
          IF (INFLAG(K) .OR. FIXV(K) .NE. 0) GO TO 200
          REDC=PROFIT(K)
          DO I=COLPT(K)+1,COLPT(K+1)
            REDC=REDC-UZERO(ACINX(I))
          ENDDO
          IF (REDC.GT.-ZL) GO TO 200
          COUNT=COUNT+1
          REDCO(COUNT)=REDC
          VIND(KEEP+COUNT)=K
200     CONTINUE
        IF (COUNT.GT.O) CALL SRTUPR( COUNT, REDCO, VIND(KEEP+1))
        LL=COUNT
        IF (LL.GT.BADD) LL=BADD
        KEEP=KEEP+LL
        TCNT=TCNT+COUNT
      ENDDO
C Augment the active A-columnstructure.
      DO K=1,KEEP
        DO I=COLPT(VIND(K))+1,COLPT(VIND(K)+1)
          NZAC=NZAC+1
          ACTCOF(NZAC)=1
          ACTRIX(NZAC)=ACINX(I)
        ENDDO
        NAMES(NACT+K)=VIND(K)
        ACNAM(VIND(K))=NACT+K
        INFLAG(VIND(K))=.TRUE.
        APROFT(NACT+K)=PROFIT(VIND(K))
        ACTPT(NACT+K+1)=NZAC
      ENDDO
C Update basis info.
      DO K=NACT+1, NACT+KEEP
        COSTA(K)=0
      ENDDO
      NACT=NACT+KEEP
      WRITE(6,*)'ADDVAR: Found',TCNT,' cols. Added',KEEP,' cols.',
     *                                          CPUTIM()-TO,' s.'
      RETURN
      END
C-----------------------------------------------------------------
      SUBROUTINE ADDROW(MACT, NACT,    NZAC,   NVAR,   MXNZ, MXM, NQ,
     *                        NZA,    NBLO,   MXADD,  TO, ZL, XCTOL,
     *                        ARINX,  ROWPT,  ACINX,  COLPT,ACTRIX,ACTCOF,
     *                        ACTPT,  NAMES,  ROSTA,  BLPT, PLANO, BLKNO,
     *                                        IND,    XSOL, XVIOL)
      INTEGER MACT,NACT,NZAC,NVAR,MXNZ,MXM,NQ,NBLO,NZF,MXADD,NZA
      REAL*8  ZL,XCTOL
      INTEGER ARINX(MXNZ),ROWPT(MXM+1),ACINX(MXNZ),COLPT(NVAR+1)
      INTEGER ACTRIX(MXNZ),ACTPT(MXM+1),NAMES(MACT),IND(MXM)
      INTEGER ROSTA(MXM),BLPT(NQ+1),PLANO(NZF),BLKNO(NZF),ACTCOF(MXNZ)
      REAL*8  XSOL(NVAR),XVIOL(NVAR)
      INTEGER NZ,MCO,MIQ1,MIQ2,MIQ3,NQSQ,J,K,L,P,Q,OLNZ,BSIZ,ADD,COL
      REAL*8  VIOL
      REAL    TO,CPUTIM
      NZ=0
      MCO=0
      MIQ1=0
      MIQ2=0
      MIQ3=0
      NQSQ=NQ*NQ
      BSIZ=NQ*(NQ-1)/2
C "String out" XSOL.
      DO K=1,NVAR
        XVIOL(K)=0
      ENDDO
      DO K=1,NACT
        XVIOL(NAMES(K))=XSOL(K)
      ENDDO
      DO K=1,NVAR
```

```
              XSOL(K)=XVIOL(K)
            ENDDO
C Check the inequs (1.11). We use ACTRIX, ACTPT as scratch here.
            DO K=1,NQ
              IF (BLPT(K+1).GT.BLPT(K)) THEN
                DO J=1,NQ-1
                  DO L=J+1,NQ
                    OLNZ=NZ
                    NZ=NZ+1
                    ACTRIX(NZ)=(K-1)*NQ+J
                    VIOL=-XSOL(ACTRIX(NZ))
                    NZ=NZ+1
                    ACTRIX(NZ)=(K-1)*NQ+L
                    VIOL=VIOL-XSOL(ACTRIX(NZ))
                    DO P=BLPT(K)+1,BLPT(K+1)
                      Q=BLKNO(P)
                      NZ=NZ+1
                      ACTRIX(NZ)=NQSQ+(Q-1)*BSIZ+(J-1)*NQ-J*(J-1)/2 + L - J
                      VIOL=VIOL+XSOL(ACTRIX(NZ))
                    ENDDO
                    IF (VIOL.GT.XCTOL) THEN
                      MCO=MCO+1
                      XVIOL(MCO)=-VIOL
                      ACTPT(MCO+1)=NZ
                      IF (NZ+2*NQ.GE.MXNZ-NZA .OR. MCO.GE.MXM-MACT) THEN
                        MIQ1=MCO
                        GO TO 10
                      ENDIF
                    ELSE
                      NZ=OLNZ
                    ENDIF
                  ENDDO
                ENDDO
              ENDIF
            ENDDO
            MIQ1=MCO
C Check the inequs (1.11a).
            DO K=1,NQ
              IF (BLPT(K+1).GT.BLPT(K)) THEN
                DO J=1,NQ-1
                  DO L=J+1,NQ
                    OLNZ=NZ
                    VIOL=0.0
                    DO Q=BLPT(K)+1,BLPT(K+1)
                      P=PLANO(Q)
                      IF (P.LT.K) THEN
                        NZ=NZ+1
                        ACTRIX(NZ)=(P-1)*NQ+J
                        VIOL=VIOL-XSOL(ACTRIX(NZ))
                      ENDIF
                    ENDDO
                    NZ=NZ+1
                    ACTRIX(NZ)=(K-1)*NQ+J
                    VIOL=VIOL-XSOL(ACTRIX(NZ))
                    DO Q=BLPT(K)+1,BLPT(K+1)
                      P=PLANO(Q)
                      IF (P.GT.K) THEN
                        NZ=NZ+1
                        ACTRIX(NZ)=(P-1)*NQ+J
                        VIOL=VIOL-XSOL(ACTRIX(NZ))
                      ENDIF
                    ENDDO
                    DO P=BLPT(K)+1,BLPT(K+1)
                      Q=BLKNO(P)
                      NZ=NZ+1
                      ACTRIX(NZ)=NQSQ+(Q-1)*BSIZ+(J-1)*NQ-J*(J-1)/2 + L - J
                      VIOL=VIOL+XSOL(ACTRIX(NZ))
                    ENDDO
                    IF (VIOL.GT.XCTOL) THEN
                      MCO=MCO+1
                      XVIOL(MCO)=-VIOL
                      ACTPT(MCO+1)=NZ
                      IF (NZ+2*NQ.GE.MXNZ-NZA .OR. MCO.GE.MXM-MACT) THEN
                        MIQ2=MCO-MIQ1
                        GO TO 10
                      ENDIF
                    ELSE
                      NZ=OLNZ
                    ENDIF
                  ENDDO
                ENDDO
              ENDIF
            ENDDO
            MIQ2=MCO-MIQ1
C Check the inequs (1.11b).
            DO K=1,NQ
```

```fortran
      IF (BLPT(K+1).GT.BLPT(K)) THEN
        DO J=1,NQ-1
          DO L=J+1,NQ
            OLNZ=NZ
            VIOL=0.0
            DO Q=BLPT(K)+1,BLPT(K+1)
              P=PLANO(Q)
              IF (P.LT.K) THEN
                NZ=NZ+1
                ACTRIX(NZ)=(P-1)*NQ+L
                VIOL=VIOL-XSOL(ACTRIX(NZ))
              ENDIF
            ENDDO
            NZ=NZ+1
            ACTRIX(NZ)=(K-1)*NQ+L
            VIOL=VIOL-XSOL(ACTRIX(NZ))
            DO Q=BLPT(K)+1,BLPT(K+1)
              P=PLANO(Q)
              IF (P.GT.K) THEN
                NZ=NZ+1
                ACTRIX(NZ)=(P-1)*NQ+L
                VIOL=VIOL-XSOL(ACTRIX(NZ))
              ENDIF
            ENDDO
            DO P=BLPT(K)+1,BLPT(K+1)
              Q=BLKNO(P)
              NZ=NZ+1
              ACTRIX(NZ)=NQSQ+(Q-1)*BSIZ+(J-1)*NQ-J*(J-1)/2 + L - J
              VIOL=VIOL+XSOL(ACTRIX(NZ))
            ENDDO
            IF (VIOL.GT.XCTOL) THEN
              MCO=MCO+1
              XVIOL(MCO)=-VIOL
              ACTPT(MCO+1)=NZ
              IF (NZ+2*NQ.GE.MXNZ-NZA .OR. MCO.GE.MXM-MACT) THEN
                MIQ3=MCO-(MIQ1+MIQ2)
                GO TO 10
              ENDIF
            ELSE
              NZ=OLNZ
            ENDIF
          ENDDO
        ENDDO
      ENDIF
      MIQ3=MCO-(MIQ1+MIQ2)
10    DO K=1,MCO
        IND(K)=K
      ENDDO
      CALL SRTUPR( MCO, XVIOL, IND)
      IF (MCO.GT.MXADD) ADD=MXADD
      ADD=MCO
      DO K=1,ADD
        L=IND(K)
        DO P=ACTPT(L)+1,ACTPT(L+1)
          NZA=NZA+1
          ARINX(NZA)=ACTRIX(P)
        ENDDO
        MACT=MACT+1
        ROWPT(MACT+1)=NZA
      ENDDO
      CALL AIXCHG( ROWPT, COLPT, ARINX, ACINX, NZA, MACT, NVAR)
C Rebuild the LP active colstructure.
      NZAC=0
      DO L=1,MACT
        COL=NAMES(L)
        IF (COL.LE.NQSQ) THEN
          DO K=COLPT(COL)+1,COLPT(COL+1)
            NZAC=NZAC+1
            ACTRIX(NZAC)=ACINX(K)
            IF (ACINX(K).LE.2*NQ-1) THEN
              ACTCOF(NZAC)=1
            ELSE
              ACTCOF(NZAC)=-1
            ENDIF
          ENDDO
        ELSE
          DO K=COLPT(COL)+1,COLPT(COL+1)
            NZAC=NZAC+1
            ACTRIX(NZAC)=ACINX(K)
            ACTCOF(NZAC)=1
          ENDDO
        ENDIF
        ACTPT(L+1)=NZAC
      ENDDO
C Reconstitute XSOL array.
```

```
            DO K=1,MACT
              XVIOL(K)=XSOL(NAMES(K))
            ENDDO
            DO K=1,MACT
              XSOL(K)=XVIOL(K)
            ENDDO
C Basis info for simplex.
            DO K=MACT-ADD+1,MACT
              ROSTA(K)=1
            ENDDO
      WRITE(6,*)'ADDROW: Found',MCO,', added',ADD,', most viol cons.',
     *                       ',CPUtim:',CPUTIM()-TO,' s.'
      RETURN
      END
C-----------------------------------------------------------------------
      SUBROUTINE DRPROW( MACT,  NACT,  NZAC,MEQ,NZA,NVAR,ZL, NQ,NWMAC,
     *                   ARINX, ROWPT, UZERO, ROSTA,   NAMES,   PROFIT,
     *                   ACINX, COLPT,ACTRIX, ACTCOF,NQ,NWMAC,NWNZA
     *                                                         APROFT)
      INTEGER MACT,NACT,NZAC,MEQ,NZA,NVAR,NQ,NWMAC,NWNZA
      INTEGER ARINX(NZA),ROWPT(MACT+1),ROSTA(MACT),NAMES(NACT)
      INTEGER ACINX(NZA),COLPT(NVAR+1),ACTRIX(NZAC),ACTPT(NACT+1)
      INTEGER PROFIT(NVAR),APROFT(NACT),ACTCOF(NZAC)
      REAL*8  ZL,UZERO(MACT)
      INTEGER DROP,SHIFT,ROW,L,NZ,COL,K,NQSQ
      DROP=0
      SHIFT=0
      ROW=MEQ
      ROW=ROW+1
      IF (ROW.GT.MACT) THEN
        WRITE(6,*)
     *    'DRPROW: Dropped',DROP,' inequalities from active set.'
        NWMAC=MACT-DROP
        ROWPT(NWMAC+1)=ROWPT(MACT+1)-SHIFT
        NWNZA=ROWPT(NWMAC+1)
        IF (DROP.EQ.0) RETURN
        DO K=NWMAC+1,MACT
          ROSTA(K)=0
        ENDDO
        GO TO 20
      ENDIF
      IF (ROSTA(ROW).EQ.0) THEN
        IF (DROP.EQ.0) GO TO 10
        DO L=ROWPT(ROW)+1,ROWPT(ROW+1)
10
                ARINX(L-SHIFT)=ARINX(L)
              ENDDO
              ROWPT(ROW-DROP)=ROWPT(ROW)-SHIFT
              UZERO(ROW-DROP)=UZERO(ROW)
              ROSTA(ROW-DROP)=ROSTA(ROW)
            ELSE
              DROP=DROP+1
              SHIFT=SHIFT+ROWPT(ROW+1)-ROWPT(ROW)
            ENDIF
            GO TO 10
20          CALL AIXCHG( ROWPT, COLPT, ARINX, ACINX, NWNZA, NWMAC, NVAR)
C Rebuild the LP active colstructure.
            NZ=0
            NQSQ=NQ*NQ
            DO L=1,NACT
              COL=NAMES(L)
              IF (ABS(PROFIT(COL)-APROFT(L)) .GT. ZL) THEN
                WRITE(6,*)' Data structure messed up in DRPROW.'
                STOP
              ENDIF
              IF (COL.LE.NQSQ) THEN
                DO K=COLPT(COL)+1,COLPT(COL+1)
                  NZ=NZ+1
                  ACTRIX(NZ)=ACINX(K)
                  IF (ACINX(K).LE.2*NQ-1) THEN
                    ACTCOF(NZ)=1
                  ELSE
                    ACTCOF(NZ)=-1
                  ENDIF
                ENDDO
              ELSE
                DO K=COLPT(COL)+1,COLPT(COL+1)
                  NZ=NZ+1
                  ACTRIX(NZ)=ACINX(K)
                  ACTCOF(NZ)=1
                ENDDO
              ENDIF
              ACTPT(L+1)=NZ
            ENDDO
            RETURN
            END
C-----------------------------------------------------------------------
```

```
      SUBROUTINE DRPVAR(NACT, NZAC, NQ, NVAR, ZL, CUTOFF, NWNAC,
     *         ACTRIX,ACTCOF, ACTPT, COSTA, NAMES,INFLAG,
     *         ACHAM,  FIXV, REDCO,  PROF, XSOL)
      INTEGER NACT,NZAC,NQ,NVAR,CUTOFF,NWNAC,DROP,SHIFT,COL,L
      INTEGER ACTRIX(NZAC),ACTCOF(NACT+1),COSTA(NACT),NAMES(NACT)
      INTEGER ACHAM(NVAR),ACTCOF(NZAC),PROF(NACT),FIXV(NVAR)
      LOGICAL INFLAG(NVAR)
      REAL*8 ZL,REDCO(NACT),XSOL(NACT)
      DROP=0
      SHIFT=0
      COL=NQ*NQ
C Do not inactivate the zero-one vars.
 10   COL=COL+1
      IF (COL.GT.NACT) THEN
          WRITE(6,*) 'DRPVAR: Dropped',DROP
     *         , ' cols from the active set. Cutoff=',CUTOFF
          NWNAC=NACT-DROP
          ACTPT(NWNAC+1)=ACTPT(NACT+1)-SHIFT
          DO L=NWNAC+1,NACT
              COSTA(L)=0
          ENDDO
          RETURN
      ENDIF
      IF (FIXV(NAMES(COL)).EQ.1 .OR.
     *    COSTA(COL).GT.0 .OR. REDCO(COL).LT.CUTOFF) THEN
          IF (DROP.EQ.0) GO TO 10
          NAMES(COL-DROP)=NAMES(COL)
          ACHAM(NAMES(COL-DROP))=ACNAM(NAMES(COL))
          COSTA(COL-DROP)=COSTA(COL)
          PROF(COL-DROP)=PROF(COL)
          REDCO(COL-DROP)=REDCO(COL)
          XSOL(COL-DROP)=XSOL(COL)
          DO L=ACTPT(COL)+1,ACTPT(COL+1)
              ACTRIX(L-SHIFT)=ACTRIX(L)
              ACTCOF(L-SHIFT)=ACTCOF(L)
          ENDDO
          ACTPT(COL-DROP)=ACTPT(COL)-SHIFT
      ELSE
          DROP=DROP+1
          SHIFT=SHIFT+ACTPT(COL+1)-ACTPT(COL)
          INFLAG(NAMES(COL))=.FALSE.
          ACHAM(NAMES(COL))=0
      ENDIF
      GO TO 10
      END
C-----------------------------------------------------------
      SUBROUTINE FIXRCO(NACT,  MACT,  NZAC, NVAR,    NQ,  NBLO,   NZF,
     *             NZA,    ZL, ZDIFF, FIXO1,
     *          BLKNO, PLANO, BLPT,INFLAG,  FIXV,    B,
     *          ARINX, ROWPT, ACINX, COLPT,ACTRIX,ACTCOF, ACTPT,
     *          NAMES, ACNAM, COSTA, XSOL, UZERO, REDCO,  PROF,
     *                                PROFIT, APOS,  ANEG)
      INTEGER NACT,MACT,NZAC,NVAR,NQ,NBLO,NZF,NZA
      INTEGER ACTRIX(NZAC),ACTPT(NVAR+1),NAMES(NVAR),ACTCOF(NZAC)
      INTEGER ARINX(NZA),ROWPT(MACT+1),ACINX(NZA),COLPT(NVAR+1),B(MACT)
      INTEGER PROFIT(NVAR),APOS(MACT),ANEG(MACT),COSTA(NVAR),PROF(NVAR)
      INTEGER FIXV(NVAR),BLKNO(NZF),PLANO(NZF),BLPT(NQ+1),ACNAM(NVAR)
      LOGICAL INFLAG(NVAR),FIXO1
      REAL*8   REDCO(NVAR),UZERO(MACT),XSOL(NVAR),ZL,ZDIFF,REDC
      INTEGER NQSQ,DROP,SHIFT,COL,L,I,K
      INTEGER ITER,BSIZ,NQ2,ZRO,FXAC,FXIA,TOTFX
      ITER=0
      FXAC=0
      FXIA=0
      NQ2=2*NQ-1
      NQSQ=NQ*NQ
      BSIZ=NQ*(NQ-1)/2
      ZRO=0
      COL=0
C Fix zero-one vars as well if possible. DO'NT in the inner loop.
      DROP=0
      SHIFT=0
      TOTFX=FXAC+FXIA
      ITER=ITER+1
 10   COL=COL+1
      IF (COL.GT.NACT) THEN
          ACTPT(NACT-DROP+1)=ACTPT(NACT+1)-SHIFT
          DO K=NACT-DROP+1,NACT
              COSTA(K)=0
          ENDDO
          NACT=NACT-DROP
          GO TO 20
      ENDIF
      IF (REDCO(COL).LT.ZDIFF+ZL .OR.
     *   (NAMES(COL).LE.NQSQ .AND. .NOT.FIXO1)) THEN
          IF (ITER.EQ.1 .AND. DABS(REDCO(COL)).LT.ZL) ZRO=ZRO+1
```

```
            IF (DROP.EQ.0) GO TO 10
            PROF(COL-DROP)=PROF(COL)
            REDC0(COL-DROP)=REDC0(COL)
            XSOL(COL-DROP)=XSOL(COL)
            COSTA(COL-DROP)=COSTA(COL)
            NAMES(COL-DROP)=NAMES(COL)
            ACNAM(NAMES(COL))=ACNAM(NAMES(COL))-DROP
            DO L=ACTPT(COL)+1,ACTPT(COL+1)
               ACTRIX(L-SHIFT)=ACTRIX(L)
               ACTCOF(L-SHIFT)=ACTCOF(L)
            ENDDO
            ACTPT(COL-DROP)=ACTPT(COL)-SHIFT
         ELSE
            DROP=DROP+1
            SHIFT=SHIFT+ACTPT(COL+1)-ACTPT(COL)
            INFLAG(NAMES(COL))=.FALSE.
            FIXV(NAMES(COL))=-1
            ACNAM(NAMES(COL))=0
            IF (ITER.EQ.1) FXAC=FXAC+1
         ENDIF
         GO TO 10
 20      IF (ITER.GT.1) GO TO 60
C Fix inactive cols.
         DO 25 K=1,NQSQ
            IF (INFLAG(K) .OR. FIXV(K).NE.0) GO TO 25
            REDC=PROFIT(K)
            DO I=COLPT(K)+1,COLPT(K+1)
               IF (ACINX(I).LE.NQ2) THEN
                  REDC=REDC-UZERO(ACINX(I))
               ELSE
                  REDC=REDC+UZERO(ACINX(I))
               ENDIF
            ENDDO
            IF (DABS(REDC).LT.ZL) ZRO=ZRO+1
            IF (REDC.LT.ZDIFF+ZL) GO TO 25
            FXIA=FXIA+1
            FIXV(K)=-1
 25      CONTINUE
         DO 30 K=NQSQ+1,NVAR
            IF (INFLAG(K) .OR. FIXV(K).NE.0) GO TO 30
            REDC=PROFIT(K)
            DO I=COLPT(K)+1,COLPT(K+1)
               REDC=REDC-UZERO(ACINX(I))
            ENDDO
            IF (DABS(REDC).LT.ZL) ZRO=ZRO+1
            IF (REDC.LT.ZDIFF+ZL) GO TO 30
            FXIA=FXIA+1
            FIXV(K)=-1
 30      CONTINUE
C Fix flow vars by logical implications.
 60      DO K=1,MACT
            APOS(K)=0
            ANEG(K)=0
         ENDDO
         DO K=1,NQSQ
            REDC0(K)=0
            IF (FIXV(K).GE.0) THEN
               DO L=COLPT(K)+1,COLPT(K+1)
                  IF (ACINX(L).LE.NQ2) THEN
                     APOS(ACINX(L))=APOS(ACINX(L))+1
                  ELSE
                     ANEG(ACINX(L))=ANEG(ACINX(L))+1
                  ENDIF
               ENDDO
            ENDIF
         ENDDO
         DO K=NQSQ+1,NVAR
            REDC0(K)=0
            IF (FIXV(K).GE.0) THEN
               DO L=COLPT(K)+1,COLPT(K+1)
                  APOS(ACINX(L))=APOS(ACINX(L))+1
               ENDDO
            ENDIF
         ENDDO
         DO I=1,MACT
            IF (ABS(B(I)).LT.ZL .AND. ANEG(I).EQ.0) THEN
               DO K=ROWPT(I)+1,ROWPT(I+1)
                  IF (INFLAG(ARINX(K))) THEN
                     IF (FIXV(ARINX(K)).GE.0) THEN
                        REDC0(ACNAM(ARINX(K)))=ZDIFF+1.0
                        FIXV(ARINX(K))=-1
                        FXAC=FXAC+1
                     ENDIF
                  ELSE
                     IF (FIXV(ARINX(K)).EQ.0) THEN
                        FIXV(ARINX(K))=-1
```

```
      SUBROUTINE SETMIP(NACT,  MEQ,   MACT,   NZAC,   NVAR,    NQ,  NBLO,
     *                  NZF,   NZA,    MXM,   MXNZ,     ZL,    NI,
     *                 BLKNO, PLANO,   BLPT, INFLAG,   FIXV,     B,
     *                 ARINX, ROWPT,  ACINX,  COLPT, ACTRIX,ACTCOF,ACTPT,
     *                 NAMES, ACNAM,  COSTA,  ROSTA,  APRFT,PROFIT, MARK)
      INTEGER NACT,MEQ,MACT,NZAC,NVAR,NQ,NBLO,NZF,NZA,MXM,MXNZ,NI
      REAL*8  ZL
      INTEGER ACTRIX(MXNZ),ACTPT(MXM+1),NAMES(NVAR),MARK(MXM),B(MXM)
      INTEGER ARINX(MXNZ),ROWPT(MXM+1),ACINX(MXNZ),COLPT(NVAR+1)
      INTEGER PROFIT(NVAR),COSTA(NVAR),ROSTA(MXM),ACTCOF(MXNZ)
      INTEGER FIXV(NVAR),BLKNO(NZF),PLANO(NZF),BLPT(NQ+1),ACNAM(NVAR)
      INTEGER APRFT(NVAR)
      LOGICAL INFLAG(NVAR)
      INTEGER NQSQ,COL,L,I,J,K,BSIZ,NQ2,ROW,MCO,ANF,NZ,P,Q,M,N,LEN
      LEN=0
      DO K=1,NVAR
        IF (FIXV(K).EQ.-1) LEN=LEN+1
      ENDDO
      WRITE(6,*)'SETMIP:',NACT,' Avars,',MACT,' cons,',
     *           NZAC,' nzs,',NVAR,' vars,',LEN,' fixed.'
      NQ2=2*NQ-1
      NQSQ=NQ*NQ
      BSIZ=NQ*(NQ-1)/2
C Generate the inequs (1.11). We use ACTRIX, ACTPT as scratch here.
      NZ=0
      MCO=0
      DO K=1,NQ
        IF (BLPT(K+1).GT.BLPT(K)) THEN
          DO J=1,NQ-1
            DO L=J+1,NQ
              ACTRIX(NZ)=(K-1)*NQ+J
              NZ=NZ+1
              ACTRIX(NZ)=(K-1)*NQ+L
              NZ=NZ+1
              DO P=BLPT(K)+1,BLPT(K+1)
                Q=BLKNO(P)
                NZ=NZ+1
                ACTRIX(NZ)=NQSQ+(Q-1)*BSIZ+(J-1)*NQ-J*(J-1)/2 + L - J
              ENDDO
              MCO=MCO+1
              ACTPT(MCO+1)=NZ
              IF (NZ+2*NQ .GE. MXNZ .OR. MCO .GE. MXM) THEN
                WRITE(6,*)' Not enough storage for inequs (1.11).'
```

```
                FXIA=FXIA+1
              ENDIF
            ENDDO
          ENDDO
          IF (FXAC+FXIA.GT.TOTFX) GO TO 1
 100      WRITE(6,*)'FIXRCO: Reduced prob:',NACT,' active cols.',
     *               ZRO,' zero reduced cost cols.'
          WRITE(6,*)
     *      ' FIXRCO: Fixed ',FXAC,' active, ',FXIA,' inactive cols.'
          RETURN
          END
C------------------------------------------------------------------
          SUBROUTINE CALRCO(  NQ,   MACT,    MACT,     NZAC,
     *                      PROFIT, UZERO,  REDCO,    ACINX,   COLPT)
          INTEGER NQ,MACT,MACT,NZAC
          INTEGER PROFIT(MACT),ACINX(NZAC),COLPT(MACT+1)
          REAL*8  UZERO(MACT),REDCO(MACT),REDC
          INTEGER NQSQ,NQ2,K,I
          NQSQ=NQ*NQ
          NQ2=2*NQ-1
          DO K=1,NQSQ
            REDC=PROFIT(K)
            DO I=COLPT(K)+1,COLPT(K+1)
              IF (ACINX(I).LE.NQ2) THEN
                REDC=REDC-UZERO(ACINX(I))
              ELSE
                REDC=REDC+UZERO(ACINX(I))
              ENDIF
            ENDDO
            REDCO(K)=REDC
          ENDDO
          DO K=NQSQ+1,NACT
            REDC=PROFIT(K)
            DO I=COLPT(K)+1,COLPT(K+1)
              REDC=REDC-UZERO(ACINX(I))
            ENDDO
            REDCO(K)=REDC
          ENDDO
          RETURN
          END
C------------------------------------------------------------------
```

```
            WRITE(6,*)' Increase MXNZA and/or MXM.'
            STOP
          ENDIF
        ENDDO
      ENDDO
C Check every constraint for containment in active row set.
      DO 20 ROW=1,MCO
        MARK(ROW)=0
        ANF=ACTPT(ROW)
        LEN=ACTPT(ROW+1)-ANF
        DO 10 I=MEQ+1,MACT
          IF (LEN.NE.ROWPT(I+1)-ROWPT(I)) GO TO 10
          K=0
          DO L=ROWPT(I)+1,ROWPT(I+1)
            K=K+1
            IF (ARINX(L).NE.ACTRIX(ANF+K)) GO TO 10
          ENDDO
          MARK(ROW)=1
          GO TO 20
   10   CONTINUE
   20 CONTINUE
C Check for blatantly inactive rows and fix some more vars.
      DO 25 K=1,MCO
        IF (MARK(ROW).EQ.1) GO TO 25
        P=0
        Q=0
        DO L=ACTPT(ROW)+1,ACTPT(ROW+1)
          IF (FIXV(ACTRIX(L)).GE.O) THEN
            IF (ACTRIX(L).LE.NQSQ) Q=Q+1
            IF (ACTRIX(L).GT.NQSQ) P=P+1
          ENDIF
        ENDDO
        IF (P.EQ.O) THEN
          MARK(ROW)=1
          GO TO 25
        ENDIF
        IF (Q.GT.O) GO TO 25
        DO L=ACTPT(ROW)+1,ACTPT(ROW+1)
          FIXV(ACTRIX(L))=-1
        ENDDO
        MARK(ROW)=1
   25 CONTINUE
C Fill in the rowstructure.
      M=MACT
      DO 30 ROW=1,MCO
        IF (MARK(ROW).EQ.1) GO TO 30
        M=M+1
        IF (M.GT.MXM .OR. NZA+ACTPT(ROW+1)-ACTPT(ROW).GT.MXNZ) THEN
          WRITE(6,*),' Not enough storage. Increase MXM and MXNZA.'
          STOP
        ENDIF
        DO L=ACTPT(ROW)+1,ACTPT(ROW+1)
          NZA=NZA+1
          ARINX(NZA)=ACTRIX(L)
        ENDDO
        ROWPT(M+1)=NZA
   30 CONTINUE
      CALL AIXCHG( ROWPT, COLPT, ARINX, ACINX, NZA, M, NVAR)
C Rebuild the LP active colstructure.
      NZAC=0
      DO L=1,NACT
        COL=NAMES(L)
        IF (COL.LE.NQSQ) THEN
          DO K=COLPT(COL)+1,COLPT(COL+1)
            NZAC=NZAC+1
            ACTRIX(NZAC)=ACINX(K)
            IF (ACINX(K).LE.2*NQ-1) THEN
              ACTCOF(NZAC)=1
            ELSE
              ACTCOF(NZAC)=-1
            ENDIF
          ENDDO
        ELSE
          DO K=COLPT(COL)+1,COLPT(COL+1)
            NZAC=NZAC+1
            ACTRIX(NZAC)=ACINX(K)
            ACTCOF(NZAC)=1
          ENDDO
        ENDIF
        ACTPT(L+1)=NZAC
      ENDDO
C Add the nonfixed inactive colums.
      N=NACT
      DO 50 K=1,NVAR
```

```fortran
            IF (INFLAG(K) .OR. FIXV(K).EQ.-1) GO TO 50
            N=N+1
            DO L=COLPT(K)+1,COLPT(K+1)
               NZAC=NZAC+1
               ACTRIX(NZAC)=ACINX(L)
               ACTCOF(NZAC)=1
            ENDDO
            NAMES(N)=K
            INFLAG(K)=.TRUE.
            ACNAM(K)=N
            APRFT(N)=PROFIT(K)
            ACTPT(N+1)=NZAC
 50      CONTINUE
C Update the basis info.
         DO K=NACT+1,N
            COSTA(K)=0
         ENDDO
         NACT=N
         DO K=NACT+1,M
            ROSTA(K)=1
         ENDDO
         MACT=M
         NI=0
         DO K=1,NACT
            IF (NAMES(K).LE.NQSQ) NI=NI+1
         ENDDO
         WRITE(6,*)'MIPproblem:',NI,' 0-1 vars,',NACT,
     *             ' vars,',MACT,' rows,',NZAC,' nzs.'
         RETURN
         END
C--------------------------------------------------------------
C SRTUPR: sorts A in ascending order and updates B.
         SUBROUTINE SRTUPR( N, A, B)
         INTEGER N,TEMP1,I,L,M,K
         REAL*8  A(N),TEMP
         INTEGER B(N)
         I=N/2
         L=I
         M=N
         DO K=1,L
            CALL HEAPR(I,M,A,B)
            I=I-1
         ENDDO
         DO K=2,N
            TEMP=A(1)
            TEMP1=B(1)
            A(1)=A(M)
            B(1)=B(M)
            A(M)=TEMP
            B(M)=TEMP1
            M=M-1
            CALL HEAPR(1,M,A,B)
         ENDDO
         RETURN
         END
C--------------------------------------------------------------
         SUBROUTINE HEAPR( I, N, A, B)
         INTEGER  N,I,J,K,L,TEMP1
         REAL*8   A(N),TEMP
         INTEGER  B(N)
         J=I
 100     K=2*J
         L=K+1
         IF (L.LE.N .AND. (A(J).LT.A(K) .OR. A(J).LT.A(L))) GOTO 200
         GOTO 500
 200     IF (A(K).GT.A(L)) GOTO 300
         GOTO 400
 300     TEMP=A(J)
         TEMP1=B(J)
         A(J)=A(K)
         B(J)=B(K)
         J=K
         A(K)=TEMP
         B(K)=TEMP1
         GOTO 100
 400     TEMP=A(J)
         TEMP1=B(J)
         A(J)=A(L)
         B(J)=B(L)
         A(L)=TEMP
         B(L)=TEMP1
         J=L
         GOTO 100
 500     IF (K.EQ.N .AND. A(J).LT.A(K)) GOTO 600
         GOTO 700
 600     TEMP=A(J)
```

```
          TEMP1=B(J)
          A(J)=A(N)
          B(J)=B(N)
          A(N)=TEMP
          B(N)=TEMP1
700   RETURN
      END
C-----------------------------------------------------------------
C SRTUPI: sorts A in ascending order and updates B.
      SUBROUTINE SRTUPI( N, A, B )
      INTEGER  N,TEMP,TEMP1,I,L,M,K
      INTEGER  A(N),B(N)
      I=N/2
      L=I
      M=N
      DO K=1,L
        CALL HEAPI(I,M,A,B)
        I=I-1
      ENDDO
      DO K=2,N
        TEMP=A(1)
        TEMP1=B(1)
        A(1)=A(M)
        B(1)=B(M)
        A(M)=TEMP
        B(M)=TEMP1
        M=M-1
        CALL HEAPI(1,M,A,B)
      ENDDO
      RETURN
      END
C-----------------------------------------------------------------
      SUBROUTINE HEAPI( I, N, A, B )
      INTEGER  N,TEMP,I,J,K,L,TEMP1
      INTEGER  A(N),B(N)
      J=I
100   K=2*J
      L=K+1
      IF (L.LE.N .AND. (A(J).LT.A(K) .OR. A(J).LT.A(L))) GOTO 200
      GOTO 500
200   IF (A(K).GT.A(L)) GOTO 300
      GOTO 400
300   TEMP=A(J)
          TEMP1=B(J)
          A(J)=A(K)
          B(J)=B(K)
          J=K
          A(K)=TEMP
          B(K)=TEMP1
          GOTO 100
400       TEMP=A(J)
          TEMP1=B(J)
          A(J)=A(L)
          B(J)=B(L)
          A(L)=TEMP
          B(L)=TEMP1
          J=L
          GOTO 100
500   IF (K.EQ.N .AND. A(J).LT.A(K)) GOTO 600
      GOTO 700
600   TEMP=A(J)
      TEMP1=B(J)
      A(J)=A(N)
      B(J)=B(N)
      A(N)=TEMP
      B(N)=TEMP1
700   RETURN
      END
C-----------------------------------------------------------------
C FUNCTION CPUTIM : returns the accumulated CPU time in seconds.
      REAL FUNCTION CPUTIM()
      REAL TIME(2)
      REAL ETIME
      CPUTIM = ETIME(TIME)
      RETURN
      END
C-----------------------------------------------------------------
      SUBROUTINE ZQAPHE( NQ, MXNZA )
      INTEGER NQ,MXNZA
      INTEGER  M01,M02,M03,M04,M05,M06,M07,M08,M09,M10
      COMMON/PT/ M01,M02,M03,M04,M05,M06,M07,M08,M09,M10
      M01=1
      M02=M01+NQ*NQ
      M03=M02+NQ*NQ
```

193

```
          M04=M03+NQ*NQ
          M05=M04+NQ
          M06=M05+NQ
          M07=M06+NQ
          M08=M07+NQ
          M09=M08+NQ
          M10=M09+NQ
          IF (M10.GT.MXNZA) THEN
             WRITE(6,*)' Not enough workspace. Increase Mxnza to',M10
             STOP
          ENDIF
          RETURN
          END
C----------------------------------------------------------------------
          SUBROUTINE QAPHEU(NVAR, NQ,NBLO,NIO,NIN,  NZF, MXITR,   OBJ,
      *                     VAROUT, PLANO, BLKNO,  BLPT,  CAUX,  DAUX,
      *                       COST, FLOW,  DIST, RAPLA, RALOC,  ASPL,
      *                       ASLO, AUX1,  AUX2,  HEAP,  OBJV, RATIO)

          INTEGER NQ,NBLO,NIO,NIN,NZF,MXITR,NVAR,OBJ
          INTEGER COST(NQ,NQ),FLOW(NQ,NQ),DIST(NQ,NQ),BLPT(NQ+1)
          INTEGER RALOC(NQ),RAPLA(NQ),ASPL(NQ),ASLO(NQ),PLANO(NZF)
          INTEGER AUX1(NQ),AUX2(NQ),BLKNO(NZF),VAROUT(MXITR*(NQ+NBLO))
          INTEGER HEAP(MXITR*NQ),OBJV(MXITR),CAUX(NQ,NQ),DAUX(NQ,NQ)
          REAL*8 RATIO(NQ)
          INTEGER COST1,DIFF,MIN,MAX,BIG,K,TRY,NOAS,TWO
          INTEGER I,NXTL,NXTP,LOB,TRYB,KEY,PL,LO,VARCNT,IND,BLO
          INTEGER LI,LK,NQSQ,BSIZ,L,K1,LI,ITER,SPIEL,ALPHA,KRED
          BIG=2**30
          DO K=1,NQ
             CAUX(I,K)=COST(I,K)
             DAUX(I,K)=DIST(I,K)
          ENDDO
          SPIEL=0
          ITER=0
    1     ITER=ITER+1
          IF (ITER.GT.MXITR) GO TO 150
          K=1+(ITER-(MXITR/2)*SPIEL)/2
C Establish a "reasonable" ranking of plants and locations.
          CALL RANKPL( K, NQ, BIG, RAPLA, RALOC, CAUX, FLOW, DAUX,
      *                                          AUX1, AUX2,RATIO)
C
C Main loop.
```

```
             TWO=1
             OBJ=0
    2        NOAS=0
   10        NOAS=NOAS+1
             IF (NOAS.GT.NQ) GO TO 50
             NXTP=RAPLA(NOAS)
             ASPL(NOAS)=NXTP
             NXTL=RALOC(NOAS)
             ASLO(NOAS)=NXTL
             CALL COSTPL(COST1,NXTP,NXTL,NOAS,NQ,ASPL,ASLO,CAUX,DAUX,FLOW)
             OBJ=OBJ+COST1
             KEY=1
C Do a 2-exchange on the partial assignment.
             MAX=0
             TRY=NOAS
   20        TRY=TRY-1
             IF (TRY.EQ.0) GO TO 30
             CALL CO2EXC( DIFF, TRY,NOAS,NOAS,   NQ,ASPL,ASLO,CAUX,FLOW,DAUX)
             IF (DIFF.GT.MAX) THEN
                MAX=DIFF
                TRYB=TRY
                LOB=ASLO(TRY)
             ENDIF
             GO TO 20
   30        IF (MAX.GT.0) THEN
                OBJ=OBJ-MAX
                ASLO(TRYB)=NXTL
                ASLO(NOAS)=LOB
                NXTL=LOB
                IF (KEY.EQ.2) GO TO 10
             ENDIF
C Greedy step.
             IF (NOAS.GE.NQ-1) GO TO 10
             MAX=-BIG
             DO K=NOAS+1,NQ
                IF (FLOW(RAPLA(K),NXTP).GT.MAX) THEN
                   MAX=FLOW(RAPLA(K),NXTP)
                   PL=RAPLA(K)
                   K1=K
                ENDIF
             ENDDO
             IF (MAX.LE.0) GO TO 10
             MIN=BIG
```

194

```
              DO K=NOAS+1,NQ
                 I=RALOC(K)
                 CALL COSTPL(COST1,  PL,   I,NOAS,NQ,ASPL,ASLO,CAUX,DAUX,FLOW)
                 IF (COST1.LT.MIN) THEN
                    MIN=COST1
                    L0=I
                    L1=K
                 ENDIF
              ENDDO
              NOAS=NOAS+1
C     Fix the rankings.
              RAPLA(K1)=RAPLA(NOAS)
              RALOC(L1)=RALOC(NOAS)
              ASPL(NOAS)=PL
              NXTP=PL
              ASLO(NOAS)=L0
              NXTL=L0
              OBJ=OBJ+MIN
              KEY=2
              TRY=NOAS
              MAX=0
              GO TO 20
C     Now do a complete 2-opt exchange on the solution.
 50           TRY=0
 70           TRY=TRY+1
              IF (TRY.GE.NQ) GO TO 100
              MAX=0
 80           NOAS=TRY
              NOAS=NOAS+1
              IF (NOAS.GT.NQ) GO TO 90
              CALL CO2EXC( DIFF, TRY,NOAS,  NQ, NQ,ASPL,ASLO,CAUX,FLOW,DAUX)
              IF (DIFF.GT.O) THEN
                 MAX=MAX+DIFF
                 OBJ=OBJ-DIFF
                 LOB=ASLO(NOAS)
                 ASLO(NOAS)=ASLO(TRY)
                 ASLO(TRY)=LOB
              ENDIF
              GO TO 80
 90           IF (MAX.GT.O) GO TO 50
              GO TO 70
C     Set the variables for output.
 100          CALL SRTUPI( NQ, ASPL, ASLO)
              IF (ITER.EQ.1) THEN
                 OBJV(1)=OBJ
                 DO K=1,NQ
                    HEAP(K)=ASLO(K)
                 ENDDO
              ELSE
                 DO K=ITER-1,1,-1
                    IF (OBJ.GE.OBJV(K)) GO TO 110
                 ENDDO
                 K=0
 110             DO L=ITER-1,K+1,-1
                    OBJV(L+1)=OBJV(L)
                 ENDDO
                 OBJV(K+1)=OBJ
                 DO L=(ITER-1)*NQ,K*NQ+1,-1
                    HEAP(NQ+L)=HEAP(L)
                 ENDDO
                 DO L=1,NQ
                    HEAP(K*NQ+L)=ASLO(L)
                 ENDDO
              ENDIF
              IF (ITER.EQ.MXITR/2) GO TO 120
              IF (TWO.EQ.2) GO TO 1
              TWO=2
              DO K=1,NQ
                 RAPLA(K)=ASLO(K)
                 RALOC(K)=K
              ENDDO
              ITER=ITER+1
              GO TO 2
C     Now the whole spiel with an asymmetrically reduced distance matrix.
 120          SPIEL=SPIEL+1
              DO K=1,NQ
                 AUX1(K)=0
              ENDDO
              DO K=1,NQ
                 DO I=1,NQ
                    AUX1(K)=AUX1(K)+FLOW(I,K)
                 ENDDO
              ENDDO
              MAX=-BIG
 130          DO K=1,NQ
                 DO I=1,NQ
```

```
              IF (I.NE.K) THEN
                 IF (DAUX(I,K).GT.MAX) MAX=DAUX(I,K)
              ENDIF
            ENDDO
            DO K=1,NQ
              RAPLA(K)=MAX+1
              RALOC(K)=MAX+1
            ENDDO
            DO I=1,NQ
              IF (I.NE.K) THEN
                 IF (DAUX(K,I).LT.RAPLA(K)) RAPLA(K)=DAUX(K,I)
              ENDIF
            ENDDO
            DO I=1,NQ
              IF (I.NE.K) THEN
                 IF (DAUX(K,I).LT.RALOC(I)) RALOC(I)=DAUX(K,I)
              ENDIF
            ENDDO
          ENDDO
          IF (RAPLA(K).GT.0 .OR. RALOC(K).GT.0) THEN
             ALPHA=RAPLA(K)+RALOC(K)
             KRED=K
             GO TO 135
          ENDIF
          GO TO 1
 135      DO K=1,NQ
            IF (K.NE.KRED) THEN
               DAUX(KRED,K)=DAUX(KRED,K)-RAPLA(KRED)
               DAUX(K,KRED)=DAUX(K,KRED)-RALOC(KRED)
               CAUX(K,KRED)=CAUX(K,KRED)+ALPHA*AUX1(K)
            ELSE
               CAUX(K,KRED)=CAUX(K,KRED)+ALPHA*AUX1(K)
            ENDIF
          ENDDO
          GO TO 130
C Set the var indices (with replication) etc.
 150      VARCNT=0
          NQSQ=NQ*NQ
          BSIZ=NQ*(NQ-1)/2
          DO TRY=1,ITER-1
            DO I=1,NQ
              IND=(I-1)*NQ+HEAP((TRY-1)*NQ+I)
              VARCNT=VARCNT+1
              VAROUT(VARCNT)=IND
            ENDDO
            DO I=1,NQ-1
              LI=HEAP((TRY-1)*NQ+I)
              DO 200 K=I+1,NQ
                IF (FLOW(I,K).EQ.0) GO TO 200
                LK=HEAP((TRY-1)*NQ+K)
                DO L=BLPT(I)+1,BLPT(I+1)
                  IF (PLANO(L).EQ.K) THEN
                     BLO=BLKNO(L)
                     IF (LI.LT.LK) THEN
                        IND=NQSQ+(BLO-1)*BSIZ+(LI-1)*NQ-LI*(LI-1)/2+LK-LI
                     ELSE
                        IND=NQSQ+(BLO-1)*BSIZ+(LK-1)*NQ-LK*(LK-1)/2+LI-LK
                     ENDIF
                     VARCNT=VARCNT+1
                     VAROUT(VARCNT)=IND
                     GO TO 200
                  ENDIF
                ENDDO
 200          CONTINUE
            ENDDO
          ENDDO
          IF (TRY.EQ.1) NIO=VARCNT
        ENDDO
        RATIO(1)=0
        DO K=1,ITER-1
          RATIO(1)=RATIO(1)+OBJV(K)
        ENDDO
        NIN=VARCNT
        RATIO(1)=RATIO(1)/(DFLOAT(ITER-1))
        WRITE(6,*) 'QAPHEU: Iters=',ITER-1,' No. vars found=',NIN,
     *                            ' No. in sol=',NIO, ObjValues:'
        WRITE(6,*) (OBJV(K),K=1,ITER-1)
        WRITE(6,*) 'Average solution value=',RATIO(1)
        OBJ=OBJV(1)
        RETURN
        END
```

```fortran
C-----------------------------------------------------------------------
      SUBROUTINE RANKPL( ITER, NQ, BIG, RAPLA, RALOC, COST, FLOW, DIST,
     *                                  AUX1, AUX2,RATIO)
      INTEGER ITER,NQ,BIG,I,K,COUNT,F,D,CNT,MAX,MIN,NXTP,NXTL
      INTEGER RAPLA(NQ),RALOC(NQ),COST(NQ,NQ),FLOW(NQ,NQ),DIST(NQ,NQ)
      INTEGER AUX1(NQ),AUX2(NQ)
      REAL*8  RATIO(NQ),RMIN,AF,RMAX
      DO K=1,NQ
        AUX1(K)=0
        AUX2(K)=0
      ENDDO
      IF (ITER.GT.1) GO TO 2
C Rank by increasing total dist and decreasing total flow for each k.
      DO K=1,NQ
        DO I=1,NQ
          AUX1(K)=AUX1(K)+DIST(I,K)
          AUX2(K)=AUX2(K)-FLOW(I,K)
        ENDDO
        RALOC(K)=K
        RAPLA(K)=K
      ENDDO
      CALL SRTUPI( NQ, AUX1, RALOC)
      CALL SRTUPI( NQ, AUX2, RAPLA)
      RETURN
 2    IF (ITER.GT.2) GO TO 3
C Rank by decreasing total dist and increasing total flow for each k.
      DO K=1,NQ
        DO I=1,NQ
          AUX1(K)=AUX1(K)-DIST(I,K)
          AUX2(K)=AUX2(K)+FLOW(I,K)
        ENDDO
        RALOC(K)=K
        RAPLA(K)=K
      ENDDO
      CALL SRTUPI( NQ, AUX1, RALOC)
      CALL SRTUPI( NQ, AUX2, RAPLA)
      RETURN
 3    IF (ITER.GT.3) GO TO 4
C Rank locs by increasing dist, plants by decreasing no of connects.
      DO K=1,NQ
        DO I=1,NQ
          AUX1(K)=AUX1(K)+DIST(I,K)
          IF (FLOW(I,K).GT.0) AUX2(K)=AUX2(K)-1
        ENDDO
        RALOC(K)=K
        RAPLA(K)=K
      ENDDO
      CALL SRTUPI( NQ, AUX1, RALOC)
      CALL SRTUPI( NQ, AUX2, RAPLA)
      RETURN
 4    IF (ITER.GT.4) GO TO 5
C Rank locs by decreasing dist, plants by increasing no of connects.
      DO K=1,NQ
        DO I=1,NQ
          AUX1(K)=AUX1(K)-DIST(I,K)
          IF (FLOW(I,K).GT.0) AUX2(K)=AUX2(K)+1
        ENDDO
        RALOC(K)=K
        RAPLA(K)=K
      ENDDO
      CALL SRTUPI( NQ, AUX1, RALOC)
      CALL SRTUPI( NQ, AUX2, RAPLA)
      RETURN
 5    IF (ITER.GT.5) GO TO 6
C Rank locs and plants in their natural order.
      DO K=1,NQ
        RAPLA(K)=K
        RALOC(K)=K
      ENDDO
      RETURN
 6    IF (ITER.GT.6) GO TO 7
C Rank plants 1,...,Nq, locations Nq,....,2,1.
      DO K=1,NQ
        RAPLA(K)=K
        RALOC(K)=NQ+1-K
      ENDDO
      RETURN
 7    IF (ITER.GT.7) GO TO 8
C Rank locs by increasing dist, plants by decreasing average flow.
      DO K=1,NQ
        COUNT=0
        DO I=1,NQ
          AUX1(K)=AUX1(K)+DIST(I,K)
          AUX2(K)=AUX2(K)+FLOW(I,K)
          IF (FLOW(I,K).GT.0) COUNT=COUNT+1
        ENDDO
        RALOC(K)=K
        RAPLA(K)=K
      ENDDO
      CALL SRTUPI( NQ, AUX1, RALOC)
      CALL SRTUPI( NQ, AUX2, RAPLA)
      RETURN
```

```
              RALOC(K)=K
              RAPLA(K)=K
              RATIO(K)=0
              IF (COUNT.GT.0) RATIO(K)=-DFLOAT(AUX2(K))/DFLOAT(COUNT)
           ENDDO
           CALL SRTUPI( NQ, AUX1, RALOC)
           CALL SRTUPR( NQ,RATIO, RAPLA)
           RETURN
    8      IF (ITER.GT.8) GO TO 9
C Rank locs by decreasing dist, plants by increasing average flow.
           COUNT=0
           DO K=1,NQ
              AUX1(K)=AUX1(K)-DIST(I,K)
              AUX2(K)=AUX2(K)+FLOW(I,K)
              IF (FLOW(I,K).GT.0) COUNT=COUNT+1
           ENDDO
           RALOC(K)=K
           RAPLA(K)=K
           RATIO(K)=0
           IF (COUNT.GT.0) RATIO(K)=DFLOAT(AUX2(K))/DFLOAT(COUNT)
        ENDDO
        CALL SRTUPI( NQ, AUX1, RALOC)
        CALL SRTUPR( NQ,RATIO, RAPLA)
        RETURN
    9   IF (ITER.GT.9) GO TO 10
C Now do a greedy on total flow and dist.
        COUNT=0
   90   COUNT=COUNT+1
        MAX=-BIG
        DO K=1,NQ
           IF (AUX1(K).EQ.0) THEN
              F=0
              DO I=1,NQ
                 IF (COUNT.EQ.1) THEN
                    IF (FLOW(I,K).GT.F) F=FLOW(I,K)
                 ELSE
                    IF (AUX1(I).GT.0) F=F+FLOW(I,K)
                 ENDIF
              ENDDO
              IF (F.GT.MAX) THEN
                 MAX=F
                 NXTP=K
              ENDIF
           ENDIF
        ENDDO
        MIN=BIG
        DO K=1,NQ
           IF (AUX2(K).EQ.0) THEN
              D=0
              IF (COUNT.EQ.1) D=BIG
              DO I=1,NQ
                 IF (COUNT.EQ.1) THEN
                    IF (K.NE.I .AND. DIST(K,I).LT.D) D=DIST(K,I)
                 ELSE
                    IF (AUX2(I).GT.0) D=D+DIST(K,I)
                 ENDIF
              ENDDO
              D=D+COST(NXTP,K)
              IF (D.LT.MIN) THEN
                 MIN=D
                 NXTL=K
              ENDIF
           ENDIF
        ENDDO
        AUX1(NXTP)=NXTL
        AUX2(NXTL)=NXTP
        RAPLA(COUNT)=NXTP
        RALOC(COUNT)=NXTL
        IF (COUNT.LT.NQ) GO TO 90
        RETURN
   10   IF (ITER.GT.10) GO TO 11
C Same as 9, but with average flows and average distances.
        COUNT=0
  100   COUNT=COUNT+1
        RMAX=-BIG
        DO K=1,NQ
           IF (AUX1(K).EQ.0) THEN
              F=0
              CNT=0
              DO I=1,NQ
                 IF (COUNT.EQ.1) THEN
                    IF (FLOW(I,K).GT.F) F=FLOW(I,K)
                    CNT=1
                 ELSE
                    IF (AUX1(I).GT.0) THEN
```

```
            IF (FLOW(I,K).GT.0) CNT=CNT+1
            F=F+FLOW(I,K)
          ENDIF
        ENDDO
        AF=0
        IF (CNT.GT.0) AF=DFLOAT(F)/DFLOAT(CNT)
        IF (AF.GT.RMAX) THEN
          RMAX=F
          NXTP=K
        ENDIF
      ENDIF
    ENDDO
    RMIN=BIG
    DO K=1,NQ
      IF (AUX2(K).EQ.0) THEN
        D=0
        IF (COUNT.EQ.1) D=BIG
        CNT=0
        DO I=1,NQ
          IF (COUNT.EQ.1) THEN
            CNT=1
            IF (K.NE.I .AND. DIST(K,I).LT.D) D=DIST(K,I)
          ELSE
            IF (AUX2(I).GT.0) THEN
              D=D+DIST(K,I)
              IF (DIST(K,I).NE.0) CNT=CNT+1
            ENDIF
          ENDIF
        ENDDO
        AF=0
        IF (CNT.GT.0) AF=DFLOAT(D)/DFLOAT(CNT)
        AF=AF+COST(NXTP,K)
        IF (AF.LT.RMIN) THEN
          RMIN=AF
          NXTL=K
        ENDIF
      ENDIF
    ENDDO
    AUX1(NXTP)=NXTL
    AUX2(NXTL)=NXTP
    RAPLA(COUNT)=NXTP
    RALOC(COUNT)=NXTL
    IF (COUNT.LT.NQ) GO TO 100
11  RETURN
    END
C----------------------------------------
    SUBROUTINE COSTPL(COSTX,  PL,    LO,NOAS,   NQ,
    *                          ASPL,ASLO,COST,DIST,FLOW)
    INTEGER COSTX,PL,LO,NOAS,NQ,I,FLO
    INTEGER ASPL(NQ),ASLO(N),COST(NQ),DIST(NQ,NQ),FLOW(NQ,NQ)
    COSTX=COST(PL,LO)
    DO I=1,NOAS
      FLO=FLOW(PL,ASPL(I))
      IF (FLO.NE.0) THEN
        COSTX=COSTX+(DIST(LO,ASLO(I))+DIST(ASLO(I),LO))*FLO
      ENDIF
    ENDDO
    RETURN
    END
C----------------------------------------
    SUBROUTINE CO2EXC( DIFF, P1, P2,  N1,   NQ, ASPL, ASLO,
    *                                COST, FLOW, DIST)
    INTEGER DIFF,P1,P2,N1,NQ
    INTEGER ASPL(N1),ASLO(N1),COST(NQ),FLOW(NQ,NQ),DIST(NQ,NQ)
    INTEGER PL1,LO1,PL2,LO2,I,F,D
    IF (P1.LT.P2) THEN
      PL1=ASPL(P1)
      LO1=ASLO(P1)
      PL2=ASPL(P2)
      LO2=ASLO(P2)
    ELSE
      PL1=ASPL(P2)
      LO1=ASLO(P2)
      PL2=ASPL(P1)
      LO2=ASPL(P1)
    ENDIF
    DIFF=COST(PL1,LO1)+COST(PL2,LO2)-COST(PL1,LO2)-COST(PL2,LO1)
    DO I=1,P1-1
      F=FLOW(PL1,ASPL(I))-FLOW(PL2,ASPL(I))
      IF (F.NE.0) THEN
        D=DIST(ASLO(I),LO1)-DIST(ASLO(I),LO2)+
    *        DIST(LO1,ASLO(I))-DIST(LO2,ASLO(I))
        IF (D.NE.0) DIFF=DIFF+F*D
      ENDIF
    ENDDO
```

199

```
          DO I=P1+1,P2-1
            F=FLOW(PL1,ASPL(I))-FLOW(PL2,ASPL(I))
            IF (F.NE.0) THEN
              D=DIST(ASLO(I),LO1)-DIST(ASLO(I),LO2)+
     *          DIST(LO1,ASLO(I))-DIST(LO2,ASLO(I))
              IF (D.NE.0) DIFF=DIFF+F*D
            ENDIF
          ENDDO
          DO I=P2+1,N1
            F=FLOW(PL1,ASPL(I))-FLOW(PL2,ASPL(I))
            IF (F.NE.0) THEN
              D=DIST(ASLO(I),LO1)-DIST(ASLO(I),LO2)+
     *          DIST(LO1,ASLO(I))-DIST(LO2,ASLO(I))
              IF (D.NE.0) DIFF=DIFF+F*D
            ENDIF
          ENDDO
      ENDDO
      RETURN
      END
C-----------------------------------------------------------
      SUBROUTINE QAPLOW( NQ,    OBJ,      ZL, BOUND,
     *              COST,    FLOW,    DIST,    TK,   CAUX,
     *         ACTRIX,ACTCOF,ACTPT,    B,PROFIT, REDCO, UZERO,  XSOL)
      INTEGER NQ,OBJ
      INTEGER COST(NQ,NQ),FLOW(NQ,NQ),DIST(NQ,NQ),ACTRIX(3*NQ*NQ)
      INTEGER ACTCOF(3*NQ*NQ),ACTPT(NQ*NQ+1),PROFIT(NQ*NQ),B(2*NQ+1)
      INTEGER DJ(NQ),TK(NQ),CAUX(NQ,NQ) NQ2,MEQ,ECHO,COUNT
      REAL*8 REDCO(NQ*NQ),UZERO(2*NQ+1),XSOL(NQ*NQ),BOUND,ZL,Z
      INTEGER BIG,NQSQ,I,J,K,CNT,PRIMAL,PERTU,OLDBA,TERMIN,L
      REAL    T0,CPUTIM
      BIG=2**30
      NQSQ=NQ*NQ
      NQ2=2*NQ
      DO J=1,NQ
        DJ(J)=BIG
        TK(J)=0
        DO K=1,NQ
          TK(J)=TK(J)+FLOW(K,J)
          IF (J.NE.K) THEN
            IF (DIST(J,K).LT.DJ(J)) DJ(J)=DIST(J,K)
          ENDIF
        ENDDO
      ENDDO
      CNT=0
      ACTPT(1)=0
      DO I=1,NQ
        DO J=1,NQ
          CNT=CNT+1
          PROFIT(CNT)=COST(I,J)+DJ(J)*TK(I)
          ACTCOF(2*CNT-1)=1
          ACTRIX(2*CNT-1)=I
          ACTCOF(2*CNT)=1
          ACTRIX(2*CNT)=NQ+J
          ACTPT(CNT+1)=2*CNT
        ENDDO
        B(I)=1
        B(NQ+I)=1
      ENDDO
      PRIMAL=1
      OLDBA=0
      PERTU=0
      ECHO=0
C Solve the linear program.
      CALL LPSOLV(NQSQ,NQ2,NQ2, BOUND,PRIMAL,TERMIN,   Z,  OLDBA,
     *                PERTU,  ECHO,PROFIT,    B, ACTPT,ACTRIX,ACTCOF,
     *                                 XSOL, UZERO, REDCO,    TK,   DJ)
      OBJ=Z+ZL
      WRITE(6,*) 'LOWbnd=',OBJ
C Gilmore-Lawler bound.
      CNT=0
      ACTPT(1)=0
      MEQ=NQ2+1
      DO I=1,NQ
        DO J=1,NQ
          CNT=CNT+1
          ACTCOF(3*CNT-2)=1
          ACTRIX(3*CNT-2)=I
          ACTCOF(3*CNT-1)=1
          ACTRIX(3*CNT-1)=NQ+J
          ACTCOF(3*CNT)   =0
          ACTRIX(3*CNT)   =MEQ
          ACTPT(CNT+1)=3*CNT
        ENDDO
      ENDDO
      B(MEQ)=1
      T0=CPUTIM()
```

```fortran
      CNT=0
      ECH0=0
      DO I=1,NQ
        COUNT=0
        DO J=1,NQ
          COUNT=0
          DO K=1,NQ
            DO L=1,NQ
              COUNT=COUNT+1
              PROFIT(COUNT)=FLOW(I,K)*DIST(J,L)
            ENDDO
          ENDDO
          CNT=CNT+1
          ACTCOF(3*CNT)=1
          CALL LPSOLV(NQSQ,MEQ,MEQ,BOUND,PRIMAL, TERMIN ,    Z, OLDBA,
     *                PERTU, ECHO,PROFIT,    B, ACTPT,ACTRIX,ACTCOF,
     *                                 XSOL,UZERO, REDCO,    TK,    DJ)
          ACTCOF(3*CNT)=0
          CAUX(I,J)=COST(I,J) + INT(Z+ZL)
        ENDDO
      ENDDO
      CNT=0
      DO I=1,NQ
        DO J=1,NQ
          CNT=CNT+1
          PROFIT(CNT)=CAUX(I,J)
          ACTCOF(2*CNT-1)=1
          ACTRIX(2*CNT-1)=I
          ACTCOF(2*CNT)=1
          ACTRIX(2*CNT)=NQ+J
          ACTPT(CNT+1)=2*CNT
        ENDDO
      ENDDO
C Solve the  linear program.
      CALL LPSOLV(NQSQ,NQ2, BOUND,PRIMAL,TERMIN ,    Z, OLDBA,
     *             PERTU,   ECHO,PROFIT,    B, ACTPT,ACTRIX,ACTCOF,
     *                                XSOL,UZERO, REDCO,    TK,    DJ)
      K=Z+ZL
      WRITE(6,*)'Gilmore-Lawler LB=',K,' CPUtim:',CPUTIM()-TO,' s.'
      IF (K.GT.OBJ) OBJ=K
      RETURN
      END
```

```c
/* LPSOLV: Interface QAPMIP and CPLEX
------------------------------------------------------------------*/
/* Include CPlex definitions */
#include "/usr/local/lib/cplex3.0/cpxdefs.inc"
#include <stdio.h>
#include <strings.h>
void
lpsolv_(nvar,meq,m,bound,primal,termin, z,oldba, echo,ndflt,
        profit,b,colpnt,acinx,accof,xsol,uzero,redco,rosta,costa)
int *primal,*termin,*nvar,*m,*meq,*oldba,*ndflt,*echo;
int *colpnt,*acinx,*rosta,*costa,*accof,*profit,*b;
double *bound,*z,*xsol,*uzero,*redco;
{
extern char *realloc();
extern struct cpxlp *loadprob();
extern int setscr_ind(), setitfoind(), lpwrite(), loadbase();
extern int setprind(), optimize(), dualopt(), hybbaropt();
extern int setdprind(),solution(), getitc(),getbase(), setedlimu();
extern int setperind(), setepmrk(), setepopt(), setreinv();
extern void freeprob();
/*---- CPLEX variables ---*/
struct cpxlp *lp;
int    mac, mar, objsen, *matbeg, *matcnt, *matind, pri_ind;
int    macsz, marsz, matsz, cplexstat, per_ind, re_inv, ca_list;
int    *cstat, *rstat, *rhsx, *matval, *bdl, *bdu, *x,*x,*piout,*slack,*dj;
double *objx, *rhsx, *matval, *bdl, *bdu, obj, *x,*x,*piout,*slack,*dj;
double ep_mrk, ep_opt, toosmall, toobig;
char   probname[16], *senx;
char   *dataname = (char *)NULL;
char   *objname  = (char *)NULL;
char   *rhsname  = (char *)NULL;
char   *rngname  = (char *)NULL;
char   *bndname  = (char *)NULL;
char   **cname  = (char **)NULL;
char   *cstore  = (char *)NULL;
char   **rname  = (char **)NULL;
char   *rstore  = (char *)NULL;
char   **ename  = (char **)NULL;
char   *estore  = (char *)NULL;
/*--- Local variables ---*/
int    iterat,iterd,i, j, k;
char   fname[10];
/*--- Set CPlex dimensions ---*/
```

201

```c
    mac = *nvar;
    mar = *m;
    macsz = *nvar;
    marsz = *m;
    matsz = colpnt[*nvar];
    /*--- Load CPLEX data structures (Part I) ---*/
    objx = (double *) malloc(macsz * sizeof(double));
    for (i=0;i<*nvar;i++) objx[i] = (double) profit[i];
    matbeg = (int *) malloc((macsz + 1) * sizeof(int));
    for (i=0;i<*nvar+1;i++) matbeg[i] = colpnt[i];
    matcnt = (int *) malloc(macsz * sizeof(int));
    for (i=0;i<*nvar;i++) matcnt[i] = matbeg[i+1] - matbeg[i];
    matind = (int *) malloc(matsz * sizeof(int));
    for (i=0;i<matsz;i++) matind[i] = acinx[i] - 1;
    matval = (double *) malloc(matsz * sizeof(double));
    for (i=0;i<matsz;i++) matval[i] = (double) accof[i];
    /*--- Initialize CPlex data structures (Part II) ---*/
    strcpy(probname,"LP");
    objsen = 1;
    rhsx = (double *) malloc(marsz * sizeof(double));
    for (i=0;i<*m;i++) rhsx[i] = (double) b[i];
    senx = (char *) malloc(marsz * sizeof(char));
    for (i=0;i<*meq;i++) senx[i] = 'E';
    for (i=*meq;i<*m;i++) senx[i] = 'L';
    bdl = (double *) malloc(macsz * sizeof(double));
    bdu = (double *) malloc(macsz * sizeof(double));
    for (i=0;i<*nvar;i++) {
        bdl[i] = 0;
        bdu[i] = *bound;
    }
    /* Output to screen */
    setscr_ind(*echo);
    setitfoind(*echo,&ptoosmall,&ptoobig);
    /*--- Load the problem ---*/
    lp = loadprob(probname, mac, mar, 0,
            objsen, objx, rhsx, senx,
            matbeg, matcnt, matind, matval, bdl, bdu,
            NULL, NULL, NULL, NULL, NULL, rhsname, rngname,
            dataname, objname, cname, cstore, rname, rstore, ename, estore,
            bndname, cname, cstore, rname, rstore, ename, estore,
            macsz, marsz, matsz,
            0, 0, (unsigned)0, (unsigned)0, (unsigned)0 );
    /*---Load old optimal basis---*/
    cstat = (int *) malloc( macsz * sizeof(int));
    rstat = (int *) malloc( marsz * sizeof(int));
    if (*oldba) {
        for (i=0;i<*m;i++)
            rstat[i] = rosta[i];
        for (i=0;i<*nvar;i++)
            cstat[i] = costa[i];
        loadbase(lp,cstat,rstat);
    }

    if (*ndflt) {
        ep_mrk = 0.02;
        setepmrk( ep_mrk, &toosmall, &toobig);
        ep_opt = 0.0001;
        setepopt( ep_opt, &toosmall, &toobig);
        re_inv = 75;
        setreinv( re_inv,&ptoosmall,&ptoobig);
        ca_list = 0;
        setedlimu(ca_list,&ptoosmall,&ptoobig);
        per_ind = 1;
        setperind(per_ind);
    }
    /*--- Solve LP ---*/
    if (*primal) {
        pri_ind = 2;
        setpprriind(pri_ind,&ptoosmall,&ptoobig);
        optimize(lp);
    }
    else {
        pri_ind = 2;
        setdprind(pri_ind,&ptoosmall,&ptoobig);
        dualopt(lp);
    }

    /*--- Return results---*/
    x    = (double *) malloc(macsz * sizeof(double));
    piout = (double *) malloc(marsz * sizeof(double));
    slack = (double *) malloc(marsz * sizeof(double));
    dj    = (double *) malloc(macsz * sizeof(double));
    solution(lp, &cplexstat, &obj, x, piout, slack, dj);
    /*--- Objective function value and solution status ---*/
    *z = obj;
    *termin = cplexstat;
    /*---Primal and dual solutions---*/
    for (i=0;i<*nvar;i++)
```

```c
/* MIPSOL: Interface QAPMIP and mipoptimize of CPLEX                  ---*/

/* Include CPlex definitions */
#include "/usr/local/lib/cplex3.0/cpxdefs.inc"
#include <stdio.h>
#include <strings.h>
void
mipsol_(nint,nvar,meq,m,first,termin,z,oldba,upbnd,bound,
        profit,b,colpnt,acinx,accof,xsol,rosta,costa)
int    *first,*termin,*nint,*nvar,*m,*meq,*oldba;
int    *colpnt,*acinx,*rosta,*costa,*accof,*profit,*b;
double *bound, *upbnd, *z, *xsol;
{
extern char *realloc();
extern struct cpxlp *loadmprob();
extern int setscr_ind(), setitfoind(), loadbase();
extern int mipoptimize(), getstat(), getmobjval(), setsossind();
extern int getndc(), getmx(), setheuristic(), setcutup();
extern void freeprob();
/*--- CPLEX variables  ---*/
struct cpxlp *lp;
int    mac, mar, objsen, *matbeg, *matcnt, *matind;
int    macsz, marsz, matsz, cplexstat, h_val, sos_ind;
int    *cstat, *rstat, ptoosmall, ptoobig;
double *objx, *rhsx, *matval, *bdl, *bdu, obj,*x, toosmall, toobig;
double cut_up;
char   probname[16], *senx, *ctype;
char   *dataname = (char *)NULL;
char   *objname  = (char *)NULL;
char   *rhsname  = (char *)NULL;
char   *rngname  = (char *)NULL;
char   *bndname  = (char *)NULL;
char   **cname   = (char **)NULL;
char   *cstore   = (char *)NULL;
char   **rname   = (char **)NULL;
char   *rstore   = (char *)NULL;
char   **ename   = (char **)NULL;
char   *estore   = (char *)NULL;
/*--- Local variables ---*/
int    iterat,i, j, k;
char   fname[10];
/*--- Set CPlex dimensions ---*/
mac   = *nvar;
mar   = *m;
macsz = *nvar;
marsz = *m;
matsz = colpnt[*nvar];
/*--- Load CPLEX data structures (Part I)  ---*/
objx   = (double *) malloc(macsz * sizeof(double));
for (i=0;i<*nvar;i++) objx[i] = (double) profit[i];
matbeg = (int *) malloc((macsz + 1) * sizeof(int));
for (i=0;i<*nvar+1;i++) matbeg[i] = colpnt[i];
matcnt = (int *) malloc(macsz * sizeof(int));
for (i=0;i<*nvar;i++) matcnt[i] = matbeg[i+1] - matbeg[i];
matind = (int *) malloc(matsz * sizeof(int));
for (i=0;i<matsz;i++) matind[i] = acinx[i] - 1;
```

```c
          xsol[i]=x[i];
      for (i=0;i<*m;i++)
          uzero[i] = piout[i];
      for (i=0;i<*nvar;i++)
          redcof[i] = dj[i];
/*---Remember basis info---*/
      getbase(lp,cstat,rstat);
      for (i=0;i<*m;i++)
          rosta[i] = rstat[i];
      for (i=0;i<*nvar;i++)
          costa[i] = cstat[i];
/*--- Unload the problem and free all malloced space ---*/
      freeprob(&lp);
      free(objx);
      free(rhsx);
      free(senx);
      free(matbeg);
      free(matcnt);
      free(matind);
      free(matval);
      free(bdl);
      free(bdu);
      free(x);
      free(piout);
      free(slack);
      free(cstat);
      free(rstat);
}/*lpsolv_*/
```

203

```c
        matval = (double *) malloc(matsz * sizeof(double));
        for (i=0;i<matsz;i++) matval[i] = (double) accof[i];
    /*--- Initialize CPlex data structures (Part II) ---*/
        strcpy(probname,"LP");
        objsen = 1;
        rhsx = (double *) malloc(marsz * sizeof(double));
        for (i=0;i<*m;i++) rhsx[i] = (double) b[i];
        senx = (char *) malloc(marsz * sizeof(char));
        for (i=0;i<*meq;i++) senx[i] = 'E';
        for (i=*meq;i<*m;i++) senx[i] = 'L';
        ctype = (char *) malloc(macsz * sizeof(char));
        for (i=0;i<*nint;i++) ctype[i] = 'B';
        for (i=*nint;i<*nvar;i++) ctype[i] = 'C';
        bdl = (double *) malloc(macsz * sizeof(double));
        bdu = (double *) malloc(macsz * sizeof(double));
        for (i=0;i<*nint;i++) {
            bdl[i] = 0;
            bdu[i] = 1;
        }
        for (i=*nint;i<*nvar;i++) {
            bdl[i] = 0;
            bdu[i] = *bound;
        }
    /*--- Set CPLEX parameters ---*/
    /* Output to screen */
        setscr_ind(1);
        setitfoind(1,&ptoosmall,&ptoobig);
    /*--- Load the problem ---*/
        lp = loadmprob(probname, mac, mar, 0,
                objsen, objx, rhsx, senx,
                matbeg, matcnt, matind, matval, bdl, bdu,
                NULL, NULL, NULL, NULL, NULL, NULL, NULL,
                dataname, objname, rhsname, rngname,
                bndname, cname, cstore, rname, rstore, ename, estore,
                macsz, marsz, matsz,
                0, 0, (unsigned)0, (unsigned)0, (unsigned)0, ctype);
    /*---Load old optimal basis---*/
        cstat = (int *) malloc( macsz * sizeof(int));
        rstat = (int *) malloc( marsz * sizeof(int));
        if (*oldba) {
            for (i=0;i<*m;i++)
                rstat[i] = rosta[i];
            for (i=0;i<*nvar;i++)
                cstat[i] = costa[i];
            loadbase(lp,cstat,rstat);
        }
    /*--- Solve MIP ---*/
        h_val = -1;
        setheuristic( h_val, &ptoosmall, &ptoobig);
        sos_ind = 0;
        setsosind( sos_ind, &ptoosmall, &ptoobig);
        cut_up = *upbnd + 0.1;
        setcutup (cut_up, &toosmall, &ptoobig);
        mipoptimize( lp );
    /*--- Return results---*/
        *termin = getstat( lp );
        *first = getndc( lp );
        x = (double *) malloc(macsz * sizeof(double));
        getmx( lp, x, 0, *nvar-1 );
        getmobjval( lp, &obj);
        *z = obj;
        for (i=0;i<*nvar;i++)
            xsol[i] = x[i];
    /*--- Unload the problem and free all malloced space ---*/
        freeprob(&lp);
        free(objx);
        free(rhsx);
        free(senx);
        free(ctype);
        free(matbeg);
        free(matcnt);
        free(matind);
        free(matval);
        free(bdl);
        free(bdu);
        free(x);
        free(cstat);
        free(rstat);
}/*mipsol_*/
```

REFERENCES

[1] Ahuja, R., T. Magnanti and J. Orlin [1993] *Network Flows: Theory, Algorithms, and Applications* Prentice-Hall, Englewood-Cliffs, New Jersey.

[2] Applegate, D., R. Bixby, V. Chvátal and W. Cook [1994] "Finding Cuts in the TSP," Paper presented in 15th International Symposium on Mathematical Programming, Ann Arbor, Michigan.

[3] Araque, A., L. Hall and T. Magnanti [1990] "Capacitated trees, capacitated routing, and associated polyhedra," Operations Research Center, M.I.T. Cambridge.

[4] Armour, G. and E. Buffa [1963] "A heuristic algorithm and simulation approach to relative location of facilities," *Management Science* 9 294-309.

[5] Assad, A. and W. Xu [1985] "On lower bounds for a class of quadratic 0,1 programs," *Operations Research Letters* 4 175-180.

[6] Balas, E. and J. Mazzola [1980] "Quadratic 0-1 programming by a new linearization," TIMS/ORSA Meeting, Washington D. C., cited in [27].

[7] Balas, E. and J. Mazzola [1984] "Nonlinear 0-1 programming: Parts I & II," *Mathematical Programming* 30 1-45.

[8] Balinski, M. [1970] "On a selection problem," *Management Science* 17 230-231.

[9] Barahona, F. [1983] "The max cut problem in graphs not contractible to K_5," *Operations Research Letters* 2 107-111.

[10] Barahona, F. [1986] "A solvable case of quadratic 0-1 programming," *Discrete Applied Mathematics* 13 23-26.

[11] Barahona, F. and A. Casari [1987] "On magnetization of the ground states in two dimensional Ising spin glasses," University of Waterloo, Canada, cited in [47].

[12] Barahona, F., M. Grötschel, M. Jünger and G. Reinelt [1988] "An application of combinatorial optimization to statistical physics and circuit layout design," *Operations Research* 36 493-513.

[13] Barahona, F., M. Grötschel and R. Mahjoub [1985] "Facets of bipartite subgraph polytope," *Mathematics of Operations Research* 10 340-358.

[14] Barahona, F., M. Jünger and G. Reinelt [1989] "Experiments in quadratic 0-1 programming," *Mathematical Programming* 44 127-137.

[15] Barahona, F. and A. R. Mahjoub [1986] "On the cut polytope," *Mathematical Programming* 36 157-173.

[16] Barthélemy, J. and B. Monjardet [1981] "The median procedure in cluster analysis and social choice theory," *Mathematical Social Sciences* 1 235-267.

[17] Bazaraa, M. and A. Elshafei [1979] "An exact branch-and-bound procedure for the quadratic-assignment problem," *Naval Research Logistics Quarterly* 26 109-121.

[18] Bazaraa, M. and M. Sherali [1980] "Benders' partitioning scheme applied to a new formulation of the quadratic assignment problem," *Naval Research Logistics Quarterly* 27 29-41.

[19] Benders, J. [1962] "Partitioning procedures for solving mixed-variables programming problems," *Numerische Mathematik* 4 238-252.

[20] Bokhari, S. [1981] "A shortest tree algorithm for optimal assignment across space and time in a distributed processor system," *IEEE Transactions on Software Engineering* 7 583-589.

[21] Bokhari, S. [1987] *Assignment Problems in Parallel and Distributed Computing* Kluwer Academic Publishers, Massachussets.

[22] Boros, E., Y. Crama and P. Hammer [1990] "Upper-bounds for quadratic 0-1 linearization," *Operations Research Letters* 9 73-79.

[23] Boros, E. and P. Hammer [1993] "Cut-polytopes, Boolean quadric polytopes and nonnegative quadratic pseudo-Boolean functions," *Mathematics of Operations Research* 18 245-253.

[24] Brown, D., C. Huntley and A. Spillane [1989] "A parallel genetic heuristic for the quadratic assignment problem," Proceedings on the Third Conference on Genetic Algorithms, Arlington, Virginia, 406-415, cited in [116].

[25] Buffa, E. [1955] "Sequence analysis for functional layouts," *The Journal of Industrial Engineering* 6 12, 13, 25.

[26] Burkard, R. [1973] "Die Störungsmethode zur Lösung quadratischer Zuordnungsprobleme," *Operations Research Verfahren* 16 84-108 cited in [27].

[27] Burkard, R. [1990] "Locations with spatial interactions: the quadratic assignment problem," in: P. Mirchandani and R. Francis (eds.) *Discrete Location Theory* Wiley, Berlin, Germany, 387-437.

[28] Burkard, R. and T. Bönninger [1983] "A heuristic for quadratic Boolean programs with applications to quadratic assignment problems," *European Journal of Operational Research* 13 374-386.

[29] Burkard, R. and U. Derigs [1980] *Assignment and matching problems: solution methods with FORTRAN-programs* Springer-Verlag, New York.

[30] Burkard, R., S. Karisch and F. Rendl [1991] "QAPLIB – a quadratic assignment problem library," *European Journal of Operational Research* 55 115-119. Updated version – Feb. 1994.

[31] Burkard, R. and J. Offerman [1977] "Entwurf von Schreibmaschinentastaturen mittels quadratischer Zuordnungsprobleme," *Zeitschrift für Operations Research* 21 B121-B132.

[32] Burkard, R. and F. Rendl [1984] "A thermodynamically motivated simulation procedure for combinatorial optimization procedures," *European Journal of Operational Research* 17 169-174.

[33] Burkard, R. and K-H Stratmann [1978] "Numerical investigations on quadratic assignment problems," *Naval Research Logistics Quarterly* 25 129-148.

[34] Cannon, T. [1988] "Large-scale zero-one linear programming on distributed workstations," Ph. D. thesis, Department of Operations Research and Applied Statistics, George Mason University, Fairfax, VA.

[35] Cannon, T. and K. Hoffman [1990] "Large-scale linear integer programming on distributed workstations," *Annals of Operations Research* 22 181-217.

[36] Carlson, R. and G. L. Nemhauser [1966] "Scheduling to minimize interaction cost," *Operations Research* 14 52-58.

[37] Carraresi, P. and F. Malucelli [1992a] "A new lower bound for the quadratic assignment problem," *Operations Research* 40 S22-S27.

[38] Carraresi, P. and F. Malucelli [1992b] "A reformulation scheme and new lower bounds for the quadratic assignment problem," *Dipartimento di Informatica, Università di Pisa* Pisa, Italy.

[39] Chakrapani, J. and J. Skorin-Kapov [1994] "A constructive method for improving lower bounds for a class of quadratic assignment problems," *Operations Research* 42 837-845.

[40] Chopra, S. [1992] "The graph partitioning polytope on series-parallel and 4-wheel free graphs," Northwestern University (to appear in *SIAM Journal on Discrete Mathematics*)

[41] Chopra, S. and M. R. Rao [1989a] "The partition problem I: Formulations, dimensions and basic facets," Stern School of Business, New York University, New York.

[42] Chopra, S. and M. R. Rao [1989b] "Facets of the k-partition polytope," Stern School of Business, New York University, New York.

[43] Chopra, S. and M. R. Rao [1993] "The partition problem," *Mathematical Programming* 59 87-115.

[44] Christofides, N. and E. Benavent [1989] "An exact algorithm for the quadratic assignment problem on a tree," *Operations Research* 37 760-768.

[45] Christofides, N. and M. Gerrard [1976] "Special cases of the quadratic assignment problem," Carnegie Mellon University, cited in [27].

[46] Clausen, J. [1994] "Announcement on discrete mathematics and algorithms network (DIMANET)," cited in [142].

[47] Conforti, M., M. R. Rao and A. Sassano [1990] "The equipartition polytope: Parts I & II," *Mathematical Programming* 49 49-90.

[48] Conway, R. and W. Maxwell [1961] "A note on the assignment of facility location," *The Journal of Industrial Engineering* 12 34-36.

[49] Crowder, H., E. Johnson and M. Padberg [1983] "Solving large-scale zero-one linear programming problems," *Operations Research* 31 803-833.

[50] Crowder, H. and M. Padberg [1980] "Solving large-scale symmetric traveling slaesman problems to optimality," *Management Science* 26 393-410.

[51] Dantzig, G., D. Fulkerson and S. Johnson [1954] "Solution of a large-scale travelling salesman problem," *Operations Research* 2 393-410.

[52] Deza, M. and M. Laurent [1992] "A survey of the known facets of the cut cone," Institut für Okonometrie und Operations Research, Universität Bonn, Bonn, Germany.

[53] Drezner, Z. [1995] "Lower bounds based on linear programming for the quadratic assignment problem," *Computational Optimization and Applications* 4 159-165.

[54] De Simone, C. [1989-90] "The cut polytope and the Boolean quadric polytope," *Discrete Mathematics* 79, 71-75.

[55] Dorn, W. [1961] "On Lagrange multipliers and inequalities," *Operations Research* 9 95-104.

[56] Dutta, A., G. Koehler and A. Whinston [1982] "On optimal allocation in a distributed processing environment," *Management Science* 28 839-853.

[57] Dyer, M., A. Frieze and C. McDiarmid [1986] "On linear programs with random costs," *Mathematical Programming* 35 3-16.

[58] Edwards, C. [1980] "A branch and bound algorithm for the Koopmans-Beckmann quadratic assignment problem," *Mathematical Programming Study* 13 35-52.

[59] Elshafei, A. [1977] "Hospital lay-out as a quadratic assignment problem," *Operational Research Quarterly* 28 167-179.

[60] Finke, G., R. Burkard and F. Rendl [1987] "Quadratic assignment problems," *Annals of Discrete Mathematics* 31 61-82.

[61] Ford, L. and D. Fulkerson [1962] *Flows in Networks* Princeton University Press, Princeton, New Jersey.

[62] Fortet, R. [1959] "L'algèbre de Boole et ses applications en recherche opérationelle," *Cahiers du Centre d'Etudes de Recherche Opérationelle* 1 5-36.

[63] Fortet, R. [1960] "Applications de l'algèbre de Boole en recherche opérationelle," *Revue Francaise de Recherche Opérationelle* 4 17-26.

[64] Frieze, A. and J. Yadegar [1983] "On the quadratic assignment problem," *Discrete Applied Mathematics* 5 89-98.

[65] Garey, M. and D. Johnson [1979] *Computers and Intractability: A Guide to the Theory of NP-Completeness* W. H. Freeman and Company, New York.

[66] Gavett, J. and N. Plyter [1966] "The optimal assignment of facilities to locations by branch and bound," *Operations Research* 14 210-232.

[67] Gilmore, P. [1962] "Optimal and suboptimal algorithms for the quadratic assignment problem," *SIAM Journal on Applied Mathematics* 10 305-313.

[68] Ghosh, A. and G. Rushton [1987] *Spatial Analysis and Location-Allocation Models* Van Nostrand Reinhold Company, New York.

[69] Glaser, R. [1959] "A quasi-simplex method for designing sub-optimal packages of electronic building blocks(Burroughs 220)," General Electric Company, cited in [107].

[70] Grötschel, M. [1992] "Discrete mathematics in manufacturing," Konrad-Zuse-Zentrum für Informationstechnik, Berlin, Germany.

[71] Grötschel, M., M. Jünger and G. Reinelt [1984] "A cutting plane algorithm for the linear ordering problem," *Operations Research* 32 1195-1220.

[72] Grötschel, M., M. Jünger and G. Reinelt [1985a] "On acyclic subgraph polytope," *Mathematical Programming* 33 28-42.

[73] Grötschel, M., M. Jünger and G. Reinelt [1985b] "Facets of the linear ordering polytope," *Mathematical Programming* 33 43-60.

[74] Grötschel, M., A. Martin and R. Weissmantel [1992] "Packing Steiner trees: a cutting plane algorithm and computation results," Konrad-Zuse-Zentrum für Informationstechnik, Berlin, Germany.

[75] Grötschel, M. and G. Nemhauser [1984] "A polynomial algorithm for the max-cut problem on graphs without long odd cycles," *Mathematical Programming* 29 28-40.

[76] Grötschel, M. and M. Padberg [1985] "Polyhedral theory," in: E. L. Lawler, J. K. Lenstra, A. H. G. Rinnoy Kan and D. Shmoys (eds.) *The Traveling Salesman Problem: A Guided Tour of Combinatorial Optimization* Wiley, Chichester, 251-305.

[77] Grötschel, M. and W. Pulleyblank [1981] "Weakly bipartite graphs and the max-cut problem," *Operations Research Letters* 1 23-27.

[78] Grötschel, M. and Y. Wakabayashi [1989] "A cutting plane algorithm for a clustering problem," *Mathematical Programming* 45 59-96.

[79] Grötschel, M. and Y. Wakabayashi [1990] "Facets of the clique partitioning polytope," *Mathematical Programming* 47 367-387.

[80] Hadlock, F. [1975] "Finding a maximum cut of planar graphs in polynomial time," *SIAM Journal on Computing* 4 221-225.

[81] Hammer (Ivănescu), P. [1965] "Some network flow problems solved using pseudo-boolean programming," *Operations Research* 13 388-399.

[82] Hammer, P., P. Hansen and B. Simeone [1984] "Roof duality, complementation and persistency in quadratic 0-1 optimization," *Mathematical Programming* 28 121-155.

[83] Hardy, G., J. Littlewood and G. Polya [1952] *Inequalities* Cambridge University Press, London.

[84] Hadley, S., F. Rendl and H. Wolkowicz [1992] "A new lower bound via projection for the quadratic assignment problem," *Mathematics of Operations Research* 17(3) 727-739.

[85] Hansen, P. [1979] "Methods of nonlinear 0-1 programming," *Annals of Discrete Mathematics* 5 53-70.

[86] Heffley, D. [1977] "Assigning runners to a relay team," in: S. Ladany and R. Machol (eds.) *Optimal Strategies in Sports* North-Holland, Amsterdam, 169-171, cited in [27].

[87] Heider, C. [1972] "A computationally simplified pair exchange algorithm for the quadratic assignment problem," Center for Naval Analyses, Arlington, VA, cited in [27].

[88] Hillier, F. [1963] "Quantitative tools for plant layout analysis," *The Journal of Industrial Engineering* 14 33-40.

[89] Hillier, F. and M. Connors [1966] "Quadratic assignment problem algorithms and the location of indivisible facilities," *Management Science* 13 42-57.

[90] Hoffman, A. and H. Wielandt [1953] "The variation of the spectrum of a normal matrix," *Duke Mathematical Journal* 20 37-39 cited in [147].

[91] Hoffman, K. and M. Padberg [1985] "LP-based combinatorial problem solving," *Annals of Operations Research* 4 145-194.

[92] Hoffman, K. and M. Padberg [1993] "Solving airline crew scheduling problems by branch-and-cut," *Management Science* 39 657-682.

[93] Jünger, M. [1985] *Polyhedral Combinatorics and the Acyclic Subdigraph Problem* Heldermann Verlag, Berlin.

[94] Jünger, M., A. Martin, G. Reinelt and R. Weissmantel [1994] "Quadratic 0/1 optimization and a decomposition approach for the placement of electronic circuits," *Mathematical Programming* 63 257-279.

[95] Jünger, M., G. Reinelt and G. Rinaldi [1995] "The traveling salesman problem," in: M. Ball *et al.* (eds) *Handbook in Operations Research and Management Science* North-Holland, Amsterdam, 7 225-330.

[96] Kaufman, K. and F. Broeckx [1978] "An algorithm for the quadratic assignment problem using Benders' decomposition," *European Journal of Operational Research* 2 204-211.

[97] Kernighan, B. and S. Lin [1970] "An efficient heuristic procedure for partitioning graphs," *Bell System Technical Journal* 49 291-308.

[98] Knuth, D. [1961] "Minimizing drum latency time," *Journal of Association of Computing Machinery* 8 119-150.

[99] Kodres, U. [1959] "Geometrical positioning of circuit elements in a computer," AIEE Fall General Meeting, cited in [107].

[100] Koopmans, T. and M. Beckmann [1957] "Assignment problems and the locations of economic activities," *Econometrica* 25 53-76.

[101] Korte, B. and W. Oberhofer [1968] "Zwei Algorithmen zur Lösung eines komplexen Reihenfolgeproblems," *Unternehmensforschung* 12 217-362.

[102] Korte, B. and W. Oberhofer [1969] "Zur Triangulation von Input-Output Matrizen," *Jahrbücher für Nationalökonomie und Statistik* 182 398-433.

[103] Krarup, J. and P. Pruzan [1978] "Computer-aided layout design," *Mathematical Programming Study* 9 75-94.

[104] Land, A. [1963] "A problem of assignment with interrelated costs," *Operational Research Quarterly* 14 185-198.

[105] Laporte, G. and H. Mercure [1988] "Balancing hydraulic turbine runners: a quadratic assignment problem," *European Journal of Operational Research* 35 378-382.

[106] Lawler, E. [1960] "Notes on the quadratic assignment problem," Harvard Computation Laboratory, cited in [107].

[107] Lawler, E. [1963] "The quadratic assignment problem," *Management Science* 9 586-599.

[108] Leontief, W. [1951] *The Structure of the American Economy 1919-1939* New York, Oxford University Press.

[109] Leontief, W. [1963] "The structure of development," *Scientific American*.

[110] Leontief, W. [1966] *Input-Output Economics* New York, Oxford University Press.

[111] Lengauer, T. [1990] *Combinatorial algorithms for Integrated Circuit Layout* Wiley-Teubner, Chichester.

[112] Li, Y., P. Pardalos, K. Ramakrishnan, M. Resende [1994] "Lower bounds for the quadratic assignment problem," *Annals of Operations Research* 50 387-410.

[113] Loberman, H. and A. Weinberger [1957] "Formal procedures for connecting terminals with a minimum total wire length," *Journal of the Association of Computing Machinery* 4 428-437.

[114] Lutton, J. and E. Bonomi [1986] "The asymptotic behaviour of quadratic sum assignment problems: a statistical mechanics approach," *European Journal of Operational Research* 26 295-300.

[115] Magirou, V. and J. Milis [1989] "An algorithm for the multiprocessor assignment problem," *Operations Research Letters* 8 351-356.

[116] Mans, B., T. Mautor and C. Roucairol [1992] "A parallel depth first search branch and bound algorithm for the quadratic assignment problem," Technical report INRIA, cited in [117].

[117] Mans, B., T. Mautor and C. Roucairol [1993] "Recent exact and approximate algorithms for the quadratic assignment problems," APMOD, Budapest, Hungary, 395-402.

[118] Marcotorchino, J. and P. Michaud [1980] "Optimisation en analyse des données relationnelles," in: E. Diday et al. (eds.) *Data Analysis and Informatics* North-Holland, Amsterdam, 655-670.

[119] Marcotorchino, J. and P. Michaud [1981a] "Heuristic approach to the similarity aggregation problem," *Methods of Operations Research* 43 395-404.

[120] Marcotorchino, J. and P. Michaud [1981b] "Optimization in exploratory data analysis," *Proceedings of 5th International Symposium on Operations Research* Physica Verlag, Köln, cited in [79].

[121] Mautor, T. and C. Roucairol [1993] "A new exact algorithm for the quadratic assignment problem," (to appear in *Discrete Applied Mathematics*), cited in [117].

[122] Mirsky, L. [1956] "The spread of a matrix," *Mathematika* 3 127-130.

[123] Moore, J. [1961] "Optimal locations for multiple machines," *The Journal of Industrial Engineering* 12 307-313.

[124] Mühlenbein, H. [1989] *Parallel genetic algorithms, population genetics and combinatorial optimization in Parallelism,learning, evolution* WOPPLOT89, Springer-Verlag.

[125] Müller, R. [1993] "Bounds for Linear VLSI-Layout Problems" Doctoral Dissertation, Mathematics, Technical University, Berlin.

[126] Nugent, C., T. Vollmann and J. Ruml [1968] "An experimental comparison of techniques for the assignment of facilities to locations," *Operations Research* 16 150-173.

[127] Opitz, O. and M. Schader [1984] "Analyse qualitativer Daten: Einführung und Übersicht," *Operations Research Spektrum* 6 67-83, cited in [78].

[128] Padberg, M. [1972] "Equivalent knapsack-type formulations of bounded integer linear programs: an alternative approach," *Naval Research Logistics Quarterly* 19 699-708.

[129] Padberg, M. [1973] "On the facial structure of set packing polyhedra," *Mathematical Programming* 5 199-215.

[130] Padberg, M. [1975] "A note on zero-one programming," *Operations Research* 23 833-837.

[131] Padberg, M. [1976] "Zero-one decision problems," GBA, New York University. Published (in German) in: M. Beckmann and R. Selten (eds.) [1978] *Handwörterbuch der Mathematischen Wirtschaftswissenschaften* Gabler-Verlag, Wiesbaden, 187-229.

[132] Padberg, M. [1989] "The Boolean quadric polytope: some characteristics, facets and relatives," *Mathematical Programming* 45 139-172.

[133] Padberg, M. [1995] *Linear Optimization and Extensions* Springer-Verlag, Berlin.

[134] Padberg, M. and M. Grötschel [1985] "Polyhedral computations," in: E. L. Lawler, J. K. Lenstra, A. H. G. Rinnoy Kan and D. Shmoys (eds.) *The Traveling Salesman Problem: A Guided Tour of Combinatorial Optimization* Wiley, Chichester, 306-367.

[135] Padberg, M. and G. Rinaldi [1990] "An efficient algorithm for the minimum capacity cut problem," *Mathematical Programming* 47 19-36.

[136] Padberg, M. and G. Rinaldi [1991] "A branch-and-cut algorithm for the resolution of large-scale symmetric traveling salesman problems," *Siam Review* 33 60-100.

[137] Padberg, M. and T-Y Sung [1991] "An analytical comparison of different formulations of the travelling salesman problem," *Mathematical Programming* 52 315-357.

[138] Padberg, M., T. J. Van Roy and L. Wolsey [1985] "Valid linear inequalities for fixed charge problems," *Operations Research* 33 842-861.

[139] Padberg, M. and M. Wilczak [1993] "Boolean polynomials and set functions," *Mathematical and Computer Modelling* 17 3-6.

[140] Padberg, M. and L. Wolsey [1983] "Trees and cuts," *Annals of Discrete Mathematics* 17 511-517.

[141] Pardalos, P. and J. Crouse [1989] "A parallel algorithm for the quadratic assignment problem," Proceedings of Supercomputing 89' Conference, ACM Press 351-360.

[142] Pardalos, P., K. Ramakrishnan, M. Resende and Y. Li [1994] "Implementation of a variance based lower bound in branch and bound algorithm for the quadratic assignment problem," AT&T Bell Laboratories, Murray Hill.

[143] Picard, J. and D. Ratliff [1975] "Minimum cuts and related problems," *Networks* 5 357-370.

[144] Pierce, J. and W. Crowston [1971] "Tree-search algorithms for quadratic assignment problems," *Naval Research Logistics Quarterly* 18 1-36.

[145] Pollatschek, M., H. Gershoni and Y. Radday [1976] "Optimization of the typewriter keyboard by computer simulation," *Angewandte Informatik* 10 363-372, cited in [27].

[146] Reinelt, G. [1985] *The Linear Ordering Problem: Algorithms and Applications* Heldermann Verlag, Berlin.

[147] Rendl, F. and H. Wolkowicz [1992] "Applications of parametric programming and eigen value maximization to the quadratic assignment problem," *Mathematical Programming* 53 63-78.

[148] Resende, M., K. Ramakrishnan and Z. Drezner [1994] "Computing lower bounds for the quadratic assignment problems with an interior point algorithm for linear programming," AT&T Bell Laboratories, Murray Hill.

[149] Rhys, J. [1970] "A selection problem of shared fixed costs and network flows," *Management Science* 17 200-207.

[150] Rijal, M. [1995] "Scheduling, Design and Assignment Problems with Quadratic Costs," Ph. D. thesis, Stern School of Business, New York University, New York, NY.

[151] Roucairol, C. [1987] "A parallel branch and bound algorithm for the quadratic assignment problem," *Discrete Applied Mathematics* 18 211-225.

[152] Sahni, S. and T. Gonzalez [1976] "P-complete approximation problems," *Journal of the Association for Computing Machinery* 23 555-565.

[153] Schlegel, D. [1987] *Die Unwucht-optimale Verteilung von Turbinenschaufeln als quadratisches Zuordnungsproblem* Doctoral thesis, ETH Zürich, Switzerland, cited in [27].

[154] Skorin-Kapov, J. [1990] "Tabu search applied to the quadratic assignment problem," *ORSA Journal of Computing* 2 33-45.

[155] Steinberg, L. [1961] "The blackboard wiring problem: a placement algorithm," *Siam Review* 3 37-50.

[156] Stone, H. [1977] "Multiprocessor scheduling with the aid of network flow algorithms," *IEEE Transactions on Software Engineering* 3 85-93.

[157] Taillard, E. [1991] "Robust tabu search for the quadratic assignment problem," *Parallel Computing* 17 443-455.

[158] Thorndike, R. [1950] "The problem of classification of personnel," *Psychometrika* 15 215-235.

[159] Tüshaus, U. [1983] "Aggregation binärer Relationen in der qualitativen Datenanalyse," in: *Mathematical Systems in Economics* 82 Hain, Königstein, cited in [78].

[160] Ugi, I., J. Bauer, J. Brandt, J. Friedrich, J. Gasteiger, C. Jochum and W. Schubert [1979] "Neue Anwendungsgebiete für Computer in der Chemie," *Angewandte Chemie* 91 99-111, cited in [27].

[161] Van Roy, T. J. and L. Wolsey [1985] "Valid inequalities and separation for uncapacitated fixed charge networks," *Operations Research Letters* 4 105-112.

[162] Van Roy, T. J. and L. Wolsey [1987] "Solving mixed integer programming problems using automatic reformulation," *Operations Research* 35 45-57.

[163] Weber, A. [1909, 1929] *Über den Standort der Industrien* Tübingen. Translated as *Alfred Weber's Theory of the Location of Industries* by C. J. Friedrich, University of Chicago Press.

[164] Wimmert, R. [1958] "A mathematical method of equipment location," *The Journal of Industrial Engineering* 9 498-505.

[165] Wilhelm, M. and T. Ward [1987] "Solving quadratic assignment problems by simulated annealing," *IIE Transactions* 19 107-119.

[166] Wolsey, L. [1989] "Strong formulations for mixed integer programming: a survey," *Mathematical Programming* 45 173-191.

[167] Young, H. [1978] "On permutations and permutation polytopes," *Mathematical Programming Study* 8 128-140.

INDEX

a general model, 111
acyclic gaph, 77
acyclic subgraph problem, 38
affine hull, 12–13, 62, 122, 151, 154, 161, 164
airline crew scheduling problem, 60
Benders' decomposition, 72–74
binary search tree, 65
bipartite graph, 57, 76–77
    weakly, 77
Boolean quadric problem, 56, 79–81
bound
    Gilmore-Lawler, 66, 68–69, 168
    lower, 5–6, 16, 25, 30–31, 49, 62, 65, 67–73, 141, 168–170
    upper, 25, 66, 68, 71, 73, 168–169
branch-and-bound, 25, 60–61, 65–67, 77, 167–169
branch-and-cut, 60, 65, 167–171
branching variable, 65
branching, 65–67, 170
capacitated network problem, 60
circuit layout design problem, 39, 81, 105
class-room scheduling problem, 47
classification of BQPSs, 58
clique inequality, 147–148, 150, 157–158
clique partitioning problem, 52
clique, 52, 55, 157
cobweb, 79
combinatorial optimization, 13–14, 26–27, 31, 34–35, 51, 56–57, 59–60, 63–64, 66, 79–81, 151, 171
    formulation of, 79
    structural property of, 35
complete graph, 12, 38, 52–53, 55, 57, 136, 164
constrained Boolean quadratic problem, 57
convex hull, 11, 13, 62, 80, 87, 93, 100, 109, 115, 121, 130, 135, 144
convexification, 11
CPLEX, 24–25, 65, 168–169
cut inequality, 147, 149–150, 157–162
cutting plane algorithm
    polyhedral, 59–61, 168
    traditional, 61, 72, 74–75
dicycle covering problem, 38
discrete set, 11, 62, 82, 89, 95, 105, 111, 117, 122
distinct equation, 153
double description algorithm, 81
double star, 76
dual linear program, 14, 67, 72–74, 85–86, 92–93, 99, 108, 114, 120
    degeneracy of, 14
duality, 72, 86, 93, 99–100, 109, 114, 120
dynamic programming, 61, 77
dynamic simplex, 11, 25
eigen value, 69–72, 78
enumeration, 15, 61, 64–65
    implicit, 64–65

equation
  aggregated, 16
  disaggregated, 16
equi-partitioning problem, 56, 60
Euclidean norm, 27
exchange heuristic, 75, 168
extreme point, 73–74, 80–81, 85,
    92, 94, 99, 101, 108, 110,
    114–115, 120–121, 148, 158
Farkas' lemma, 139
feedback arc set problem, 38
fixed-charge problem, 60
floorplan, 40
formulation
  bad, 62
  better, 22, 81, 134
  compactness of, 64
  comparable, 134, 141, 143
  comparison of, 133, 136, 144
    analytical, 133, 144
    empirical, 133, 144
  data-dependent, 5
  data-independent, 17
  minimal, 7, 12–13, 17, 65, 79,
    122, 128, 131, 151, 156, 161
  mixed zero-one, 3–4, 9–11,
    15–16, 20, 24–28, 37, 59,
    61–62, 64–65, 79, 161, 167,
    169–171
FORTRAN, 25, 30, 168
genetic algorithm, 75
graph partitioning problem, 52,
    81–82
independent set problem, 56
integrality gap, 25
Koopmans-Beckmann problem, 1,
    117
  symmetric, 8, 122
Lagrangian multiplier, 49, 70
Lagrangian relaxation, 69–70
layout
  full-custom, 41
  semi-custom, 41
linear assignment polytope, 3, 7,
    32, 156
linear assignment problem, 38
linear description, 60, 62, 81,
    133–134, 136–137, 140, 143,
    151, 161
  complete and minimal, 12–13, 79
  complete, 11, 156, 161–162
  ideal, 80, 128, 154, 156, 161
    locally, 80, 86, 93, 100, 109,
    114, 120, 128
  quasi-unique, 13
linear independence, 83, 90, 97,
    102, 107–108, 110–111,
    113–114, 116, 119, 124, 139,
    144, 151, 164
linear ordering problem, 37
linearization, 5–6, 8–9, 34, 55, 58,
    61–62, 65, 67, 79–81, 87, 93,
    100, 105, 109, 114, 120, 129,
    141
  locally ideal, 87, 93, 100, 105,
    107, 109, 113–114, 118, 120,
    129, 141
Manhattan norm, 27–29
marriage problem, 4
matrix
  adjacency, 76–77
  connection, 26, 28–29
  identity, 114, 136, 164
  incidence, 12, 136, 164
  indefinite, 49
  input-output, 36–37
  nonsingular, 134, 155–156, 164
  positive semidefinite, 49, 77
  rank of, 7, 12
  spread of, 71
  trace of, 3
  triangular, 36–37, 43, 56,
    107–108, 110–111, 113, 116,
    154

max cut problem, 57, 60
max flow min cut theorem, 76
max flow problem, 76
meta-heuristic, 75
min cut problem, 56–57, 76
multi-processor assignment problem, 44, 81, 95
NP-hard problem, 34–35, 43, 46, 49, 52, 55–57, 60, 75, 81
odd cycle, 77
operations-scheduling problem, 50, 81, 88
OSL, 30, 65
permutation problem, 38
personnel assignment problem, 38
placement, 40–42, 50
planar graph, 77
plant and office layout planning, 20
polyhedral cone, 134, 137–139, 143
  extreme ray of, 138–139, 143
  generator of, 134, 138, 140–141, 143
    minimal, 134, 139
  intersection property of, 138, 143
  lineality space of, 134, 138, 143
  pointed, 138, 143
polyhedral theory, 79
polyhedron, 59–60, 94, 133–134, 142
  dimension of, 11–12, 83, 90, 96, 102, 107, 110, 113, 116, 119, 124, 133–134, 142, 144, 151, 155–156, 161, 165
  face of, 84–85, 88, 91–92, 97–98, 102–104, 108, 125–127, 145–146, 148–149
  facet of, 13, 31, 62, 81, 84–85, 87, 91, 97–98, 102–103, 108, 111, 114, 116, 119, 125–127, 130, 145–150, 155–156, 158–161, 165, 170
    distinct, 155

    local, 150
  facial structure, 59–60, 133, 144, 151, 168
  image of, 134, 140–143
  polytope, 11–13, 31, 60, 62, 79–83, 85–87, 89–90, 92–96, 99–100, 102, 105–106, 109–110, 112, 114–118, 120–123, 128, 130, 135–136, 140–144, 147–151, 156–158, 161, 165, 168, 171
primal linear program, 72
  degeneracy of, 14
projection, 112, 142–144
QAPLIB, 26, 77–78, 167, 169–170
quadratic assignment problem, 1, 31, 117
  applications of, 33
  symmetric, 122
rank, 7, 12, 14, 17, 134, 138, 151–152, 155–156, 163–164
reduction procedure, 8, 16, 67–69, 71–72
redundant equation, 70, 106–107, 112, 115, 118–119, 121, 151, 156
redundant inequality, 9, 120, 129, 140–141, 143
routing, 40–42, 50
sea-of-gates technology, 41
semi-net revenue, 2
separation problem, 60, 88, 101, 168
series-parallel graph, 76–77
set partitioning problem, 59–60
simulated annealing, 75
solution
  fractional, 4, 16, 80–81, 161
  optimal, 4–5, 15, 17, 20, 22–26, 28, 43, 48, 51, 55–56, 66, 72, 75–77, 85–86, 88, 92–93, 98–100, 108–109, 114, 120, 156, 169

alternative, 43, 48
  unique, 4, 45
 suboptimal, 23, 73, 75, 78
special ordered set, 1, 35, 57, 79
stable set problem, 56
Steinberg's wiring problem, 26
structured linear program, 17
support graph, 147
tabu search, 75, 78
task graph, 76
tournament, 38
transformation
  affine, 134
  linear, 64, 89, 106, 122, 136,
      138–144, 165
traveling salesman problem, 35,
    59–60, 62, 81, 133, 167–168
tree, 76–77
triangular inequality, 2, 4–5
triangulation problem, 36
valid equation, 12, 106, 112, 118,
    124
valid inequality, 11, 13, 84–85, 88,
    91–92, 97–98, 102–104, 106,
    112, 118, 124, 145–149, 151,
    157
violated inequality, 17, 60, 73, 75,
    88, 101
VLSI circuit layout design
    problem, 39, 81, 105
wheel-free graph, 77
wiring problem, 26